UNCONVENTIONAL METHODS IN EXPLORATION FOR PETROLEUM AND NATURAL GAS III

UNCONVENTIONAL METHODS IN EXPLORATION FOR PETROLEUM AND NATURAL GAS
III

A SYMPOSIUM UNDER THE AUSPICES OF THE
INSTITUTE FOR THE STUDY OF
EARTH AND MAN
AT
SOUTHERN METHODIST UNIVERSITY

Edited by
Martin J. Davidson and Benjamin M. Gottlieb

SMU PRESS • DALLAS

Typesetting and production by Blair Graphics,
Plano, Texas 75074
Printing through the SMU Printing Department,
Dallas, Texas 75275
Published by Southern Methodist University Press, Dallas, Texas 75275

Library of Congress Cataloging in Publication Data

Main entry under title:

Unconventional methods in exploration for petroleum and natural gas III.

1. Prospecting—Congresses. 2. Petroleum—Congresses. 3. Gas, Natural—Congresses. I. Davidson, Martin J., 1924- . II. Gottlieb, Benjamin M., 1912- . III. Southern Methodist University. Institute for the Study of Earth and Man.

TN271.P4U533 1984 622'.1828 83-8653
ISBN 0-87074-188-8

INSTITUTE FOR THE STUDY OF EARTH AND MAN

"Nature and history, the earth and man, stand in the closest relations to each other, and form only one grand harmony."

—Arnold Guyot, 1849

The Institute, under whose auspices this symposium was arranged, is organized as an affiliate of Southern Methodist University "to encourage, foster, promote and advance research and development and education in the field of the anthropological and geological sciences and other related fields of scholarship at Southern Methodist University."

Incorporated in Texas on May 11, 1966

CONTENTS

CONTENTS

PREFACE

The Institute's third symposium on Unconventional Methods in Exploration for Petroleum and Natural Gas was held on September 8 and 9, 1982. In spite of the depressed state of the petroleum economy, the symposium was attended by approximately two hundred persons, including independent and consulting geologists, university faculty members, and representatives of governmental agencies. Response to the program was enthusiastic. Particular interest was generated by the keynote address presented at the symposium luncheon by Philip Abelson, editor of *Science*.

The planning of the symposium was under the general supervision of the Institutes's symposium committee which was chaired by Benjamin M. Gottlieb and included Claude C. Albritton, Jr., Martin J. Davidson, Charles F. Dodge, William B. Heroy, Jr., W.T. Lehman, Earl R. Price, and Richard C. Weart. The detailed planning was done by a subcommittee consisting of Martin J. Davidson and Earl R. Price. Members of the ISEM staff also played key and continuous roles in staging the symposium. Included were Netta G. Blanchard, Liebe M. Purnell, Kristi H. Hunsucker, and Doris Ainsworth (public relations consultant).

After the symposium, the speakers were given opportunities to clarify or amplify the written versions of their presentations in the light of the questions and discussions that followed their talks. Also, many of the illustrations, originally drafted for viewing as slides, were redrafted for publication. All abstracts were received during the first three months of 1982.

Because all the illustrations in this volume are reproduced in black and white, the paper by John A. Ward is not included since it contained two color figures essential to the text. His paper will be available as a separate publication.

Special thanks are due Charlotte Whaley of the SMU Press for guiding the manuscript through the press and Martin J. Davidson for his extensive and significant contributions in the technical editing of this volume. We are also grateful to Ken Caldwell and the SMU Printing Department for exemplary performance in the face of difficulties of time and circumstance.

James E. Brooks

Dallas, Texas
December 28, 1983

UNCONVENTIONAL METHODS IN EXPLORATION FOR PETROLEUM AND NATURAL GAS

III

HISTORY AND FUTURE PROSPECTS OF THE SYMPOSIA ON UNCONVENTIONAL METHODS

Benjamin M. Gottlieb

VERITY OIL CO., INC.

At our last Symposium in September, 1979, many of us felt uncomfortable using the term *unconventional* to describe what we felt were very viable alternative exploration techniques. Today, however, we should rest at ease. I can think of nothing more unconventional in the oil industry than whiskey, steel, chemical, and a sundry assortment of other companies buying up our oil companies!

Approximately fourteen years ago, a small group of creative and courageous geoscientists assembled here in order to expedite the "innovative cycle" of improving our conventional exploration methods. At that time we established an interdisciplinary forum with the following objectives:

1. To assemble the most creative and authoritative thinking of speakers, panelists, and attendants of all the earth science disciplines.
2. To coordinate and document all available research data, both basic and applied.
3. To strengthen communication between the different fields of scientific endeavor (geology, geophysics, geochemistry, engineering, etc.) and to develop a team effort to evaluate and innovate unconventional exploration methods.

I have pondered over analogies which best describe and summarize the progress that we "unconventional explorers" have made through the years. In solution, I have coined the following expressions which describe the evolution of our unconventional methods. We have graduated from the "black-box" days when our revolutionary techniques were perceived as mere hocus-pocus. In the early days, black boxes were nothing more than wizardry dreamed up by scientific doodlebuggers.

After interviewing our attendants at Symposium II, I was convinced we had reached what I call the "pill-box" stage. We unconventional doodlebuggers had become unconventional medicine men; our unconventional-method approach was a sort of pill box containing tranquilizers to ease the pain of drilling a dry hole on a well-planned geological prospect.

Today, at Symposium III, it is my opinion that we have reached the "safety deposit box" stage, since many of our speakers, panelists, and attendants have applied unconventional methods, have actually found oil and gas, and have deposited their profits. We need no longer cling to the apron strings of disbelievers; we can now stand proudly next to geologists and seismologists.

It is our philosophy that the technology is available today to reduce drastically the finding cost of domestic oil and gas.

The implementation of our technology and unconventional methods is a viable solution to the problem of our domestic energy shortage. By continuing with increasingly stale and costly methods currently being employed by the industry, we can expect a general decline in the discovery of new oil and gas fields. The industry is trapped in its own rhetoric; it must break loose from its self-imposed shackles and augment its programs with unconventional exploration techniques if it is to move forward.

The hydrocarbons are there; it is our belief that there are as many or possibly more reserves to be found in the future as have been found in the past.

We sincerely hope that Symposium III will enable you to drill smarter, with more producers and fewer dry holes.

STATE OF THE ART AND THE DIRECTION OF UNCONVENTIONAL OIL AND GAS EXPLORATION

Martin J. Davidson

NORTH TEXAS STATE UNIVERSITY

It is the predominant judgement of the most eminent authorities in industry, government, and academe that we have *not* experienced any major inventive breakthrough in oil-finding methodology. As so well expressed last year by Richard Nehring (1981) of the Rand Corporation "think tank," the conventional wisdom holds that any method that might come to be considered a major breakthrough "could occur only if one assumed that *every* exploration geologist, geophysicist, and geochemist working in the United States over the past forty years had been operating with massive mental blinders, which rendered them oblivious to such a great and lucrative potential" (Nehring, 1981, p. 177). We owe it to Dr. Nehring and to the Rand Corporation to examine the possibility that explorationists may, in fact, have been operating with massive mental blinders.

From their inception fifteen years ago, the SMU symposia have been designed to sensitize the industry to changes in oil-finding thought and practice which, cumulatively, might constitute a slow-motion breakthrough. For a deeper understanding of what is happening and, more importantly, what is *not* happening in unconventional exploration, it is worth referring to Thomas S. Kuhn (1962), whose thought, however indirectly, influenced the establishment of the symposia. Kuhn is widely accepted among his peers as the outstanding contemporary figure in the philosophy of science. Examined against the backdrop of Kuhn's writing, we may find insights concerning the direction of petroleum exploration that might otherwise escape us.

Central to Kuhn's thought is what he calls "paradigms." He defines paradigms as "universally recognized scientific achievements that for a time provide model problems and solutions to a community of practitioners." We might think of paradigms as current, working conceptual frameworks.

The prevailing paradigm in scientific petroleum exploration is based exclusively on the primacy of structure, stratigraphy, and the low temperature primary migration of hydrocarbons in bulk-phase (i.e., as oil or gas) over short distances from source rock to reservoir.

A new paradigm has now arrived on the scene that is rooted in hydrogeology, the accommodation by molecular solution or colloidal suspension in water of organic derivatives, high-temperature deep-water discharge, and vertical migration from great depths occasionally accompanied by long distance lateral migration in a basin.

The new paradigm is in conflict with the prevailing paradigm, although not completely; it is not in conflict with each and every facet of the existing paradigm. The issue of changing paradigms is really one of defining a new puzzle as opposed to solving old puzzles. This is not to say that solving old puzzles is not of great value; it is. What solving old puzzles will not do, however, is to provide Dr. Nehring with the exploration breakthrough that his vision denies him.

Thomas Kuhn believes that circumstances exist, however rare, under which two paradigms can coexist peacefully, but then only in the later period of the new paradigm's existence (Kuhn, p. ix). Fundamental change in scientific thought cannot occur without conflict. Each such change in the history of science necessitated the community's rejection of one time-honored scientific theory in favor of another theory essentially incompatible with its predecessor. One defining characteristic of paradigm change is the controversy that invariably accompanies it (Kuhn, p. 6). "Competition between segments of the scientific community is the only historical process that ever actually results in the rejection of one previously accepted theory or in the adoption of another" (Kuhn, p. 8).

Kuhn also writes:

> Mopping-up operations are what engage most scientists throughout their careers. They constitute what I am here calling

> normal science. Closely examined, whether historically or in the contemporary laboratory [normal science] . . . seems an attempt to force nature into the pre-formed and relatively inflexible box [that] the [existing or prevailing] paradigm supplies. No part of the aim of normal science is to call forth new sorts of phenomena; indeed those [facts] that will not fit the box are often not seen at all. Nor do scientists normally aim to invent new theories and they are often intolerant of those invented by others. Instead, normal scientific research is directed to the articulation of those phenomena and theories that the [existing or prevailing] paradigm already supplies (Kuhn, p. 24).
>
> So long as the tools a paradigm supplies continue to prove capable of solving the problems it defines, science moves fastest and penetrates most deeply through confident employment of those tools (Kuhn, p. 76).
>
> Debates do not vanish once and for all with the appearance of a [new] paradigm (Kuhn, p. 48).

By ensuring that the existing or prevailing paradigm will not be too easily surrendered, resistance to a new paradigm guarantees that scientists will not lightly be distracted and that before switching paradigms they will penetrate existing knowledge to the core (Kuhn, p. 65). The emergence of new theories is generally preceded by a period of pronounced professional insecurity. Failure of existing rules is the prelude to a search for new ones (Kuhn, p. 67). As in manufacturing, so it is in science; retooling (or revocationalizing) is an extravagance reserved only for the critical occasion that demands it. It is only crisis that announces that the time for retooling has arrived (Kuhn, p. 76).

The new paradigm is itself a practical exploration model in that it ties vertical petroleum migration, during the origin and accumulation process, to the general probability of production in basins or in portions of basins (Munn, 1909; Rich, 1921; Culberson and McKetta, 1951; Price, 1976, 1980a, 1980b, 1981a, 1981b; Jones, 1980, 1981; Roberts, 1980, 1981).

The more specific vertical migration exploration methods (Davidson, 1982), such as surface geochemistry, radiometrics, geobotany, geomicrobiology, induced polarization, temperature, magnetoelectrics, shallow magnetic effects, etc., are *not* identical to the new hot-deep origin and accumulation paradigm, but they are easily conceivable, natural sequitors of the new paradigm. It is on these grounds that Roberts and Jones (1981) explicitly lend support to the value of near-surface geochemical prospecting, "the idea being that water ascending through covering strata would help to deliver the signals of deeper entrapment to the surface." The new paradigm will not pick a location for you, but the specific vertical migration exploration methods are intended to do just that.

In fitting our exploration industry experience to Thomas Kuhn's framework, can anyone doubt the conflict that has so often and so vigorously pursued the vertical migration methods; especially geochemistry and radiometrics? The empirical success of individual techniques, both in their positive predictions and their negative predictions, could be and has been easily brushed off by the industry (Davidson, 1982). The real significance of the new paradigm is that it is one within which the "vertical migration methods" fit comfortably.

Each exploration subject presented at this symposium tends to fall into one of three categories, which I am calling the "old-new," the "new-new," and the "new." Four papers fall into the "old-new" category. These address themselves to specific exploration methods which possess a history of empirical success but which tend to depend, for their theoretical validity, upon the new paradigm. The Symposium III papers in this "old-new" category are Curry's on total gamma radiation, Gallagher's on iodine, Hebert's on fluorescence, and Sternberg and Oehler's on induced polarization as a particular aspect of electrical methods.

The difference between the "new-new" and "old-new" categories has only to do with the existing quantity of empirical data. Both categories are rooted in the new paradigm. The methods associated with the "new-new" category have simply not been on the scene long enough to accumulate large amounts of data, or else, as in temperature, where the data exist they are known mainly to researchers rather than to line explorationists. Of the papers in this symposium, Paul Jones's on vertical subsurface fluid migrations comes the closest to a pure expression of the new paradigm. Mary Dalziel's paper and Bob Foote's paper are deeply anchored in the new paradigm.

The papers that may be placed in the "new" category fit comfortably into the existing or prevailing paradigm. The methods and insights in this category, which includes all seven of the other papers presented, do not depend upon vertical migration. The methods and insights that fall in the "new" category find little fundamental resistance on theoretical grounds and hold the promise of early and widespread utilization. Increased effectiveness within the prevailing paradigm could be of tremendous scientific value and economic importance, although often at high cost. The chances are slim, however, that any method in the "new" category will provide Dr. Nehring with the breakthrough for which he despairs.

There is no doubt that some overlap occurs among the categories. To the extent that a portion of Brian Bailey's remote sensing update might have touched upon stressed vegetation and/or tonal anomalies, it would have been related to the new paradigm and the "new-new" category. The extent to which empirical data on the use of iodine to track deep-water discharge may be widespread is itself a matter of judgment, and Gallagher's paper on iodine might well be placed in the "new-new" category.

The literature of petroleum exploration is replete with case histories and success rates on the effectiveness of surface geochemistry and radiometric methods (Horvitz, 1954; Davidson, 1963, 1967, 1982; Curry, this volume; Duchscherer, 1982; Weart and Heimberg, 1981). The marked controversy that has surrounded these applications of a new paradigm fit well into Kuhn's enunciation that fundamental change in scientific thought can only occur in an atmosphere of conflict. That the applications arrived on the scene substantially ahead of the new paradigm itself does not alter Kuhn's observation. History, in the form of experience and experiment, provides the raw material for new theory and new paradigms.

Resistances to change are anything but irrational. Intellectual and financial interests can rightfully be expected to fight to maintain their own positions. It is proper that they should. Too early an emphasis on new methods, when too little is known about their limitations, can retard the development and the application of those methods. The generally negative perception, based on limited and improper usage of surface geochemistry in the 1930s and 1940s, provided an effective barrier to its use in the 1950s and 1960s when, because of the low price of oil and gas, it was critically needed.

Expenditures on vertical migration exploration methods probably constitute less than two-tenths of one percent (.2%) of predrilling exploration expenditures. Guessing the extent of vertical migration methods in use is further complicated by the failure of the Society of Exploration Geophysicists to tabulate the findings of geochemical or radiometric crews in the field on the grounds that examining the physical chemistry of the earth is not in the field of geophysics. There are indications, however, that the SEG is becoming more liberal in recognizing that exploration geophysics is broader rather than narrower.

In an attempt to determine, with statistical accuracy, the effectiveness of some of the vertical migration methods, the Institute for the Study of Earth and Man applied for and received a grant from the Department of Energy to determine the feasibility of field testing vertical migration exploration methods. A field demonstration project was determined to be feasible, and the study provided a structure and methodology to accomplish it. The DOE has now sent the feasibility study, with its recommendation to move forward, to the Geophysics Research Board of the National Academy of Sciences, where it is now awaiting a decision on whether and how to proceed with the field demonstration.

The attempt to fit Kuhn's analysis to oil and gas exploration seems to twist off when we consider his observations that "the emergence of new theories is generally preceded by a period of pronounced professional insecurity . . . Failure of existing rules is the prelude to a search for new ones" (Kuhn, p. 67). "It is only crisis that announces that the time for retooling has arrived" (Kuhn, p. 76).

In attempting to account for the industry's continued resistance to exploration methods rooted in the new paradigm, it is my judgment that the industry has successfully insulated itself from the crisis that would force it to accept the new paradigm. The willingness to accept a low exploratory success rate and the concomitant high finding costs associated with it stems from the mistaken conviction that exploration is essentially an art and only incidentally a science. The mental image keyed to the prevailing paradigm is an utter disbelief that we can ever hope to achieve a 60 percent or better new field wildcat success rate. When a prospect is based solely upon the prevailing paradigm, too few operators would be dissuaded from drilling the prospect merely because it may be condemned under the new paradigm. Kuhn writes, "Though they may begin to lose faith and then consider alternatives, they do not renounce the paradigm that has led them into crises" (Kuhn, p. 77).

What we believe and how we react to that belief

creates the statistics that we mistake for reality. The prevailing paradigm, together with the widespread unwillingness to change, has provided the industry with formidable crisis-avoiding insulation. Tax benefits, cartel pricing, and the willingness of investors, top management, and stockholders to accept a high percentage of exploration failures and high finding costs have the effect of perpetuating the prevailing paradigm.

Though it is seldom conceived of in those terms, the belief that a new paradigm is impossible is enhanced by the conscious and even subconscious perception of loss that would be suffered by those companies and individuals who are secure economically, professionally, and intellectually in the prevailing paradigm. The persistent belief in the prevailing paradigm even after it has been found wanting effectively prohibits the new paradigm, or exploration methods based upon it, to be adequately tested, or (as we have seen) even adequately counted.

It is not improbable that low molecular surface hydrocarbon geochemistry can vastly improve our anticipation of new field wildcat dry holes. Assuming proper interpretation, which can by no means be taken for granted, surface hydrocarbon geochemistry is much better at determining where not to drill than where to drill. Positive determination is weakened because of the method's inability to determine depth and the possibility that production may exist below an exploratory well's total depth. Cosden Petroleum Company, after finding five or six seismic structures on the Parkey Ranch in Archer County, Texas, performed a surface hydrocarbon geochemical survey over the ranch prior to drilling. Only one of the seismic structures was *not* condemned by hydrocarbon geochemistry. Of the five or six seismic structures actually drilled, the single discovery was in the one structure that was also a hydrocarbon geochemical anomaly (Davidson, 1967).

There are aspects of compatibility between the two paradigms, and these should be stressed in application. Compatible aspects of the prevailing paradigm and the new paradigm provide added dimension to each other. For example, an induced polarization anomaly on a seismic high is a better prospect than either one of them alone.

Based as it is on the central role of hydrogeology, deep-water discharge, the accommodation by molecular solution or by colloidal suspension of organic derivatives in water, and the tracking of vertically rising water by any number of different methods, the new paradigm is totally incompatible with the concept of primary migration of hydrocarbons in bulk-phase as oil and gas from a source rock to a reservoir. A serious reading of Price (1976, 1981a, 1981b), Jones (1980), and Roberts (1980) renders totally extraneous the entire bulk-phase migration concept. The longer we are hung up on a bulk-phase, limited distance primary migration dependency, the longer we will be immobilized by the prevailing paradigm, and the longer we will be unable to pursue aggressively vertical migration exploration methods.

The structure and stratigraphy aspects of the prevailing paradigm are compatible with the new paradigm, although they play a much less critical and, in part, even a different role in the design and function of oil and gas traps (Roberts, 1980).

In the transition to the new paradigm, the greatest emphasis should be placed on using the full range of structural and stratigraphic tools along with the vertical migration exploration methods. Farm out or, better yet, sell the leases where a great seismic high possesses nothing but background on surface hydrocarbon geochemistry. Let those who would embrace the prevailing paradigm pay the cost in dry holes.

It is an expensive error to believe that science is merely cumulative and additive and that all we have to do is to pile knowledge on top of knowledge. Like the phlogiston theory (that fire is a released substance) and like Ptolemaic astronomy, paradigms or conceptual frameworks can outlive their usefulness and impede scientific achievement.

It would be totally wrong to claim that commercial hydrocarbons are ubiquitous. By relaxing the temperature and "proximity to source" requirements of the prevailing paradigm, however, the new paradigm does allow for a greater sense of ubiquity than the prevailing one.

If this is true for oil, it is even more true for natural gas. A natural-gas-based economy over the next fifty or sixty years is a distinct and marked probability. Public and private energy policies should begin to reflect that view.

The real cost of continuing to embrace bulk-phase, limited distance primary migration is to render impotent the new paradigm and to sustain the self-fulfilling prophecy of Dr. Nehring and his scientific advisers that meaningful petroleum exploration technology has reached the end of the line.

REFERENCES

Culberson, O.L., and McKetta, J.J., Jr., 1951, Phase equilibria in HC-water systems III. The solubility of methane in water at pressures to 10,000 psia: Petroleum Transactions, American Institute of Mining and Metallurgical Engineers, v. 192, p. 223-226.

Curry, William H. III, 1984, Evaluation of surface gamma radiation surveys for pertoleum exploration in the Deep Powder River Basin, Wyoming, pp. 25-39, this volume.

Davidson, Martin J., 1963, Geochemistry can help find oil if properly used: World Oil (July), p. 94-106.

__________, 1967, Petroleum geochemical exploration from space: a radical new technology: Houston, National Aeronautics and Space Administration, Manned Spacecraft Center, p. 1-74.

__________, 1982, Toward a general theory of vertical migration: Oil and Gas Journal (June 21), p. 288-300.

Duchscherer, William, Jr., 1982, Geochemical exploration for hydrocarbons: Oil and Gas Journal (July 5), p. 163-176.

Jones, Paul H., 1980, Role of geopressure in the hydrocarbon and water system, *in* Roberts, W.H. III, and Cordell, Robert J., eds., Problems of petroleum migration: American Association of Petroleum Geologists Studies in Geology, no. 10, p. 207-216.

__________, 1981, Hydrogeology of petroleum and natural gas, *in* Gottlieb, Benjamin M., ed., Unconventional methods in exploration for petroleum and natural gas II, Dallas: Southern Methodist University Press, p. 50-62.

Kuhn, Thomas S., 1962, The structure of scientific revolutions: Chicago, University of Chicago Press (pages in text refer to 2nd edition, 1972).

Munn, M.J., 1909, The anticlinal and hydraulic theories of oil and gas accumulation: Economic Geology, v. 4, p. 509-529.

Nehring, Richard, 1981, The discovery of significant oil and gas fields in the United States: Santa Monica, California, The Rand Corporation.

Price, L.C., 1976, Aqueous solubility of petroleum as applied to its origin and primary migration: American Association of Petroleum Geologists Bulletin, v. 60, p. 213-244.

__________, 1980a, Utilization and documentation of vertical oil migration in deep basins: Journal of Petroleum Geology, v. 2, no. 4, p. 353-387.

__________, 1980b, Shelf and shallow basin oil as related to hot-deep origin of petroleum: Journal of Petroleum Geology, v. 3, no. 1, p. 91-116.

__________, 1981a, Primary petroleum migration by molecular solution: consideration of new data: Journal of Petroleum Geology, v. 4, no. 1, p. 89-101.

__________, 1981b, Aqueous solubility of crude oil to 400° C and 2,000 bars pressure in the presence of gas: Journal of Petroleum Geology, v. 4, no. 2, p. 195-223.

Rich, J.L., 1921, Moving underground water as a primary cause of the migration and accumulation of oil and gas: Economic Geology, v. 16, no. 6, p. 347-371.

Roberts, W.H. III, 1980, Design and function of oil and gas traps, *in* Roberts and Cordell, eds., Problems of petroleum migration: American Association of Petroleum Geologists Studies in Geology, no. 10, p. 217-240.

__________, 1981, Some uses of temperature data in petroleum exploration, *in* Gottlieb, Benjamin M., ed., Unconventional methods in exploration for petroleum and natural gas II, Dallas: Southern Methodist University Press, p. 8-49.

__________, and Paul H. Jones, 1981, Research conference considers hot prospects: American Association of American Geologists Explorer, v. 3, no. 4, p. 24.

Weart, Richard C., and Heimberg, George, 1981, Exploration radiometrics: postsurvey drilling results; *in* Gottlieb, Benjamin M., ed., Unconventional methods in exploration for petroleum and natural gas II, Dallas: Southern Methodist University Press, p. 116-123.

THE FUTURE OF METHANE AS AN ENERGY SOURCE

Keynote Address

Philip H. Abelson

SCIENCE

For decades, natural gas was regarded by the oil industry as a nuisance. It was flared or sold at a trivial price. Later governmental regulations held down the price, and accordingly, there was little interest in exploring for the gas. But we have entered a new era in which methane will assume an increasingly important role in the nation's energy supplies. Both the National Petroleum Council and the American Gas Association estimate that very large amounts of gas remain to be discovered and tapped.

Methane is far more ubiquitous than petroleum, and its occurrence more widespread. For the most part, petroleum occurs in a limited window between 7,000 and 15,000 ft in depth. To create petroleum requires a special composition of source beds. These beds must contain organic matter that originally was rich in fats and other lipides. In contrast, the diagenesis of all organic matter gives rise to methane. For example, every coal seam has associated methane, a principal cause of coal mine explosions. Methane can withstand temperatures that destroy the hydrocarbons of petroleum. For example, exposure to a temperature of 400°F for a few million years cracks hydrocarbons into methane plus graphitic carbon. But appreciable amounts of methane are also formed at comparatively low temperatures. Thus, natural gas is being produced from the Devonian shales of the Appalachian Basin, where, in many instances, the burial is only a few thousand feet. The rate of production from a given well is small, but the wells are long-lived, and their cost is low.

Natural gas can easily be freed from impurities, and the product methane has chemical and physical properties that give it premium value in many applications. Natural gas of pipeline quality has practically no sulfur content. In contrast to other fuels, therefore, it is clean-burning and comparatively nonpolluting. The combustion products of natural gas make minimal contribution to acid rain, a fact that causes environmentalists to regard it favorably. Of all the fuels, methane produces the most energy per unit of carbon dioxide released. In the event that the CO_2 problem looms large, methane will have an advantage over other fuels, especially coal. Adding to the value of methane as a fuel is the development of new efficient burners that convert 95 percent of the fuel value into heat. This figure is in contrast to 60 to 65 percent for older type burners. Because the stack gas is relatively noncorrosive, all the heat can be extracted. In contrast, gases derived from sulfur-containing fuels cannot be cooled below the dew point.

As a heat generator, methane has a premium value in those industrial processes in which precise temperature control is important. The strength of the flame is highly controllable and can be almost instantly adjusted. Another advantageous role for methane as a fuel is its use in gas turbines that are employed in peak-shaving of electrical demand. The turbines have a comparatively low capital cost.

Methane is the feedstock of choice for important petrochemicals, including ammonia and methyl alcohol. The gas serves as a source of hydrogen in hydrogenation processes. Methane has other advantages that foster a long-term role for it. There are a large number of consumers served by an existing distribution network. Most areas of the country are served by pipelines. When natural gas is discovered and produced, it can be quickly purified to pipeline quality, and the well can be connected to the existing network. Ultimately, there may be competition from syngas derived from coal. Very large investments are involved, however, and the elapsed time between authorization and completion tends to be of the order of ten years.

Thus far the advantages of methane have been emphasized. But there are price limitations. In a little over half of its consumption, the gas can command a premium price, that is, above $6 per million BTU or a thousand cu ft. But, in the remainder of its use, methane is in competition with other energy sources, such as coal and residual oil. In addition, economists and others have been surprised at the extent to which higher prices have led to conservation of both oil products and natural gas.

As a member of the board of directors of the Washington Gas Light Company, I have noted some of the effects of fuel competition and conservation. Our company serves the Washington metropolitan area. Its customers are mainly residential, but it has some large users, including large buildings, apartment houses, and industrial plants.

By early 1981, prices of natural gas at the burner tip had about doubled. The fuel was still considerably cheaper than distillate or residual oil for heating, and it was also a much less expensive source of heat than electricity. Individual customers had already responded to the price increase, however, but cutting consumption by 15 percent. Fortunately, new customers took up the slack. By the end of 1981, unit costs of gas had increased, and the new year started with a particularly cold month. When customers received unprecedented high bills in early February 1982, they immediately reacted by cutting their consumption. Our experience is that part of this decrease in send-out will be restored later as construction of large office buildings continues. But the lesson should be clear. In those parts of the country where population and new construction is stagnant, higher prices will lead to lessened consumption.

The gas distribution companies face another limitation in competition from fuel oils. The residential market is secure for natural gas, but many commercial and industrial users have dual burners that give them the option of using residual oil. They can and do switch quickly when the price of heat favors one or the other fuel. Thus, the price of residual oil at the burner tip places a ceiling on what can be charged for natural gas. Henry Linden of the Gas Research Institute has emphasized this point. He has estimated that an annual 6 trillion cu ft of gas might be quickly lost to residual oil. That, of course, presupposes the availability of residual oil. The replacement of six trillion cu ft of gas would require about a billion barrels of oil a year beyond what is now used. Such a demand would impact the oil market. Beyond that, the longer term prospect for oil prices is that they will increase faster than inflation. Therefore, the competition of residual oil with natural gas is likely to be short-term rather than long-term.

For large-scale applications, such as base load electric power plants, coal is likely to replace both oil and natural gas, especially in areas close to mines. For example, in Texas the lignite belt is a source of fuel that costs in the neighborhood of only 70 cents per million BTU at the mine. It should be evident, therefore, that the long-term future for methane lies in emphasizing its premium value, rather than its large-scale use as a boiler fuel.

To maintain and increase their share of the market, gas producers and distributors need to promote expanded applications. Progress in increasing the efficiency of residential burners while cutting consumption temporarily will in the end enhance the total market. Expansion of use in refrigerators, air conditioning, and fuel cells seems likely. The big opportunity, however, is in the field of transportation. Methane has an octane rating of about 130. In an engine designed to use it, methane is a superior fuel. Efficiency is high, pollution is negligible. Today, gasoline costs about $10 per million BTU and the price is not likely to drop. Rather, it will increase. When they become available, synthetic liquids will cost considerably more than petroleum.

Use of methane as a motor fuel is not an impractical dream. Around the world, some 400,000 vehicles are already fueled by methane. Because the gas does not dilute motor oil as does gasoline, engine wear is reduced when methane is used. Methane can be used in existing automobile engines, but the efficiency is diminished somewhat. The cost of conversion, including pressure storage vessels designed for 3,000 psi, is about $1,500. In principle, a homeowner could have his own compressor and obtain the fuel from his home outlet. At the moment, the more feasible application is for fleet operation in cities. For example, a number of gas distributing companies, including Washington Gas Light, operate fleets of trucks fueled by methane. To expand the use of methane as a motor fuel will require the development of engines especially designed for it. There will also need to be distribution stations. Stations are an essential, though simple, requirement. Indeed, for only modest added cost, each vehicle could have its own compressor.

One of the questions raised about methane-fueled vehicles is the matter of safety. Actually, a high-pressure cylinder containing methane presents less of a hazard in a collision than does a thin metal tank containing gasoline. In the event of breakage, gasoline is more dangerous than methane. To insure safety, the pressure vessels are thick-walled and heavy. However, strong, lighter vessels are being designed and built. Once the various components of the methane-energy system have been perfected, other applications may arise.

In summary, prospects for finding and developing much more natural gas are excellent. There is deep gas, gas in the tight sands, Devonian Shale, and coal seams. Substantial areas of the country have not been explored. A large amount of non-associated gas will be found, and some of it will be cheap to produce.

Methane has advantages of form and composition that give it premium value in many applications. When burned in boilers merely to make heat, however, it will have price competition from residual oil and coal. But prospects are good that new large-scale premium uses for methane as motor fuel, for example, will develop. In that event, methane will become the nation's number one energy source, and it will command an excellent price.

SIGNIFICANCE OF NEAR-SURFACE MAGNETIC ANOMALIES

Robert S. Foote

GEODATA INTERNATIONAL

INTRODUCTION

Donovan, Forgey, and Roberts presented the paper "Aeromagnetic Detection of Diagenetic Magnetite over Oil Fields" in the February 1979 Bulletin of the American Association of Petroleum Geologists (Donovan et al., 1979). They discussed the presence of unusual amounts of magnetic field intensity over the Cement Oil Field, Oklahoma, where they proposed a mechanism for creation of magnetic material by reduction of nonmagnetic hematite or hydrated iron oxides to magnetic magnetite. Results of their tests showed that magnetite amounts increased toward the surface from a depth of approximately 300 m over the anticline region of the Cement Field.

The purpose of this study is to examine a large assortment of previously gathered airborne, high sensitivity, proton precession magnetic field data for contributions of magnetic material in the upper approximate 300 m. Airborne magnetic data included in this study have been taken from a surface height of approximately 100 m, with a series of computer software being developed to remove and enhance the shorter wavelength data. The evolution of data reduction involved evaluation of different gradient-measuring techniques where the contributions to the residual magnetic field from basement material and from surface contamination were removed. This led to presentation techniques which provide the results to be shown. It is not the plan of this paper to describe the computer methods developed, but to show the significance of the results related to oil and gas areas and to untested areas.

RESULTS OF DATA ANALYSES OVER OIL AND GAS FIELD AREAS

Data presented have been gathered using an airborne magnetometer mounted as a "bird" in a Douglas DC-3S aircraft. Noise levels for all data are no worse than approximately ± 1 gamma. Data having higher noise levels than ± 1 gamma provide less certain results. Much lower noise level data are now obtainable and are highly desirable. The following sets of data covering eight oil field areas have been gathered over the past several years and have been reprocessed to provide the results indicated.

Cement Oil Field, Oklahoma (1975)

The Cement Oil Field was surveyed originally in 1975 by Geodata International, Inc., for the United States Geological Survey. These data tapes have been reprocessed to remove and process the gradient data, and to provide the contoured data of Figure 1. The contour interval is 0.2 gammas, and the flight line spacing is ¼ mi. The gradient data have had the basement and surface contamination effects removed on a flight line basis prior to contouring. The oil field area over the northern section of the field includes very few individual anomalies and is not identifiable. Over the anticlines (Donovan, 1974), giving the top of the Pennsylvanian formation, the intensity of the gradient data is very large. It is suspected that the most intense and prolific oil areas are in those anticline regions which show the highest gradient results.

In any contouring process, it is necessary to average data according to a predetermined grid size. The exactness and preciseness of anomaly location is then lost in this process of averaging if puddling of small anomalies is to be prevented. It has been determined that treatment of the anomalous data on a flight line basis is a much superior approach, since precise anomaly location is preserved. Figure 2 shows the same data presented by flight line location. Anomalies are weighted and relatively numbered to indicate their intensity, with the number 1 being the lowest, and the highest being over the anticlines. The flight line data are given as a function of geographic location and oil drilling data. The anomalous clustering periphery encompasses the area given by adjacent line-to-line reinforcement of anomalous magnetic material. This clustering periphery is believed to represent the "fossil track" of the microseepage path, as first stated by

Donovan et al., and believed to be closely associated with an oil or gas region below.

ELK CITY OIL FIELD, OKLAHOMA (1981)

The Elk City Oil Field is shown to be anomalous in Figure 3, which gives the contoured map of the gradient reduced data. Flight line spacing is .707 mi, which necessitates a larger grid size for the contouring. In all cases, the computer contouring process removes the preciseness of the flight line data location because of averaging, and contour mapping by such methods has been discontinued. Figure 4 presents the Elk City data by anomaly sizes along each flight line, with a clustering periphery being shown as the region of significance. Inclusion of the residual magnetic basement contour map is also given in Figure 4. The southwest edge of the field is nearly coincident with the clustering periphery, but anomalous data extend northeast beyond the drilling edge.

PATRICK DRAW OIL FIELD, WYOMING (1980)

The Patrick Draw Oil Field is one of the GEOSAT oil and gas field study areas. The GEOSAT program, sponsored by private industry and the U.S. government, was initially set forth to study the possible relationship between LANDSAT data and stratigraphically deposited oil and gas fields. Three specific oil and gas areas were selected, these being the Patrick Draw area, Wyoming; the Coyanosa Field area, Texas; and Lost River Field area, West Virginia. Geodata surveyed the Patrick Draw Field in support of the GEOSAT program by providing eU, dTh and K results and magnetic field data. In this study, data were reprocessed to give flight line-located anomalous gradient data. The Patrick Draw area shown in Figure 5 includes the Table Rock Field to the east, Desert Springs Field to the north, and West Desert Springs to the northwest. All fields indicate to be anomalous by the clustering periphery, with the gas cap and west-central part of the Patrick Draw Field not included in the anomalous periphery. Table Rock Field is highly anomalous with two untested intense regions appearing near the Table Rock Field.

COYANOSA FIELD, PECOS CO., TEXAS (1977)

A second GEOSAT stratigraphically trapped oil and gas area is the Coyanosa Field area. These data were taken by Geodata in 1977* and recently reprocessed to give the data of Figure 6. Highly anomalous regions are observed in the ½-mi spaced flight line data which show intense anomalous regions over the Coyanosa, Waha, and Worsham oil and gas field areas. Intense anomalous areas not drill-tested also exist. Gas production is mostly from the Ellenberger formation at about 19,000 ft, and oil mostly from the Delaware formation at 6,500 ft. The most intense anomalies appear over the Waha Field area.

ISOM SPRINGS OIL FIELD, MARSHAL CO., OKLAHOMA, 1980

The Isom Springs Oil Field is a fault-trapped Arkansas novaculite field north of Lake Texoma in Marshall County, Oklahoma, as shown in Figure 7. Anomalous clustering shows existing productive areas and a few untested areas.

BELL CREEK OIL FIELD, MONTANA, 1974

The Bell Creek Oil Field data were reprocessed to give the results shown in Figure 8 as a function of flight line location and geographic location, showing drilling history and the anomalous clustering periphery around most of the Bell Creek Oil Field, with other prospective anomalies also shown. A band of more intense magnetic gradient anomalies having values of 5, 6, 7, and 8 exist near the western edge of the Bell Creek Field and within the anomaly clustering periphery. The average intensity and number of anomalies in this Montana area are much reduced.

BUTLER/GREENWOOD KANSAS STRATIGRAPHIC OIL FIELD TREND (1979)

One of the most intense anomalous trends yet observed is shown in Figure 9 as a function of ¼-mi spaced flight line and geographic location, giving the periphery outline around the dense line-to-line clustering. Flight line anomalies in this trend not only cover the existing production, but, in the areas not shown, extend up to one mile from production, with the intensity increase believed to be caused by the presence in the near surface of

*Data tapes supplied by Exxon Production Research Laboratories.

Permian redbeds which provide an abundant source of hematites.

JAY FIELD AREA, FLORIDA (1974)

Data acquired over the Jay Field area in 1974 have been reprocessed and shown in Figure 10. This survey was on a ½-mi flight line spacing. Intense anomalies are seen within sections of the clustering periphery. Since the Jay Field is rich in H_2S gas, it was suspected that any hematite conversion would be to pyrite, thus giving no near-surface magnetic anomaly. However, highly anomalous magnetic gradient results exist comparable in intensity to the Oklahoma and Kansas data included.

SUMMARY

Much data over widely varying geologic environments have been reprocessed and correlated with area drilling history. Most oil and gas fields are shown to be anomalous, and they would have been discovered if drilled as a result of the use of anomalous near-surface magnetic field gradient data. The use of the flight line presentation of computer-weighted near-surface anomalous magnetic field gradient data gives the best correlation to known oil and gas field areas yet found by any technology known to the author. The potential for locating stratigraphic oil and gas fields in shallow off-shore continental shelf environments is highly intriguing. Frontier regions can be rapidly surveyed to develop optimum areas for seismic evaluation and in turn reduce the total exploration costs.

REFERENCES

Donovan, T.J., et al., 1979, Aeromagnetic detection of diagenetic magnetite over oil fields: American Association of Petroleum Geologists Bulletin, v. 63, p. 245-248.

Donovan, T.J., 1974, Petroleum microseepage at Cement, Oklahoma—evidence and mechanism: American Association of Petroleum Geologists Bulletin, v. 58, p. 429-446.

Geodata International, Inc.
7035 JOHN W. CARPENTER FREEWAY • DALLAS, TEXAS 75247
CEMENT OIL FIELD AREA

Fig. 1

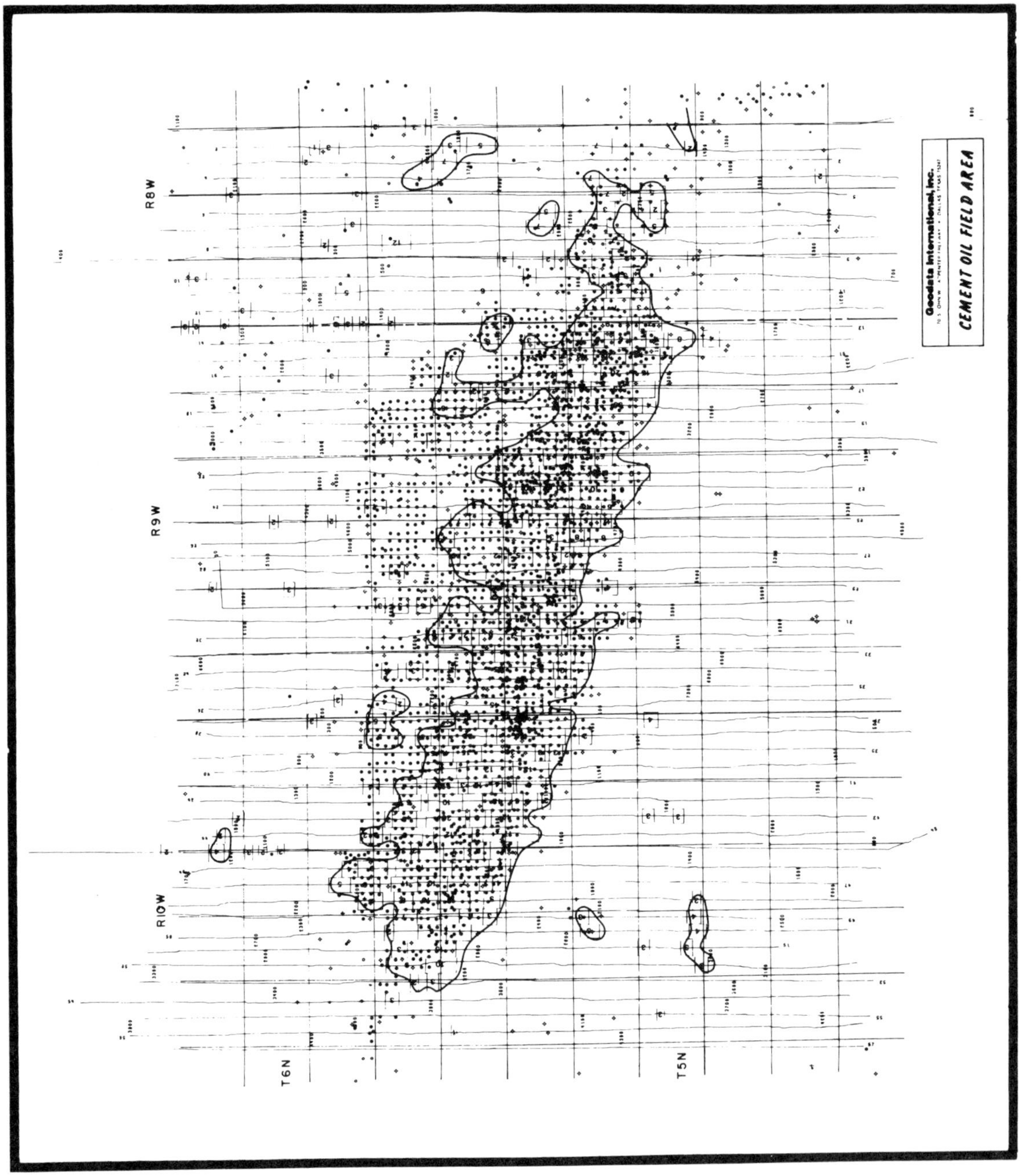
R8W
R9W
R10W
T6N
T5N
Geodata International, Inc.
CEMENT OIL FIELD AREA

Fig. 2

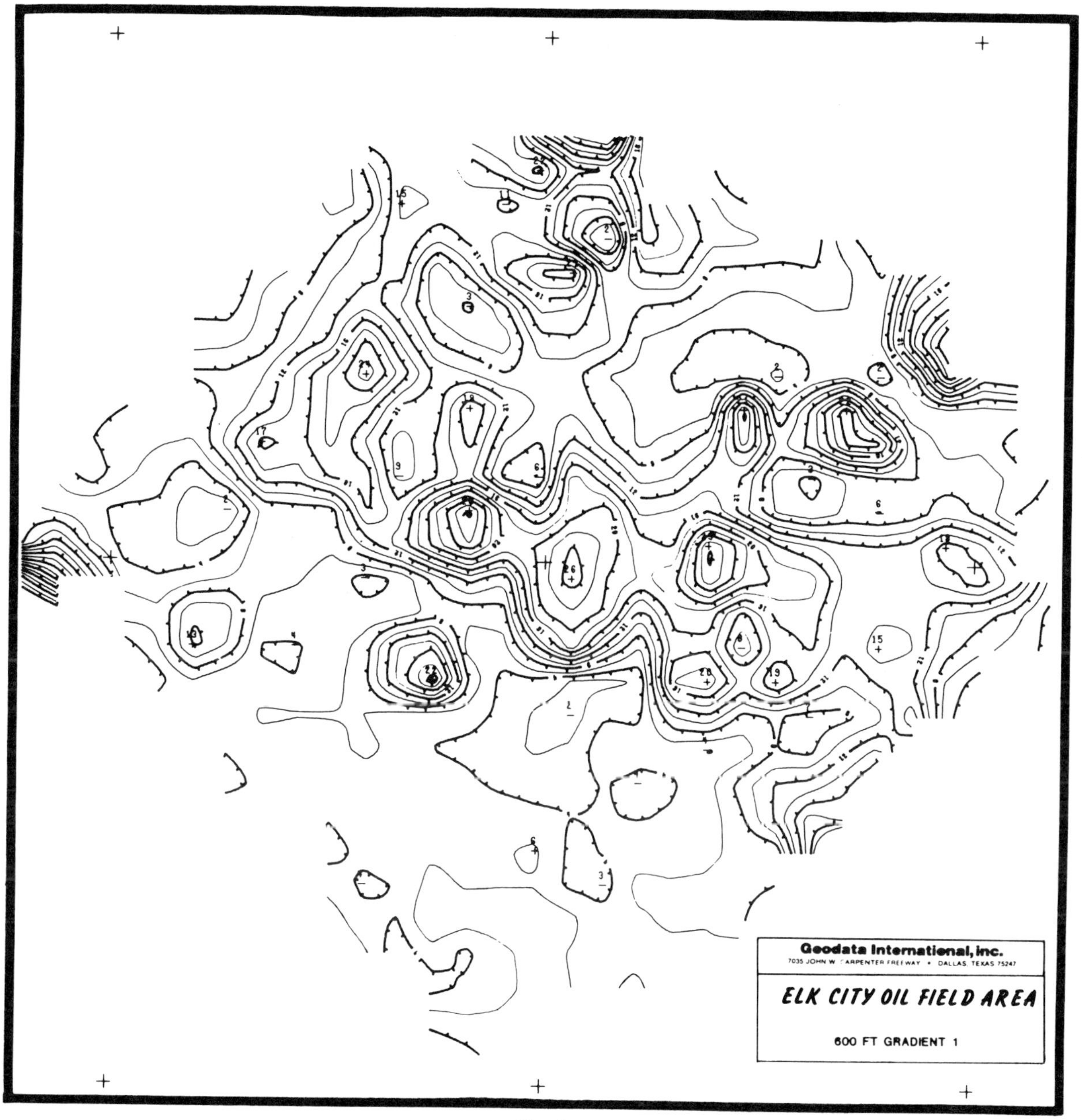
Geodata International, inc.
7035 JOHN W. CARPENTER FREEWAY • DALLAS, TEXAS 75247
ELK CITY OIL FIELD AREA
600 FT GRADIENT 1

Fig. 3

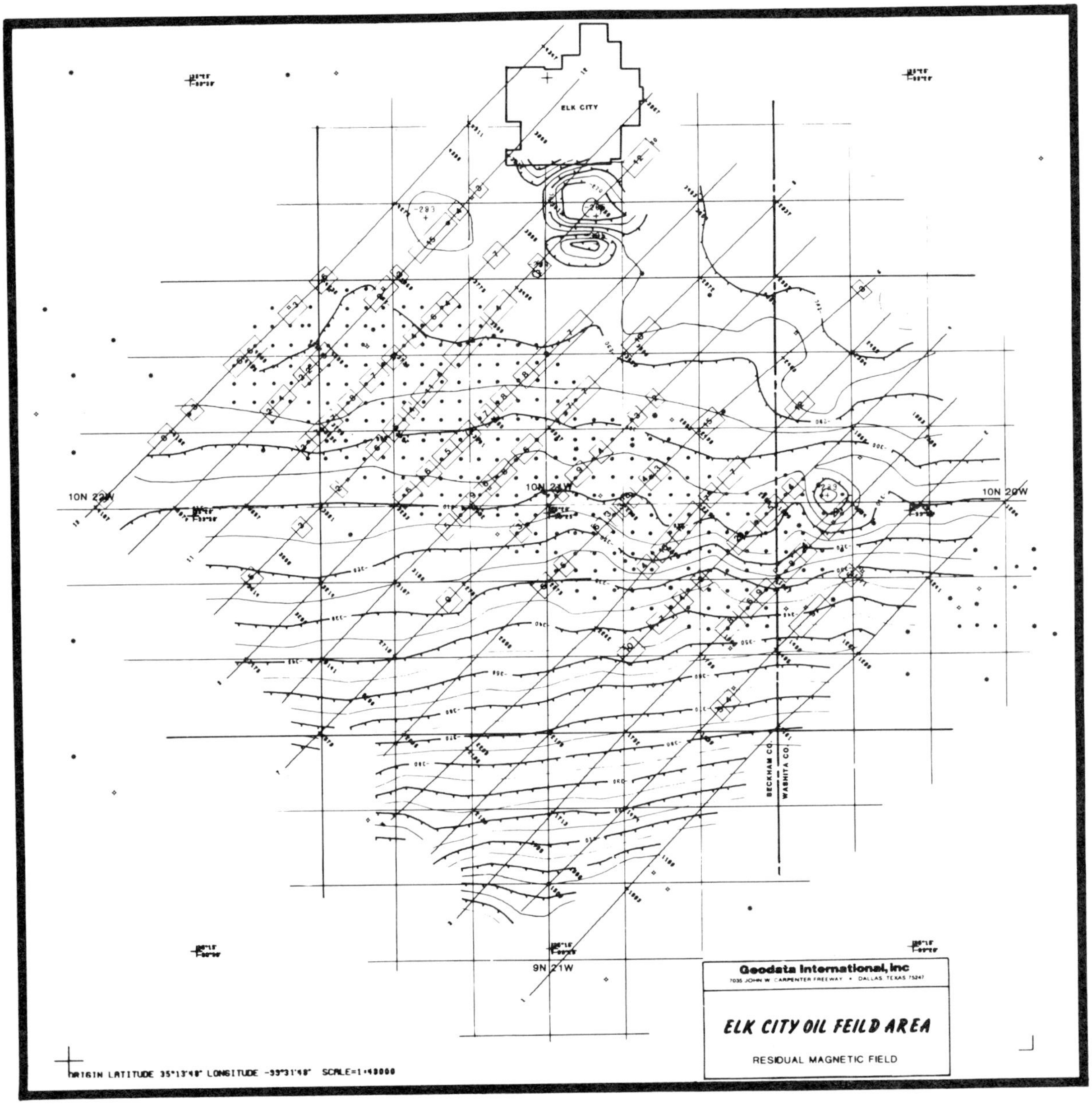
ELK CITY
10N 22W
10N 21W
10N 20W
9N 21W
BECKHAM CO.
WASHITA CO.
Geodata International, Inc
ELK CITY OIL FEILD AREA
RESIDUAL MAGNETIC FIELD

Fig. 4

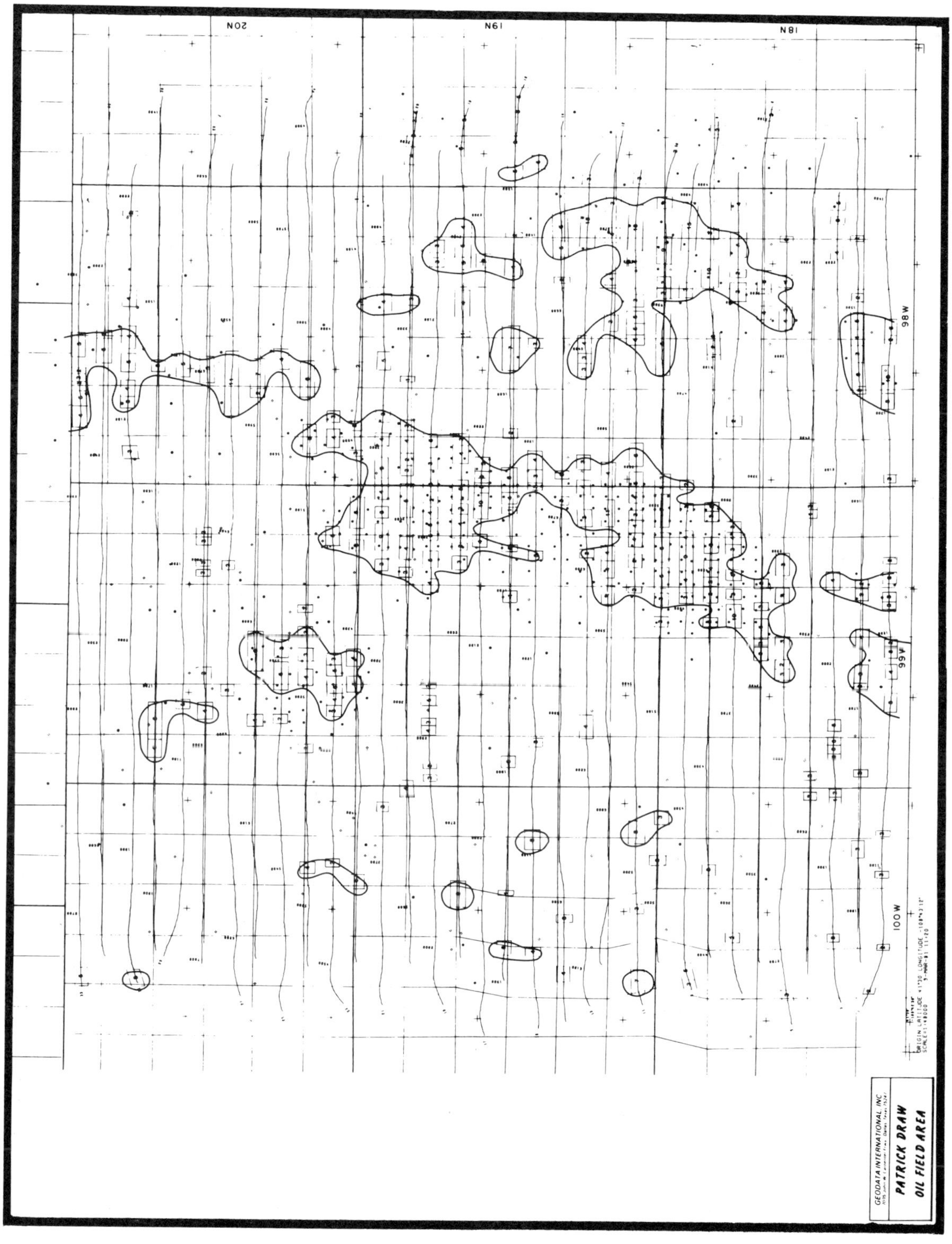

Fig. 5. Table Rock Field is 5 mi east of the Patrick Draw Oil Field.

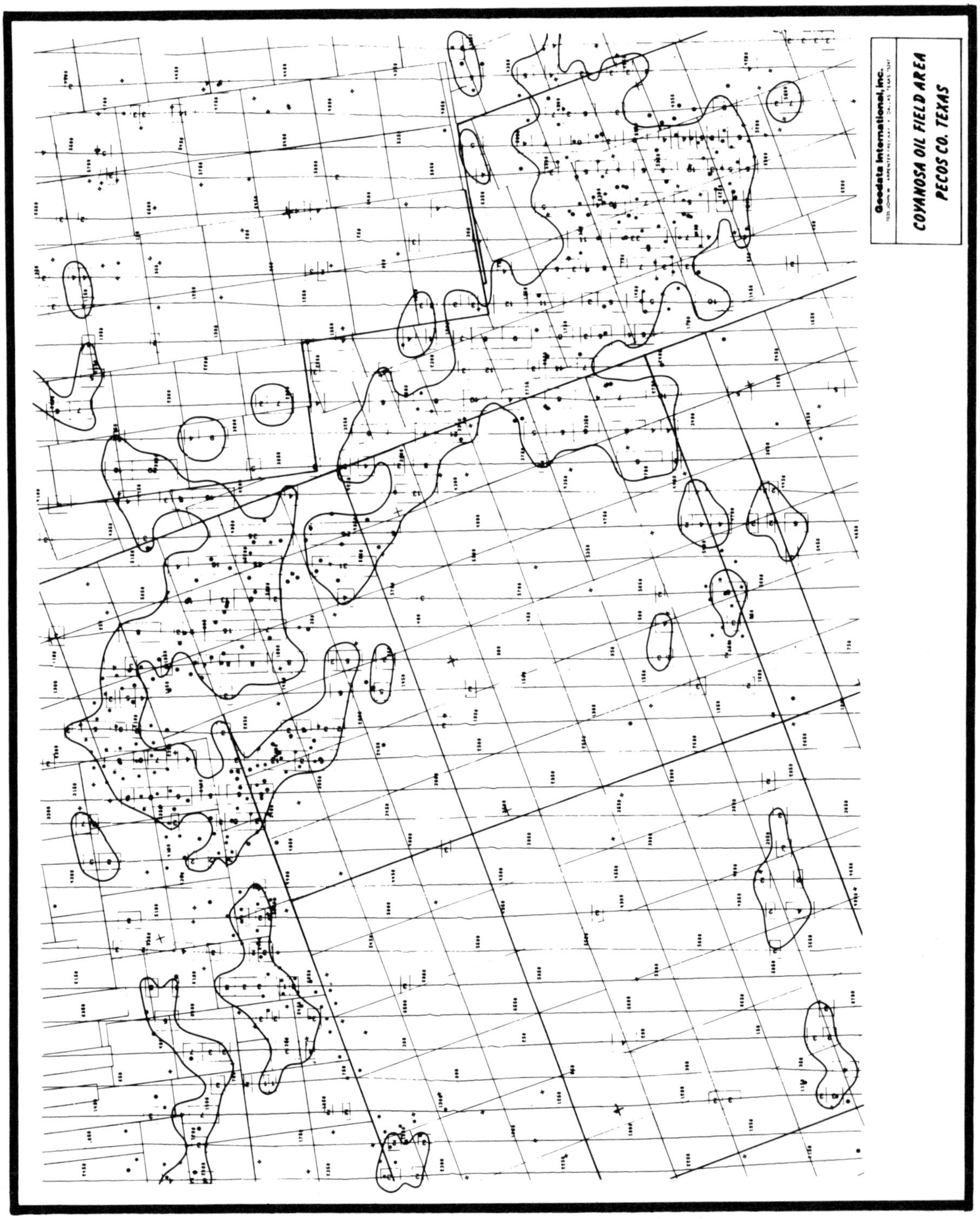
Geodata International, inc.
COYANOSA OIL FIELD AREA
PECOS CO. TEXAS

Fig. 6

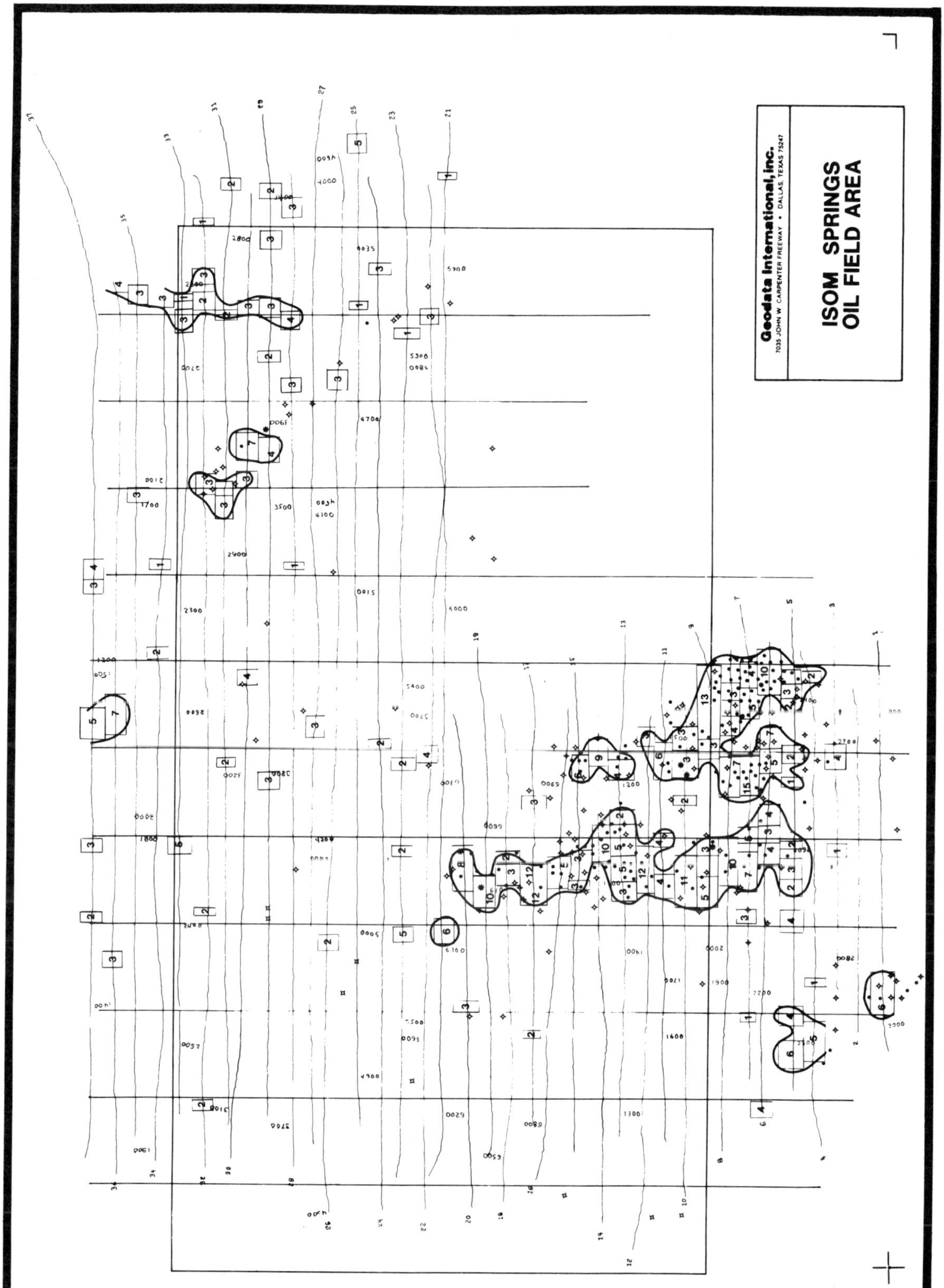
Geodata International, Inc.
7035 JOHN W CARPENTER FREEWAY • DALLAS, TEXAS 75247
ISOM SPRINGS
OIL FIELD AREA

Fig. 7

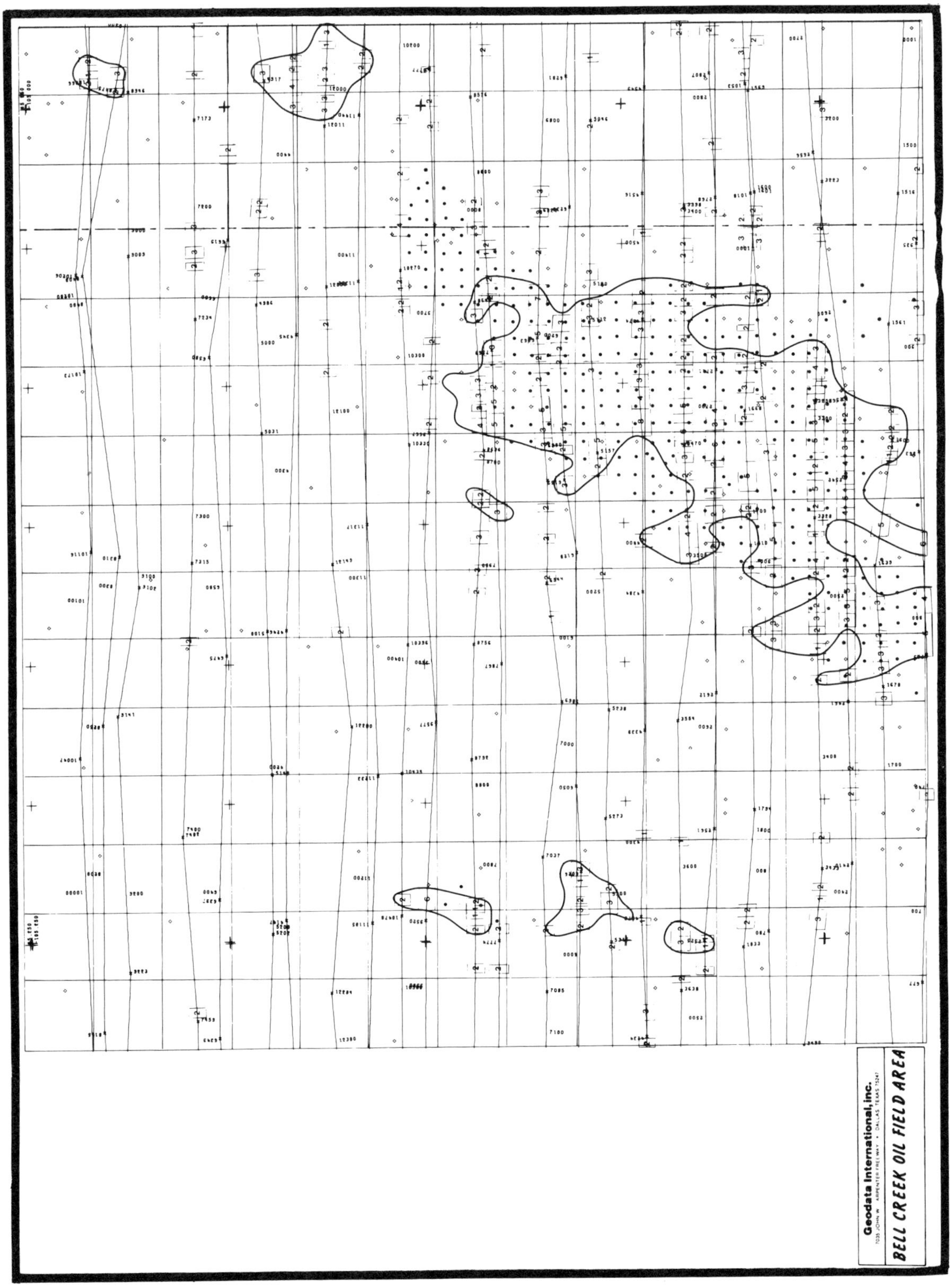

Fig. 8

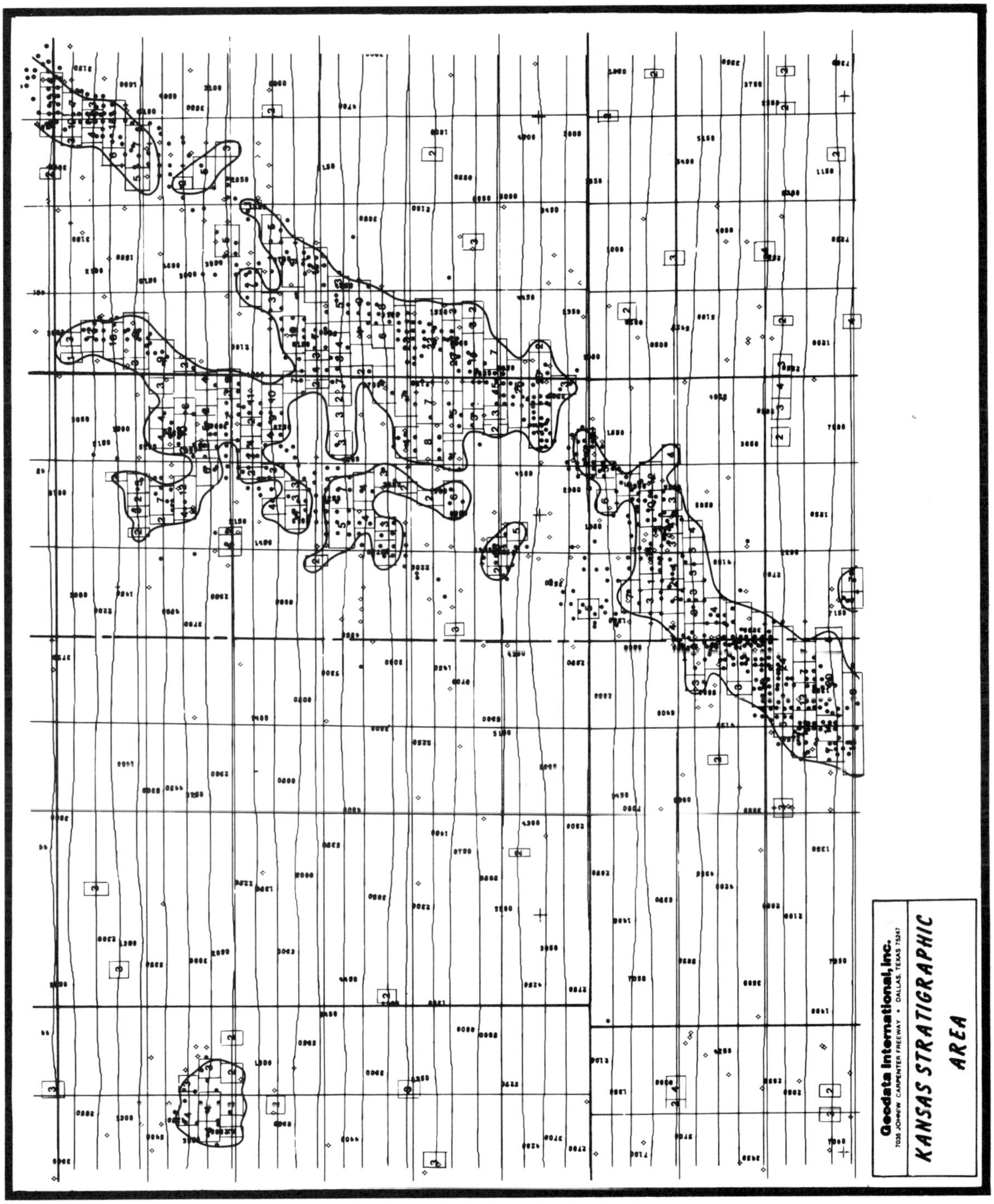

Fig. 9

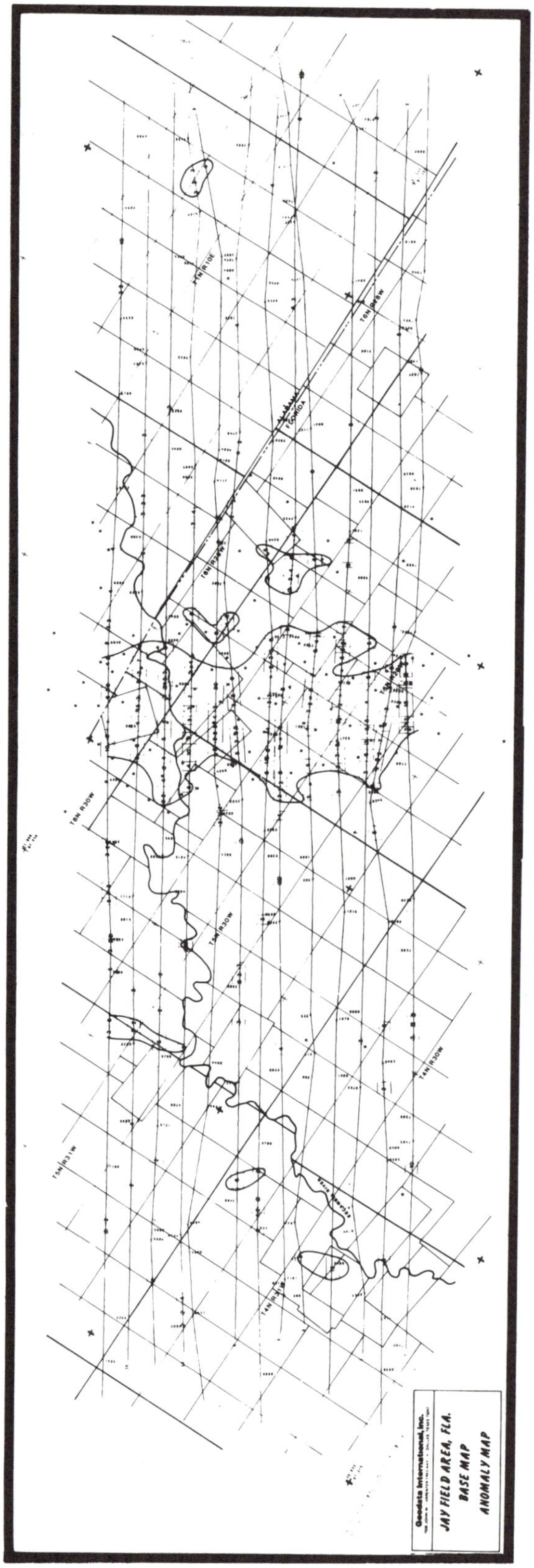
Geodata International, Inc.
JAY FIELD AREA, FLA.
BASE MAP
ANOMALY MAP
ALABAMA
FLORIDA

Fig. 10

EVALUATION OF SURFACE GAMMA RADIATION SURVEYS FOR PETROLEUM EXPLORATION IN THE DEEP POWDER RIVER BASIN, WYOMING

William H. Curry III

CONSULTING GEOLOGIST, CASPER, WYOMING

EXPLORATION PROBLEM

In 1973, an area of 4,465 sq mi in the deep part of the Powder River Basin of Wyoming was evaluated for its Cretaceous stratigraphic oil and gas potential (Fig. 1). At the start of the study, the paucity of well control needed to map prospective stratigraphic traps was immediately evident. Within the study area, only 6.38% of the sections (a section equals a 640-acre sq mi) had at least one well drilled to the marine Cretaceous rocks, and only 2.12% of the sections had been tested substantially through the Cretaceous rocks. With only two full Cretaceous penetration wells in every 100 sq mi, it was possible to map the generalized distribution of the different Cretaceous sandstone systems in the mainly marine shale sequence, but essentially impossible to locate specific stratigraphic trap prospects. Despite the paucity of test wells, 3.49% of the sections in the study area already had some Cretaceous oil and gas production, so oil and gas did appear to be relatively widespread. The problem was how to find "pure" stratigraphic traps with so little well control. The discovery of Frontier oil at Spearhead Ranch and Shannon oil at Heldt Draw during the early stages of the evaluation encouraged further consideration.

GAMMA RADIATION SURVEYS

In 1974, M.E. Douglas was retained to make a reconnaissance gamma radiation map of the study area. To make it as objective as possible, Douglas's work as an independent contractor was plotted and contoured on base maps lacking well control. The initial reconnaissance gamma radiation maps were then evaluated in December, 1974, in terms of the 282 wells that had been drilled to that date. The following statistical association between low gamma radiation anomalies and oil and gas production was indicated:

	Oil	Dry	Total	% correct
Wells drilled in middle of anomalies	36	19	55	65
Wells drilled inside edges of anomalies	21	17	38	55
Total wells drilled in anomalies	57	36	93	61
Wells drilled outside mapped anomalies	20	169	189	89
Total wells evaluated	77	205	282	80

In summary, the evaluation indicated that 61% of the wells drilled inside anomalies had been completed as oil and gas producers, and 89% of the wells drilled outside mapped anomalies had been plugged and abandoned. Accordingly, the 282 total wells were divided correctly into oil and gas producers inside anomalies and dry holes outside anomalies 80% of the time. Clearly, in the absence of adequate well control, the radiometric anomalies were welcomed as guides to exploration.

SPEARHEAD RANCH OIL FIELD

Radiometric surveys were completed at Spearhead Ranch Oil Field by August 20, 1974. Map of the status of drilling at that time (Fig. 2) shows that the main Frontier sandstone trap had been discovered and the northwest trend was evident. The radiometric map (Fig. 3) shows that the producers are located in several large radiometric anomalies. Actually, oil and gas in this anomaly is from three different sandstones—the Frontier, Muddy, and Dakota—and each of these stratigraphic traps has a different trend and geographic distribution. Therefore, the radiation anomaly reflects a complex of hydrocarbon pools, not one simple, single accumulation.

Drilling results in and around the anomaly (Fig. 4) have been:

	Oil	Dry	Total	% correct
Wells drilled in middle of anomalies	14	4	18	78
Wells drilled inside edges of anomalies	4	2	6	67
Total wells drilled in anomalies	18	6	24	75
Wells drilled outside mapped anomalies	2	7	9	78
Total wells evaluated	20	13	33	76

Several different gamma radiation instruments have been run along the same traverse across this field (Fig. 3):

1. Douglas's ionization chamber shows a distinct decrease of the gamma radiation intensity in the anomaly (Fig. 5), and his tool defines all anomalies evaluated in this report.

2. A Geometrics crystal instrument was used to traverse the same roads four times to record potassium, bismuth, thallium, and "total" radiation intensity. The potassium curve is similar to Douglas's curve and shows the production to be in the low anomaly. However, the east edges of the bismuth and thallium anomalies are shifted east from the potassium anomaly, and the west part of the potassium anomaly is high on both the bismuth and thallium curves. The "total" gamma radiation anomaly reflects all three curves, but the west edge of the "total" anomaly does not match Douglas's evaluation or the potassium curve.

3. A Mt. Sopris gamma ray detector was also compared, and its "total" gamma ray intensity curve is similar to the Geometrics "total" curve. However, the production is not in a radiometric anomaly based on the "total" gamma radiation because the high bismuth and thallium readings cancel the potassium low in the west part of the anomaly.

Although more comparisons should be made before overgeneralizing, this information tentatively suggests:

A. There appear to be several associated gamma radiation anomalies represented by the "total" anomaly, but each has somewhat different limits.

B. Douglas's instrument, evaluated in this report, seems to be most sensitive to potassium variations. The potassium anomaly is apparently caused by a decrease of K_{40} in the anomaly, and most potassium is in the illite or mixed layer clays. Therefore, it would be interesting to determine more exactly the changes in the composition of the clay minerals which cause the changes in gamma radiation intensity. A recent paper by R.E. Ferrell (1982) suggests that a rising circulation of deep, hot, formation water can and does cause an illite to Kaolinite conversion. Although Ferrell was not concerned with radiometric anomalies in his study, it seems possible that a similar vertical migration of deep, hot water might reduce K_{40} and illite in some radiometric anomalies.

At the present time, well control indicates that the Spearhead Ranch Oil Field is caused by a thin (2-6') northwest trending zone of upper Wall Creek (First Frontier) sandstone porosity (Fig. 6) at a depth of about 12,750 ft. The reservoir, with porosity over 20% and permeability over 50 md, is located in the coarsest grained uppermost and central part of an east sloping "shingle" sandstone (Fig. 7).

The 14 Frontier wells at Spearhead had average initial potential flow rates of 993 BOPD and 3231 mcgpd, many without stimulation (Fig. 6). The single Muddy well at Bear Creek had produced 463,252 BO and 2,811,788 mcf gas by December, 1980 (Fig. 2).

HELDT DRAW OIL FIELD

The Heldt Draw Oil Field was surveyed about October 1, 1974, when there were four Shannon oil producers in the field and six wells updip which lacked Shannon sandstone (Fig. 8). The radiometric map showed the four oil producers to be in one of several narrow northwest trending anomalies (Fig. 9). In general, the oil production is associated with the radiation anomaly except on the northwest end where the production gradually deflects updip from the anomaly and ends up entirely outside the anomaly in an area of well-documented, high radiation values. Despite this divergence between the stratigraphic oil accumulation and the surface radiation anomaly, drilling results in and around these anomalies (Fig. 10) have been:

	Oil	Dry	Total	% correct
Wells drilled in middle of anomalies	24	3	27	89
Wells drilled inside edges of anomalies	8	1	9	89
Total wells drilled in anomalies	32	4	36	89
Wells drilled outside mapped anomalies	11	11	22	50
Total wells evaluated	43	15	58	74

In a broader twelve-township area around this field, 205 wells were evaluated and 88% of the wells drilled inside anomalies were completed as producers and 62% of the wells outside the anomalies were dry for an overall accuracy of 79%.

The Shannon sandstone at Heldt Draw is a 1-mi-wide, 10-mi-long northwest trending "bar" sandstone at a depth of about 9,000 ft. The distribution of the sandstone reservoir is shown by the initial potential rates for oil (Fig. 11), contoured in barrels of oil per day. The reservoir is about 8 ft thick (Fig. 12) and has average porosity of about 13% and average permeability of about 20 md. Average initial production rates were 238 BOPD and 83 mcfgpd. By December, 1980, the field had produced 3,028,213 BO and 1,926,719 mcf gas.

SUBSEQUENT DRILLING AND EVALUATION

By the end of 1981, the industry had drilled marine Cretacious test wells in 21.68% of the sections in the study area; however, only 4.79% of the sections in the study area have been tested by at least one well through the Cretaceous.

As a result of the drilling effort from 1976 through 1980, oil and gas production was established in 272 sections of the study area where no previous production had existed. Currently, 428 sq mi or 9.59% of the sections in the study area have some oil and gas production within the section.

Currently, of the 535 total wells drilled inside the anomalies, 385 of the wells (72%) have been completed as oil and gas wells. However, this was not as objective an evaluation of the radiometric anomalies as the 61% success measured in 1974 because subsequent fields, which clearly demonstrated a NW-SE stratigraphic trend, influenced the 1981 contouring of the radiometric data.

Further, of the 221 mapped anomalies in the study area, 87 have been tested by at least one well, and 56 (64.4%) have produced some oil. Thirty-one of the tested anomalies lack production to date; however, virtually all of the producing anomalies also contain some dry holes.

This evaluation of the radiometric mapping of M.E. Douglas is similar to the evaluation of Douglas's work presented by Weart & Heimberg at Symposium II in which they evaluated 724 wells from a broader geographic setting and found that 85.7% of the wells drilled in anomalies were productive, 68.1% of the wells drilled outside anomalies were dry holes, and that the accuracy of the tool in predicting producers and dry holes was 79.7%.

REFERENCES

Ferrell, R.E., 1982, Clay mineral evidence for movement of high temperature subsurface fluids: Annual American Association of Petroleum Geologists Abstract, American Association of Petroleum Geologists Bulletin, v. 66, no. 5, p. 568.

Weart, R.C., and Heimberg, G., 1981, Exploration radiometrics: Postsurvey drilling results, *in* Benjamin M. Gottlieb, ed., Unconventional methods in exploration for petroleum and natural gas, II, Dallas, Southern Methodist University Press, p. 116-123.

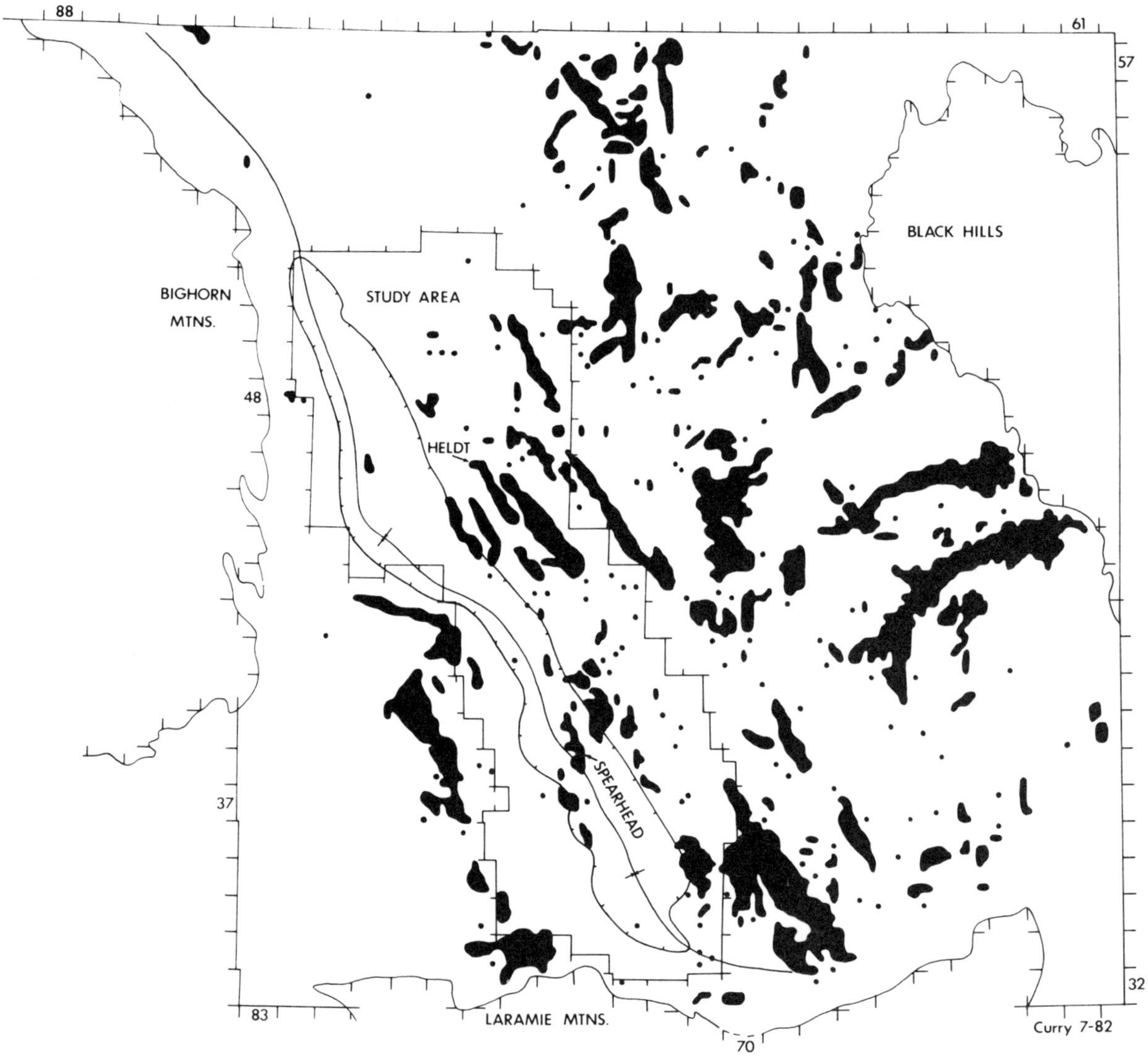

Fig. 1. Index map of the Powder River Basin of northeastern Wyoming showing the flanking mountain uplifts, the outline of the basin, deep basin axis, Cretaceous oil fields (mostly stratigraphic traps), outline of the study area, and locations of the two case studies at Spearhead Ranch and Heldt Draw.

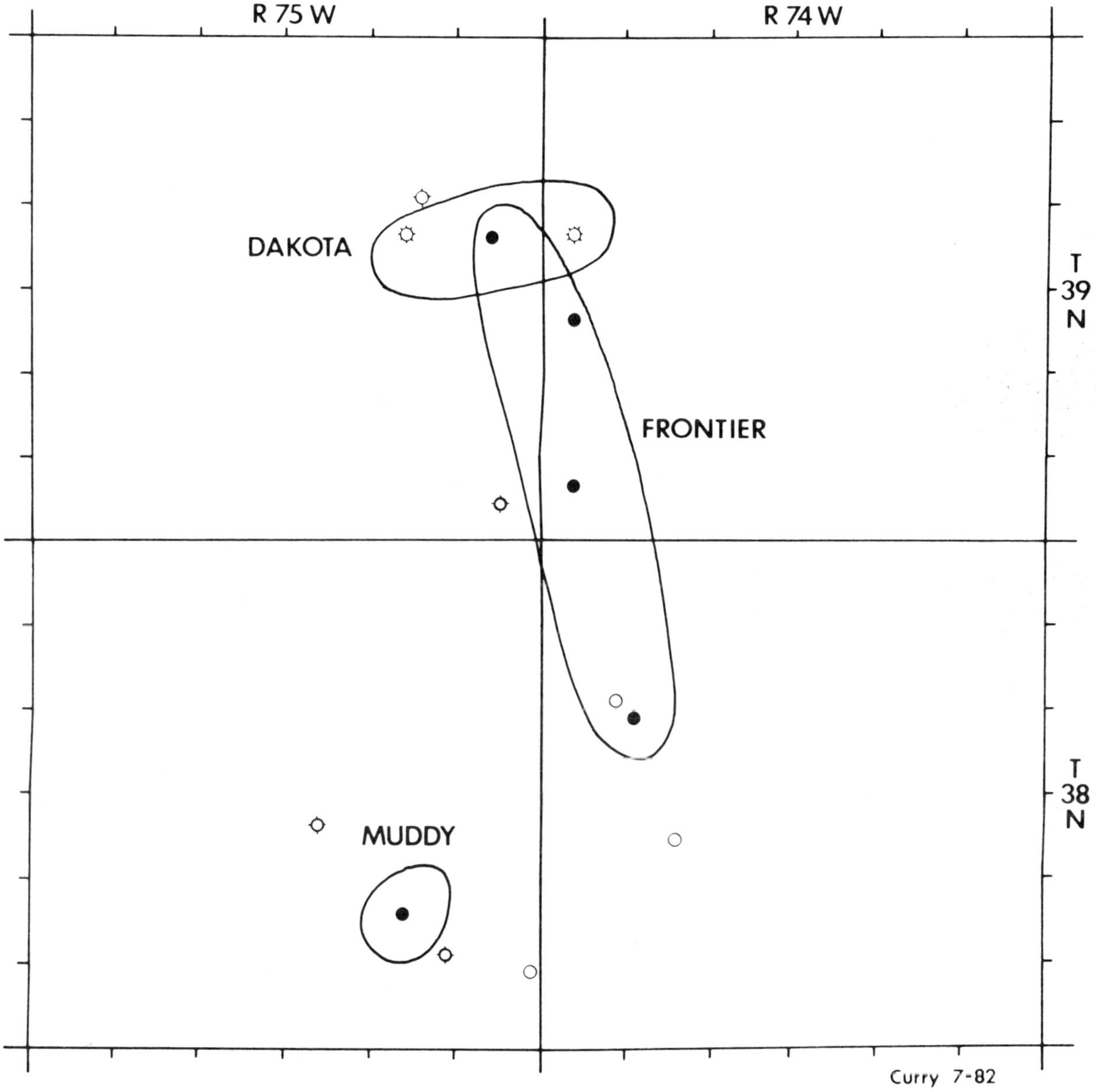

Fig. 2. Status of exploration around Spearhead Ranch Field by August 20, 1974, when gamma radiation surveys were completed. An old Muddy well at Bear Creek and two Dakota gas wells add complications to the northwest trending Frontier stratigraphic field.

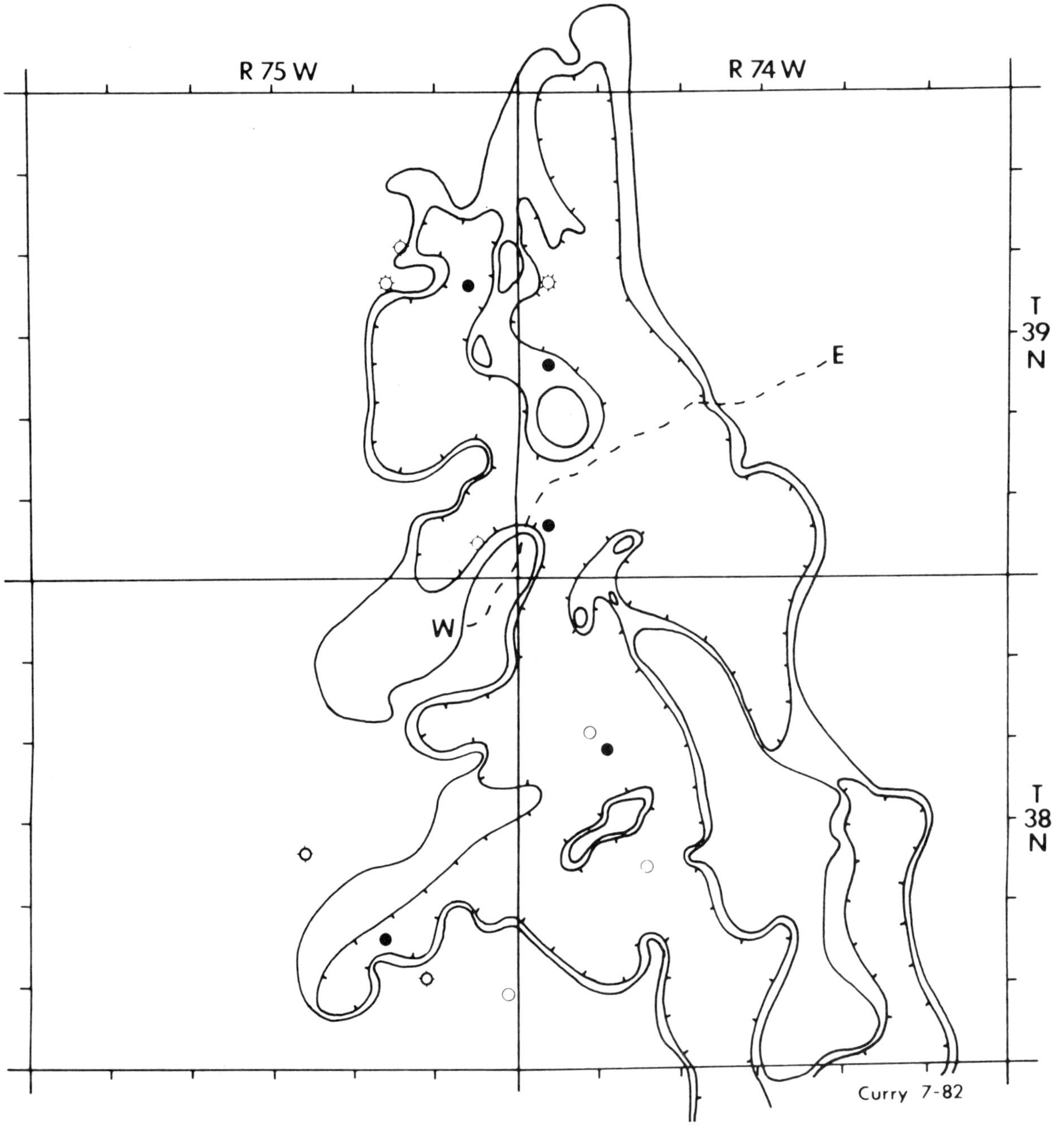

Fig. 3. Gamma radiation map completed August 20, 1974, showed the developing Frontier stratigraphic accumulation at Spearhead Ranch to be associated with several large NW trending radiometric anomalies. Also note the location of the multiple traverses of the anomaly, illustrated in Fig. 5.

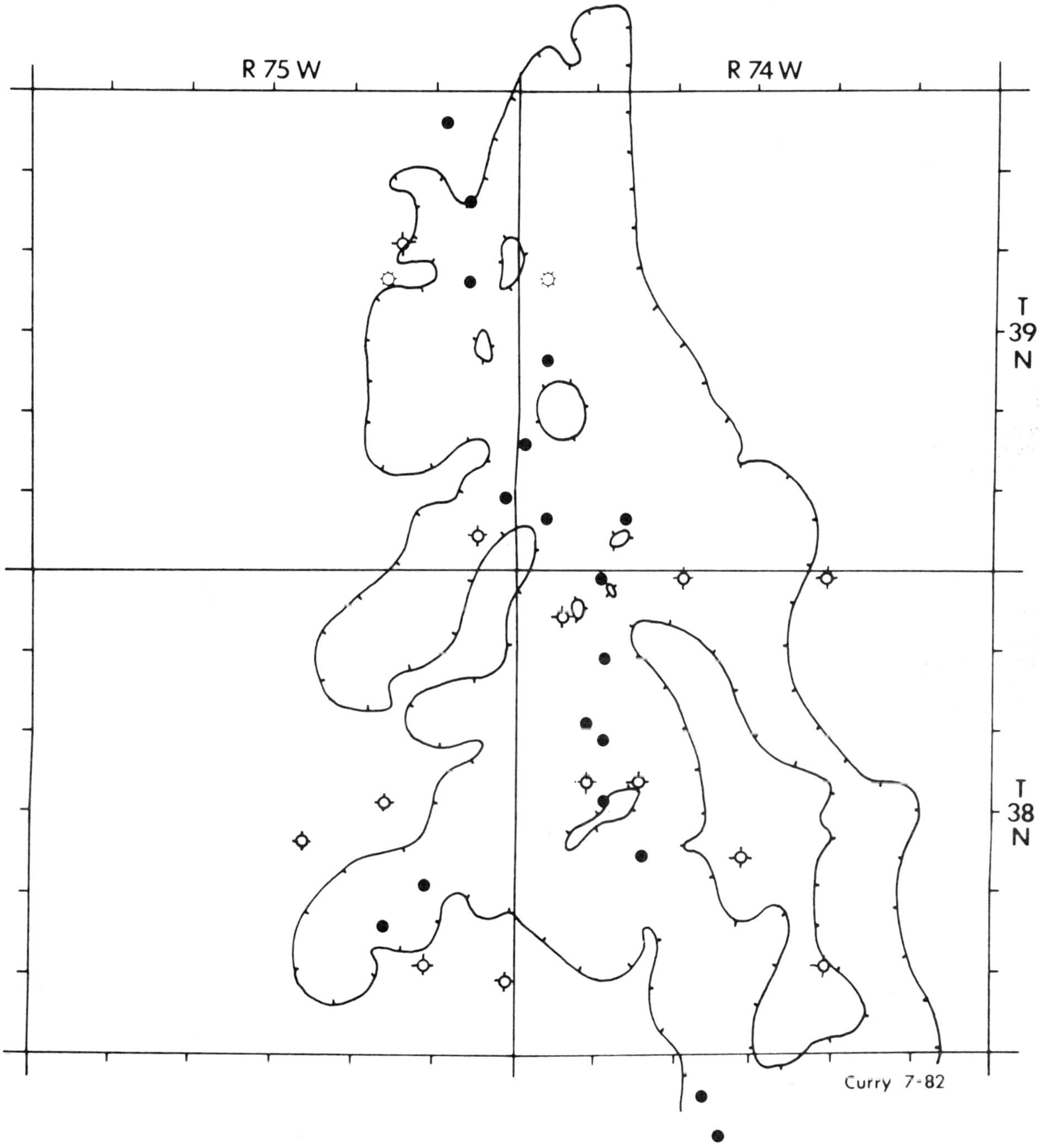

Fig. 4. Outline of gamma radiation anomaly at Spearhead Ranch showing that 75% of the wells drilled in the anomaly have been completed as oil and gas producers, and that the status of 82% of all wells in and around the anomaly were correctly predicted by the radiometric anomaly.

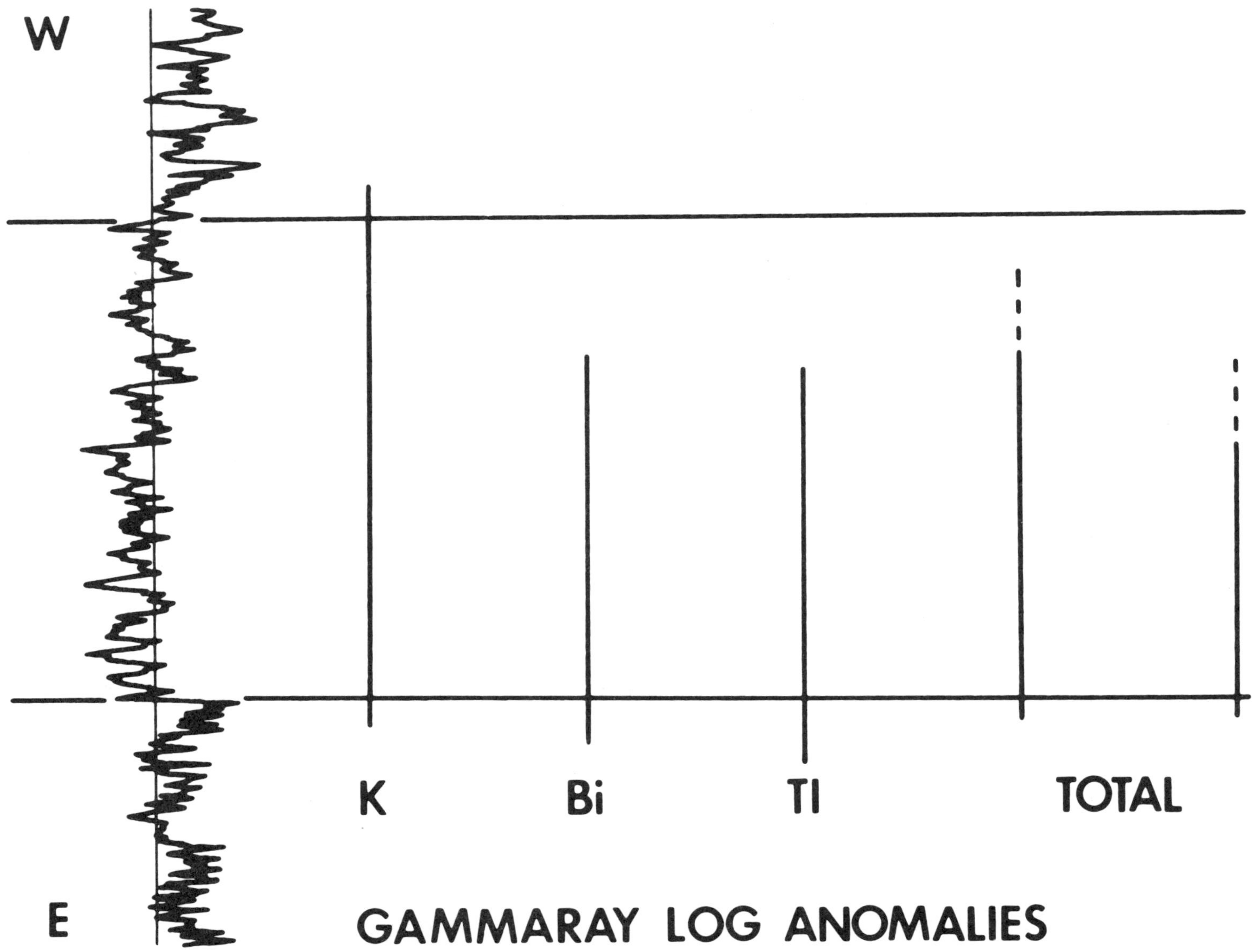

Fig. 5. Gamma ray cross section (see location Fig. 3) across the Spearhead Ranch by M. E. Douglas. Note that Douglas's record agrees with that recorded for potassium 40 with a Geometrics crystal, but that the thallium, bismuth, and "total" anomalies from the Geometrics crystal do not extend into the west part of the Douglas-potassium anomaly where production is presently located.

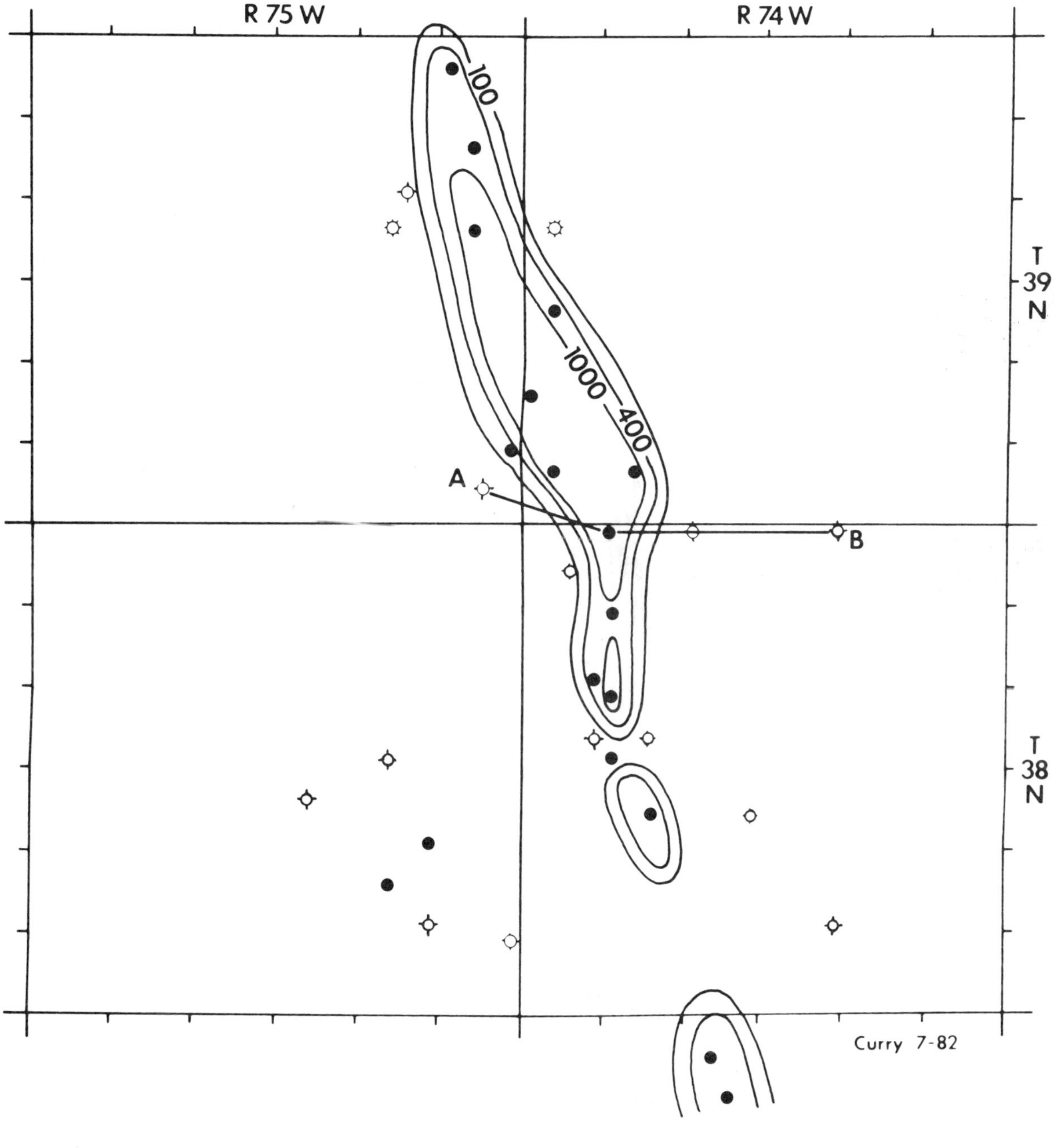

Fig. 6. The distribution of the thin (2-6 ft thick) uppermost Wall Creek (First Frontier) sandstone reservoir is shown from the initial well completion rates. Contours show initial potentials for oil of 100 BOPD, 400 BOPD, and 1,000 BOPD.

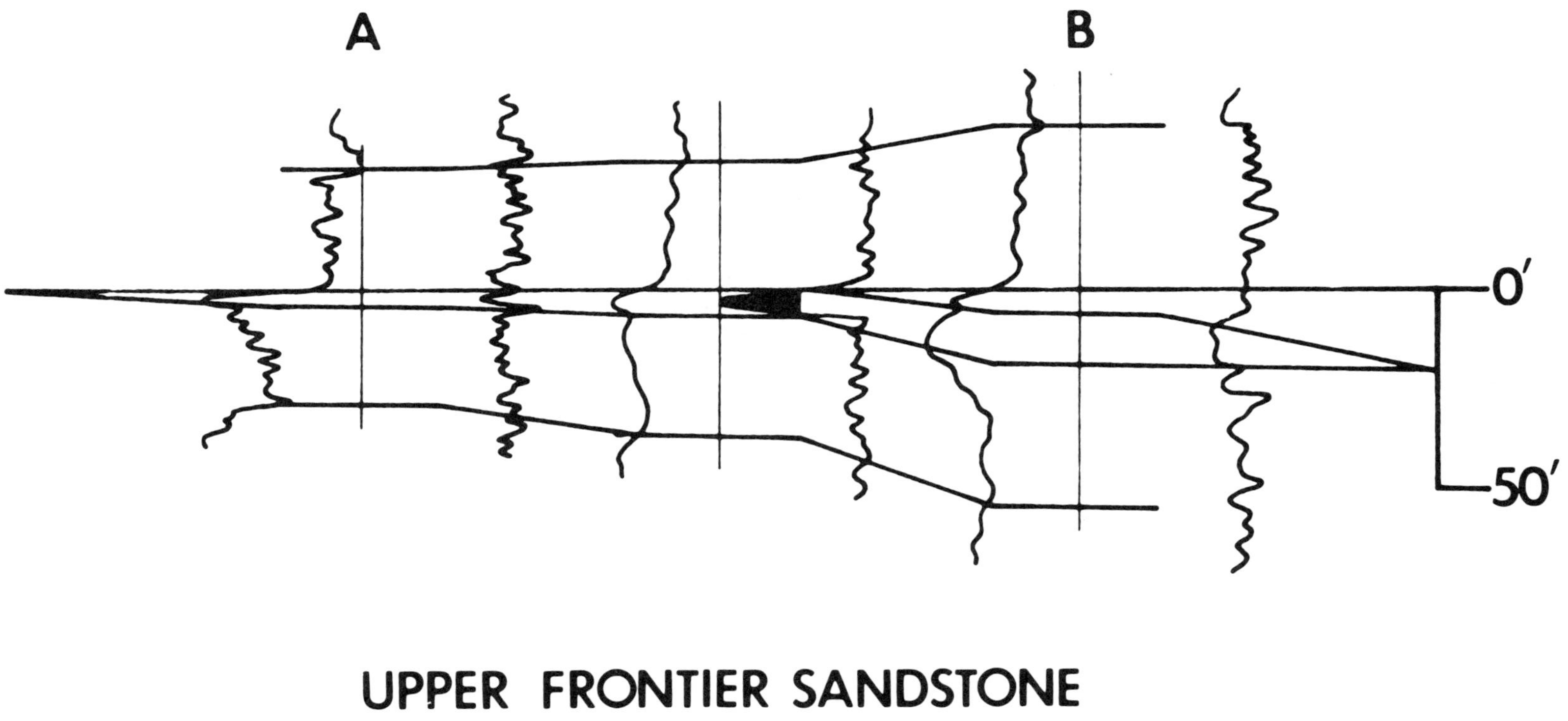

Fig. 7. Log cross section of upper Wall Creek (First Frontier) sandstone at Spearhead Ranch illustrating the east sloping depositional geometry of the "shingle" sandstone body. The reservoir is in the coarsest grained sandstone at the central top of the sandstone body.

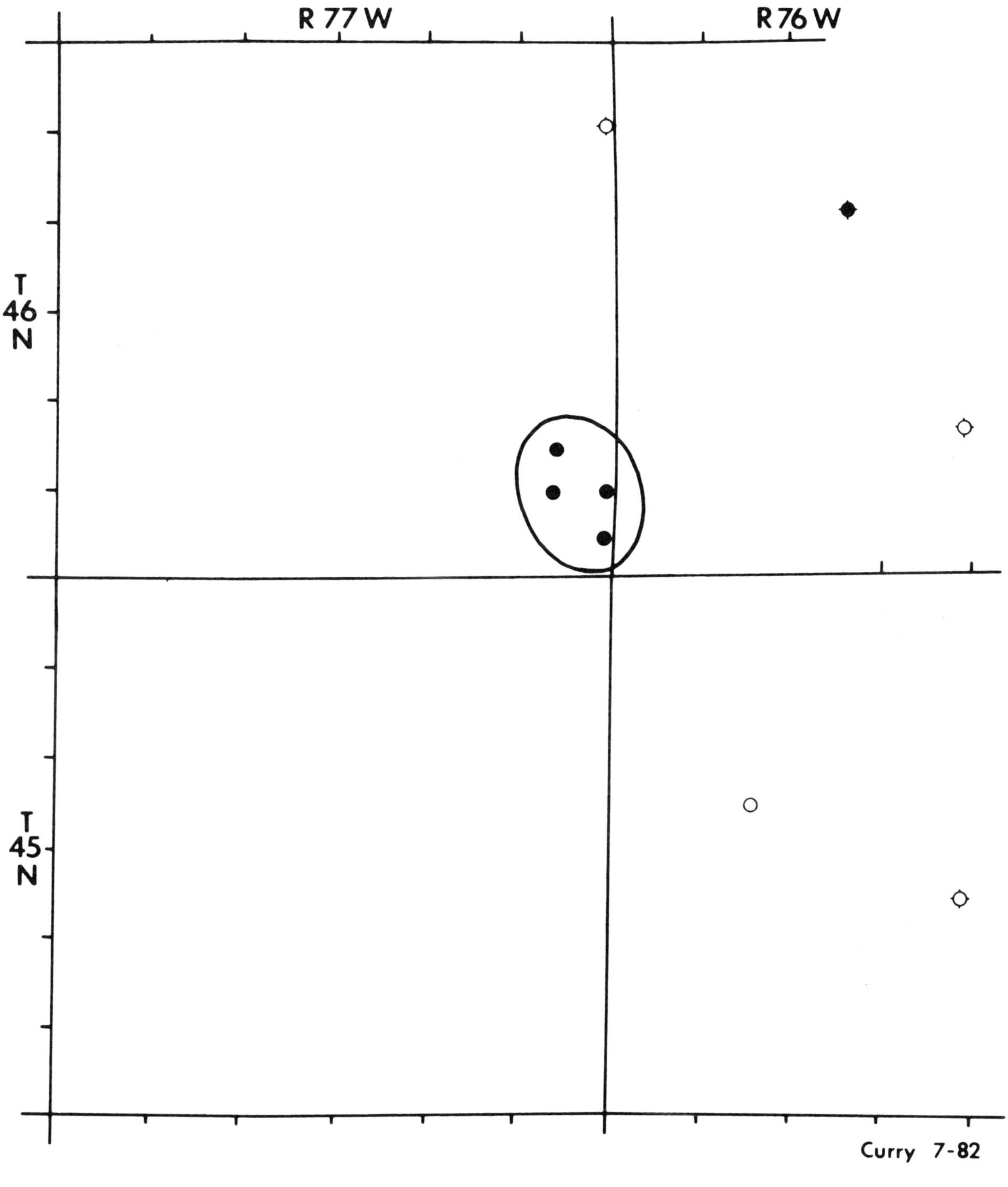

Fig. 8. Shannon oil production at Heldt Draw in October, 1974, when radiometric mapping was completed.

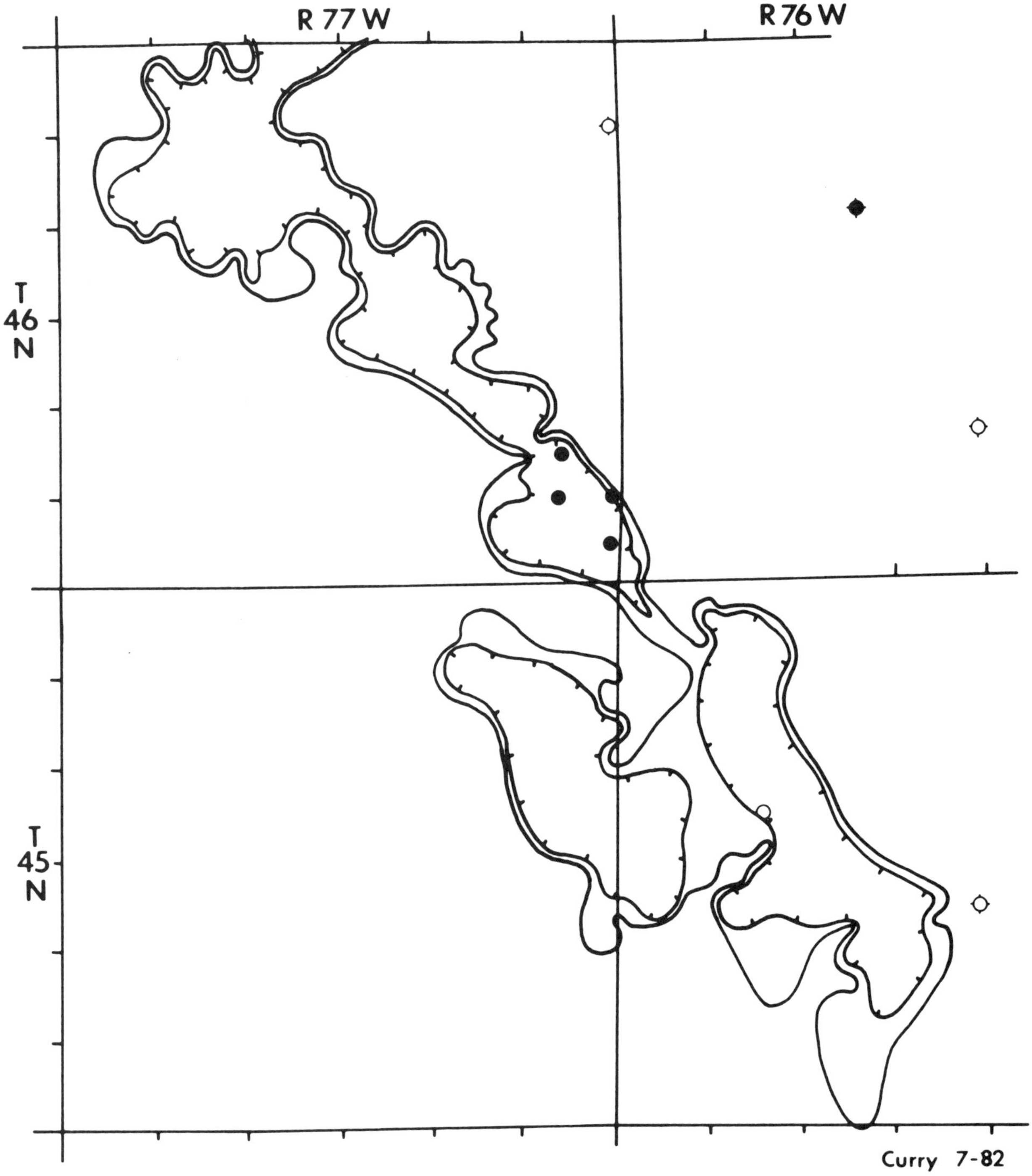

Fig. 9. Radiometric map of Heldt Draw area completed October, 1974, showing the four Shannon oil wells to be located in one of several northwest trending anomalies.

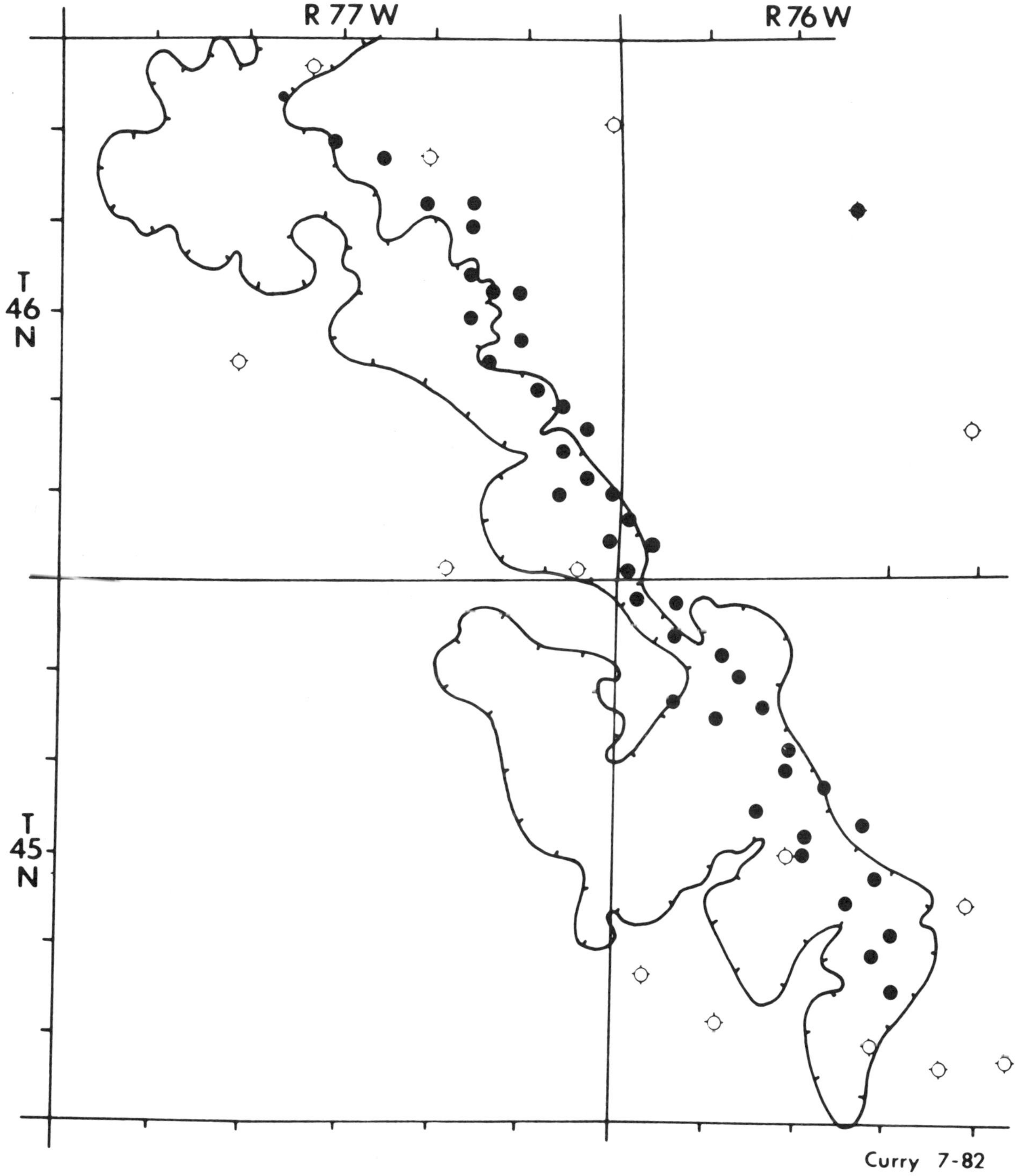

Fig. 10. Current status of drilling in and around the Heldt Draw radiometric anomaly. Despite the divergence of the field from the anomaly in the north part of the field, 89% of the wells drilled in the anomaly have been completed as oil and gas producers, but only 50% of the wells drilled outside the anomaly have been dry, for an overall accuracy of prediction of 74%.

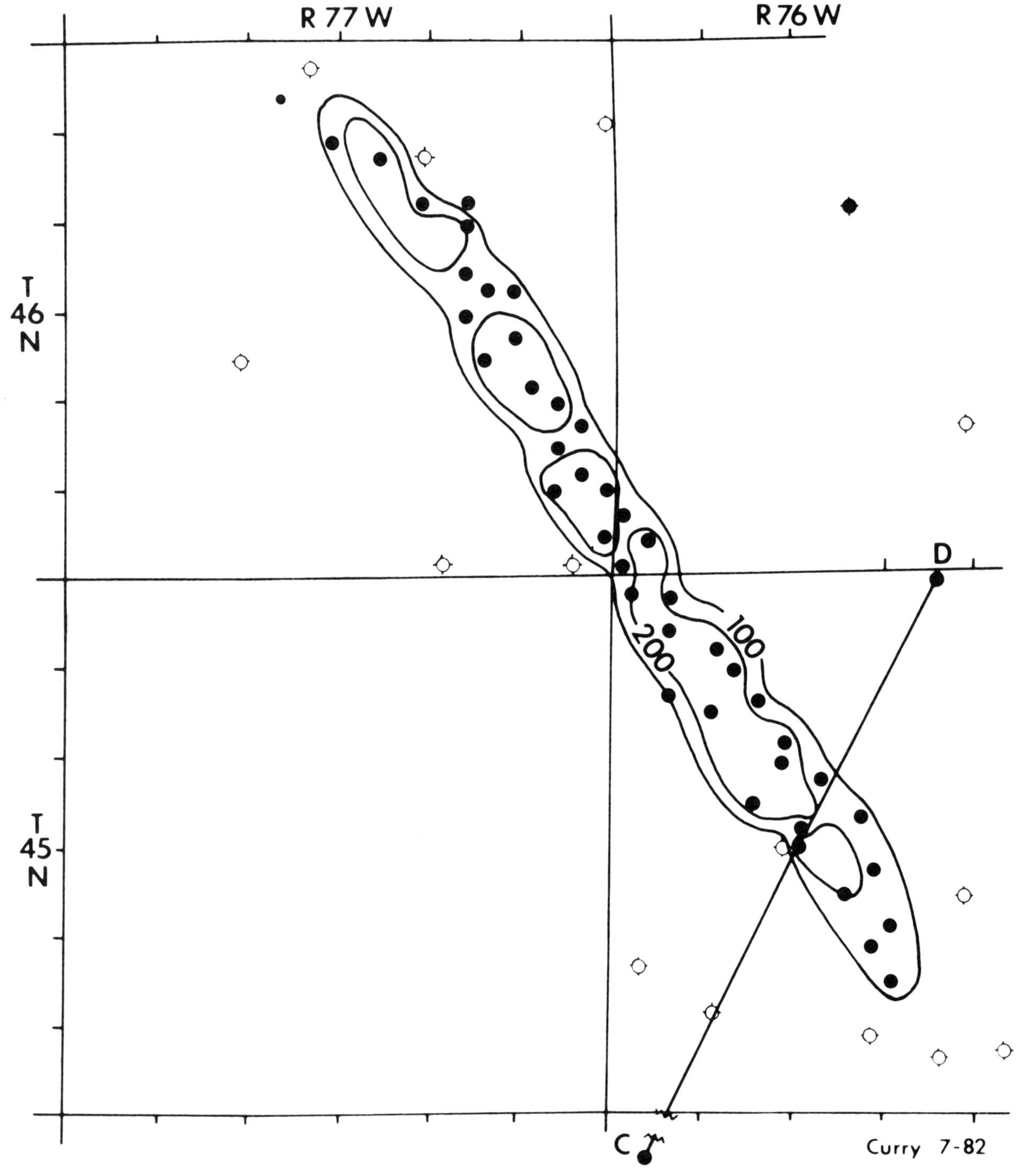

Fig. 11. Initial potential in BOPD for Shannon sandstone at Heldt Draw. Contours 100 BOPD and 200 BOPD.

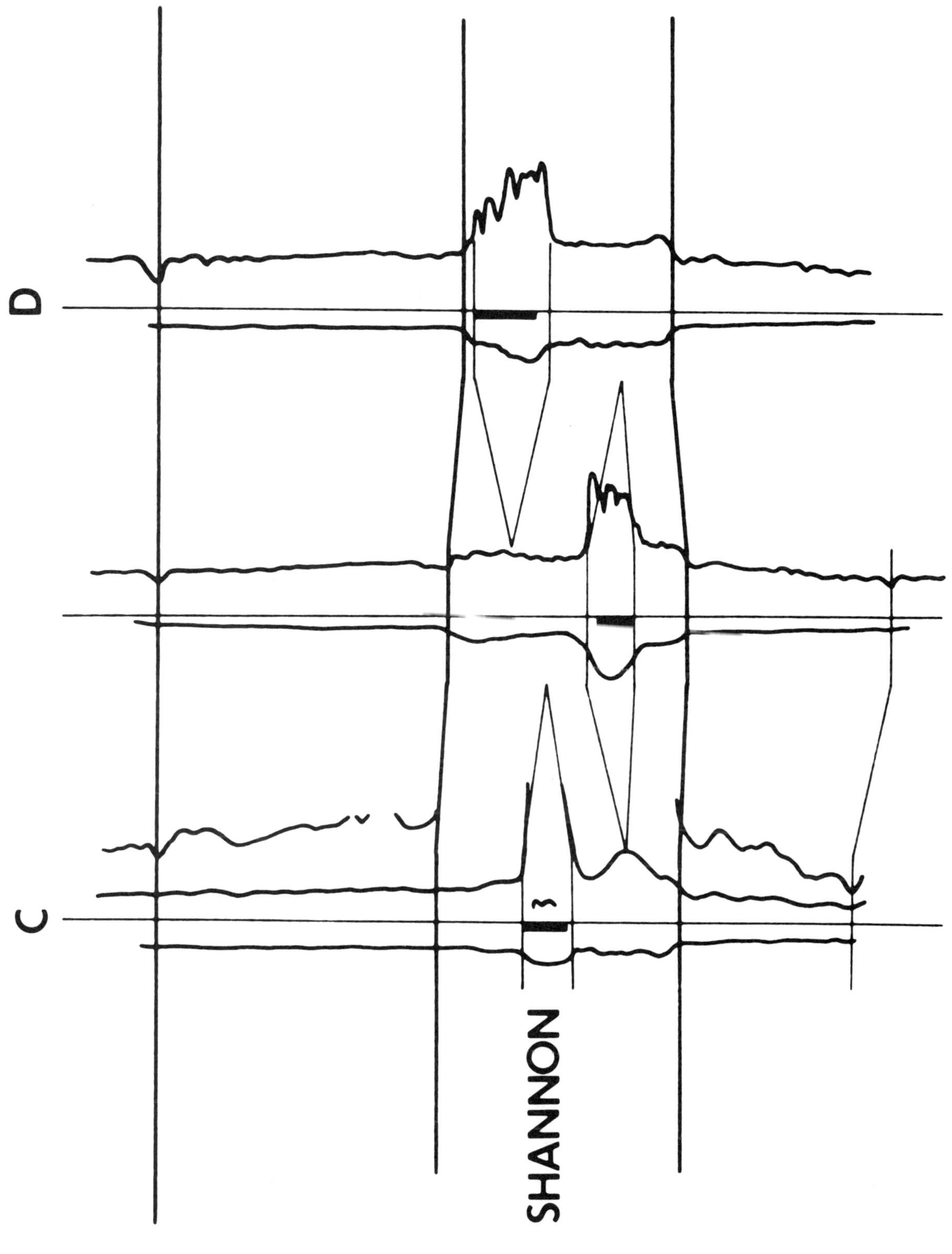

Fig. 12. Log cross section of Shannon sandstones from Table Mountain, Heldt Draw, and Hartzog Draw oil fields showing isolated productive sandstone reservoirs.

GEOCHEMICAL PROSPECTING FOR OIL AND GAS, USING HYDROCARBON FLUORESCENCE TECHNIQUES

Carroll F. Hebert

QUANTUM TECHNICAL LABORATORIES

ABSTRACT

Recent advances in instrumental technology have led to the use of several new methods of detecting geochemical anomalies in near-surface soil samples. Although the fluorescence of reservoir hydrocarbons has been known for some time, modern spectrophotofluorometers have allowed researchers to determine the concentration and type of hydrocarbons very accurately and efficiently.

A geochemical prospecting technique based on hydrocarbon fluorescence has been used to study several prospective areas. The results were analyzed statistically to provide a basis for determining the presence of an anomaly, and maps were drawn in accordance with statistical guidelines. Subsequent drilling activity was posted on these maps as it became available, in order to show the predictive capacity of this technique. Comparisons with other exploration techniques are made where they are available.

INTRODUCTION

Geochemical prospecting for oil and gas has become an established tool in the petroleum exploration industry because of a history of successful applications. Most of the work was based on variations of soil-gas techniques pioneered by Horvitz (Horvitz, 1939, 1954). More recently, the remote sensing of anomalies associated with petroleum deposits has thrived as a consequence of advances in detection, equipment, and electronics (Richers, 1982).

A prospecting system using fluorescence, first documented by the Russians (Kartsev, 1959), used the high level of detectability of fluorescence from oils and the portability of instrumentation as assets in performing field surveys. Although not based on fluorescence, a gravimetric procedure to detect soil waxes (Rosaire, 1939) and high molecular weight hydrocarbons was used in conjunction with soil-gas surveys to detect geochemical anomalies.

Using fluorescence as a tool to detect small quantities of hydrocarbons has been an established method in drilling for oil and gas since geologists first trained their black lights on oil-stained cuttings many years ago. Instrumentation in this area has evolved to a very sophisticated level and can now be used as a highly accurate indicator of hydrocarbon type and concentration (Riecker, 1962; Zierfuss, 1956; Thurston, 1971). The system that will be described uses the fluorescence technique to determine the type and quantity of hydrocarbons in near-surface soil samples to detect anomalies associated with hydrocarbon reservoirs.

SOIL SAMPLING

The highest concentrations of liquid and solid hydrocarbons left after the migration of lighter fluids are at the surface. For this reason, it was unnecessary to sample at great depth to acquire acceptable cores for testing. All sampling was done at a uniform depth throughout the survey. In order to avoid surface contamination, samples were taken at a depth of approximately six in. They were then treated with a 300 parts per million (ppm) solution of copper sulfate to retard bacterial growth and placed in sealed and numbered plastic bags.

ANALYTICAL PROCEDURE

Each sample was dried completely in a microwave oven until no further moisture was present in the soil. Upon evaporation of all adsorbed water, sufficient methanol was added to moisten the entire sample, and it was stirred to form a paste. The sample was again dried in the microwave oven, stirred again to pulverize the soil into a fine powder, and again returned to the microwave oven for a final drying cycle.

The samples were then allowed to cool in a desiccator until room temperature was reached. They were then sieved to obtain particles smaller than one hundred mesh, and a sample of approximately ten gr was weighed into an aluminum dish.

The weighed portion was placed in a tissue

homogenizer along with 50 ml of spectroquality hexane and 50 ml of methanol. After one minute in the homogenizer, the resulting suspension was poured into a beaker, and 100 ml of distilled water was added. This broke the system into two layers, the alcohol-clay, particle-water layer on the bottom and the clear hexane layer on the top.

The extract was examined in an Aminco-Bowman spectrophotofluorometer, Model SPF. The instrument had been altered to detect fluorescence in the ultraviolet region of the spectrum. The excitation wavelength was set at 265 nanometers (nm) and the spectrometer was calibrated to read 100 units when a 100 ppm solution of an oil was placed in the sample chamber. Two ml of the soil sample extract was then placed in the cuvette and the spectrum scanned from 240 to 500 nm. The extract was diluted if the readings went off-scale. A record of the scan was displayed on a Hewlett-Packard 7044A X-Y recorder.

A typical soil sample extract scan is shown in Figure 1. The first peak, at 265 nm, is simply the scattered light from the excitation lamp. The area from 300 to 350 nm is fluorescence due to light hydrocarbons, while that from 350 to 400 nm is caused by medium weight hydrocarbons. The area past 400 nm is that attributable to soil waxes. If a refined hydrocarbon, such as a petroleum base for herbicides, were in the sample, only one sharp peak would be observed. The area under the curve, less the excitation peak, is compared to that of the one hundred parts per million standard, adjusted for the instrument settings, dilution, and sample weight to calculate a relative concentration. This is the value used in all subsequent procedures.

STATISTICAL ANALYSIS

A group of samples was tested to determine the presence or absence of anomalous values in the set. First, a graph (similar to Fig. 2) showing the frequency of occurrence of each concentration interval was prepared. Then, by using the highest frequency interval, samples meeting Chauvenet's criterion (Young, 1962) were excluded, and an average background and standard deviation for the set were calculated. Then, if the set contained anomalous samples more than three standard deviations from the average background, a contrast was computed by dividing the average of the samples more than one standard deviation from background by the average background value.

In comparing Figure 2 and Figure 3, the difference between two similar sample sets can be seen. The first has anomalous values and the second does not. The first has a contrast of 4.56, which could indicate a normally pressured hydrocarbon zone as the source of the anomaly. In the area of this geochemical anomaly, six wells were drilled in 1977, each encountering a normally pressured oil reservoir; but none were deemed to be commercial. In the second area, two dry holes were drilled with no shows.

MAPPING

Maps were prepared by noting the position of the sample in the area surveyed and indicating the calculated concentration at that point. Then samples of more than one standard deviation from background were enclosed in one contour line; samples of more than two standard deviations from background were enclosed in two lines, and samples of more than three standard deviations from background were enclosed in three contour lines. This rotation produced a map showing those samples with a 66% probability of being anomalous in one contour line, ones with a 93% probability in two lines, and those with a 99% probability in three. Connecting adjacent samples of the same probability in the same lines should produce a geochemical anomaly or halo around the potential hydrocarbon reservoir. Figure 4 is a statistical analysis of the Nixon Ranch field and Figure 5 is a map based on that analysis.

The Nixon Ranch field is in an area of chalk production at about 3,000 ft, which is shown in the upper left corner of the map. This type of production is stratigraphic in nature and is intermingled with reservoirs at other depths, some of which are also stratigraphic. This group of samples shows three areas surrounded by geochemical anomalies which extend past the limits of the survey. The interpretation in these unsampled areas could be refined with additional data. In each area, one or more wells have been drilled and these are shown by the encircled well symbols on the map. Every well drilled, thus far, has encountered economical production.

SURFACE SPECTRUM SIGNIFICANCE

The relationship between reservoir and surface hydrocarbons has been studied using isotopic techniques (Stahl, 1981). With the use of fluorescence geochemistry in an area of shallow production, it is common to find that fluorescence spectra of the soil will match, or closely match, those of the shallowest production. An example of this matching was found in the Nixon Ranch field where Figure 6 shows a comparison between Olmos "B" oil and one of the soil samples from the area. Even though the common production from the area is in the chalk, those wells in the lower left of Figure 5 are producing from the Olmos "B." A sample spectrum taken from the upper right portion of the field is shown in Figure 7. The sample location is to the right of a line between the two wells in the upper right portion of Figure 5. It lies about three samples to the right of that line and shows a concentration of 71 ppm. It is obviously different from the Olmos "B" oil of Figure 6. In this area, the Olmos "B" sand is pinched out and the production encountered is from the chalk.

FALSE RIVER FIELD

Figure 8 is a statistical analysis of samples taken across what is now the False River field. Figure 9 is a map based on that analysis. The purpose of this survey was to determine whether any geochemical activity existed in association with a large geophysical anomaly. This was a reconnaissance survey and too few samples were taken to map accurately the limits of the reservoir with a geochemical halo. The fifty-three soil samples in this set provided enough data to show specific, high-contrast, geochemical activity. The high contrast was attributed to the presence of an abnormally pressured reservoir. The map in Figure 9 shows that the geochemical anomaly encircles most of the productive area. The area in the upper right was not sampled and the position of the halo was approximated from the available data. Even with such a small reconnaissance survey, the general shape and size of the field is shown. All of the drilling indicated on the map occurred after the survey. This area is now a prolific gas producer from the lower Tuscaloosa near 19,000 ft.

LITTLE RIGOLETS FIELD

Figure 10 is a map of the Little Rigolets field which now produces from the lower Tuscaloosa. Prior to the survey, one hole had been drilled and plugged in top center of the map and a gas well in the lower left-hand corner produces from a shallower horizon. All subsequent drilling is indicated by the circled well symbols. The analytical methods and the statistics used for this survey are the same as those applied to the other fields and prospects in this paper. The map shows an anomaly encircling an area with two breaks in it, one in the top center without samples and the other in the center of the map. Although drilling continues in the area, to date, there are four gas wells inside the encircling anomaly and five dry holes outside of the encircled area. The anomaly does not close in the center of the map as it might be expected to do. This configuration is probably a result of the shallow sampling depth of six in, which may be from recently deposited sediment draining into the lake.

COMPARISONS TO OTHER TECHNIQUES

Two lines of soil samples were taken at points across a gas geochemical halo in south Louisiana. The results are shown in Figure 11. The area of proven production extends from beyond the left-hand side of the graph to the arrows on the right. Several dry holes have been drilled to the right of these arrows. Samples were analyzed by using a gas extraction method and the values plotted on the upper portion of Figure 11. Fluorescence geochemistry was used to analyze samples plotted on the lower portion. It can be seen that the fluorescence background is more consistent and that the halo is much stronger, while the gas chromatograph group has a larger number of high value samples. Figure 11 also shows the highest fluorescence value to be outside the highest value of the gas anomaly. We have found that the highest value of a fluorescence halo anomaly is almost always outside of the field, and wells drilled at the highest value are rarely successful.

Two prospects were surveyed using fluorescence geochemistry and seismic techniques. In each case, the surveys were done independently, and no attempt was made to correlate the results of one survey with the other. Figures 12 and 13 show the surveys on the Algiers Prospect. Note the similarities in size and shape of the seismic and the florescence geochemical prospect. Note also that there is disagreement as to the location of the center of the prospect, the geochemistry placing it more to the right on the map. Drilling near the apex of the seismic structure but in the high value portion of the geochemical halo produced a dry hole with a show and westerly dip. This hole is indicated by the circled well symbol near the center of the map. A subsequent location to the east should prove more successful.

The Irish Bayou Prospect is a small area extensively studied, using geochemistry and seismology. The two respective maps are shown in Figures 14 and 15. Besides having similar shapes and sizes, note that both maps show the presence of additional anomalies to the north and south. Again, the surveys were entirely independent. The plugged hole which occurs inside of the surrounding geochemical anomaly, near the sample with a concentration of 29 ppm, has a thirty-foot section of gas by electric log. The other hole does not have such a show. Both holes were drilled prior to the surveys.

CONCLUSION

When the fluorescence geochemistry technique is compared with other tools used in petroleum exploration, the results show it to be a useful system in determining the presence or absence of geochemical halos associated with hydrocarbon reservoirs. The technique has been applied successfully to the deep, abnormally pressured reservoirs of south Louisiana as well as to shallower normally pressured reservoirs encountered in other areas.

In addition, it has been shown that fluorescence spectra of soil samples, especially in areas with shallow production, exhibit characteristics of the oil in those shallow reservoirs. This technique may be used to determine the shallowest accumulation that might be encountered in drilling. In areas where there is uncertainty as to the potential producing zone, these spectra can often be used to relate specific geographical areas to known production horizons.

REFERENCES

Horvitz, L., 1939, On geochemical prospecting, Geophysics, v. 4, no. 3, pp. 210-228.

__________, 1954, Near-surface hydrocarbons and petroleum accumulation at depth, Mining Engineering, pp. 3-7.

Kartsev, A., (1959), Geochemical methods of prospecting and exploration for petroleum and natural gas, trans. by P.A. Witherspoon and William D. Romney, University of California Press, Berkley, chap. 8.

Richers, D.M., Reed, R.J., Horstman, K.C., Michels, G.D., Baker, R.N., Lundell, L., and Marrs, R.W., 1982, Landsat and soil-gas geochemical study of Patrick Draw oil field, Sweetwater County, Wyoming, American Association of Petroleum Geologists Bulletin, v. 66, no. 7, pp. 903-922.

Riecker, R.E., 1962, Hydrocarbon fluorescence and migration of petroleum, American Association of Petroleum Geologists Bulletin, v. 46, no. 1, pp. 60-75.

Rosaire, E.E., (1939), The handbook of geochemical prospecting, published by the author, Houston, Texas.

Stahl, W., Faber, E., Carey, B.D., and Kirksey, D.L., 1981, Near-surface evidence of migration of natural gas from deep reservoirs and source rocks, American Association of Petroleum Geologists Bulletin, v. 65, no. 9, pp. 1543-1550.

Thurston, A.D., and Knight, R.W., 1971, Characterization of crude and residual-type oils by fluorescence spectroscopy, Environmental Science and Technology, v. 5, pp. 64-69.

Young, H.D., 1962, Statistical treatment of experimental data, McGraw-Hill, New York, pp. 78-80.

Zierfuss, H., and Coumou, D.J., 1956, Use of quantitative fluorescence measurements in drilling operations, American Association of Petroleum Geologists Bulletin, v. 40, no. 11, pp. 2724-2734.

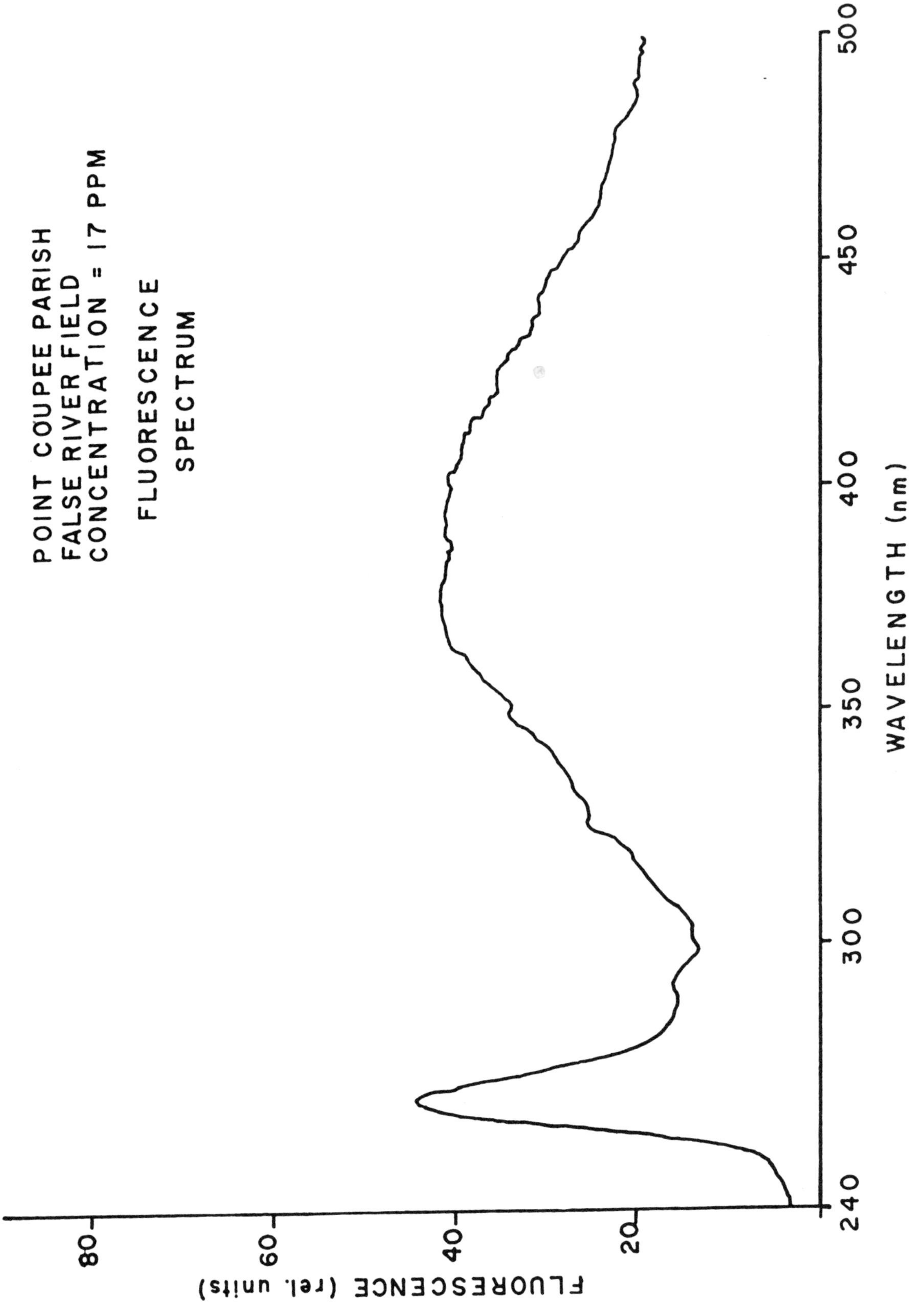

Fig. 1. Typical ultraviolet soil sample spectrum.

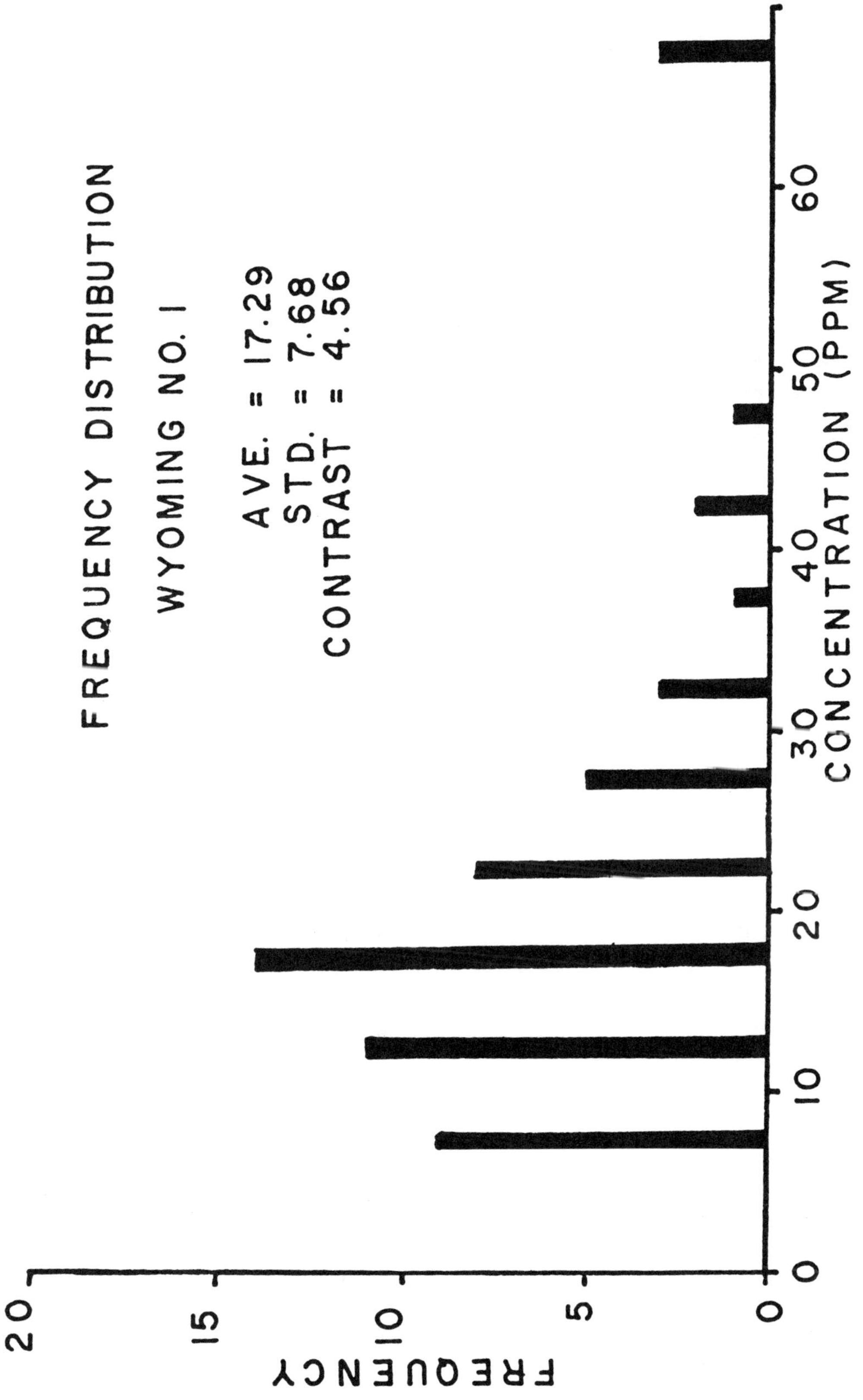

Fig. 2. Statistical analysis of a group of soil samples showing anomalous values to the right.

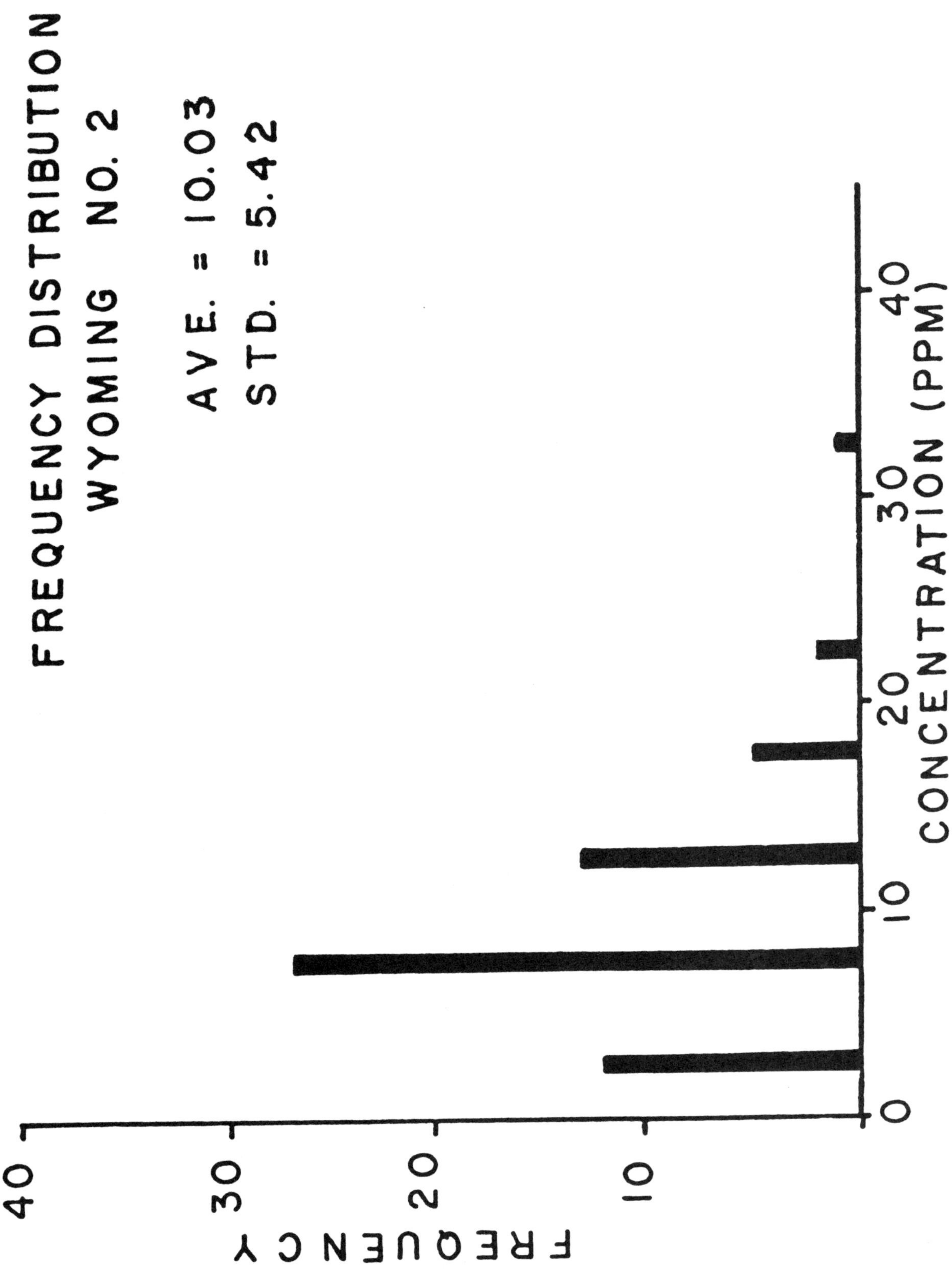

Fig. 3. Statistical analysis of a group of soil samples in an area near those of Figure 2. No anomalous values appear in this set.

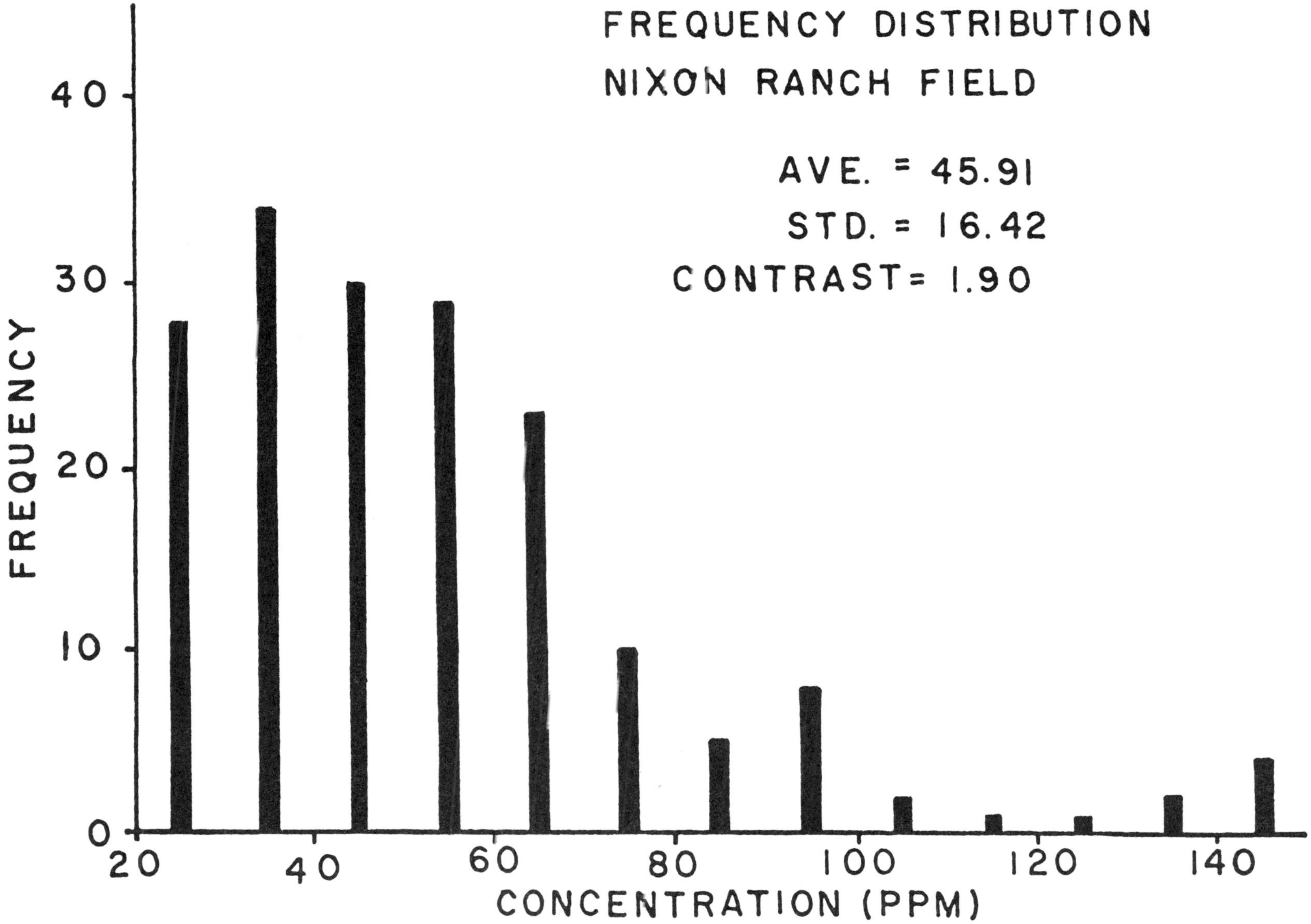

Fig. 4. Statistical analysis of a soil sample group showing low contrast, indicating a normally pressured reservoir as the source.

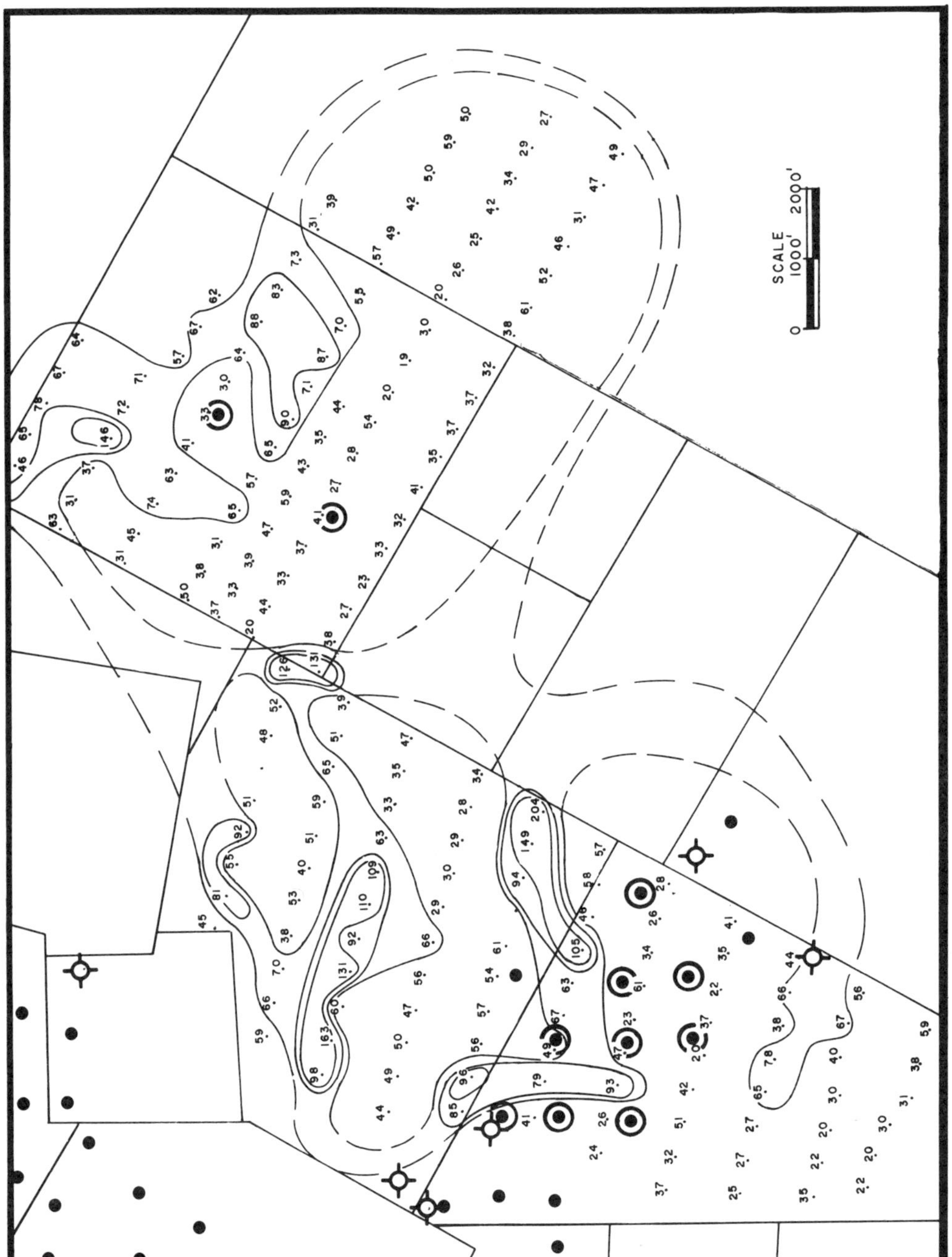

Fig. 5. Map of the Nixon Ranch field. Drilling after the survey is indicated by the circled well symbols.

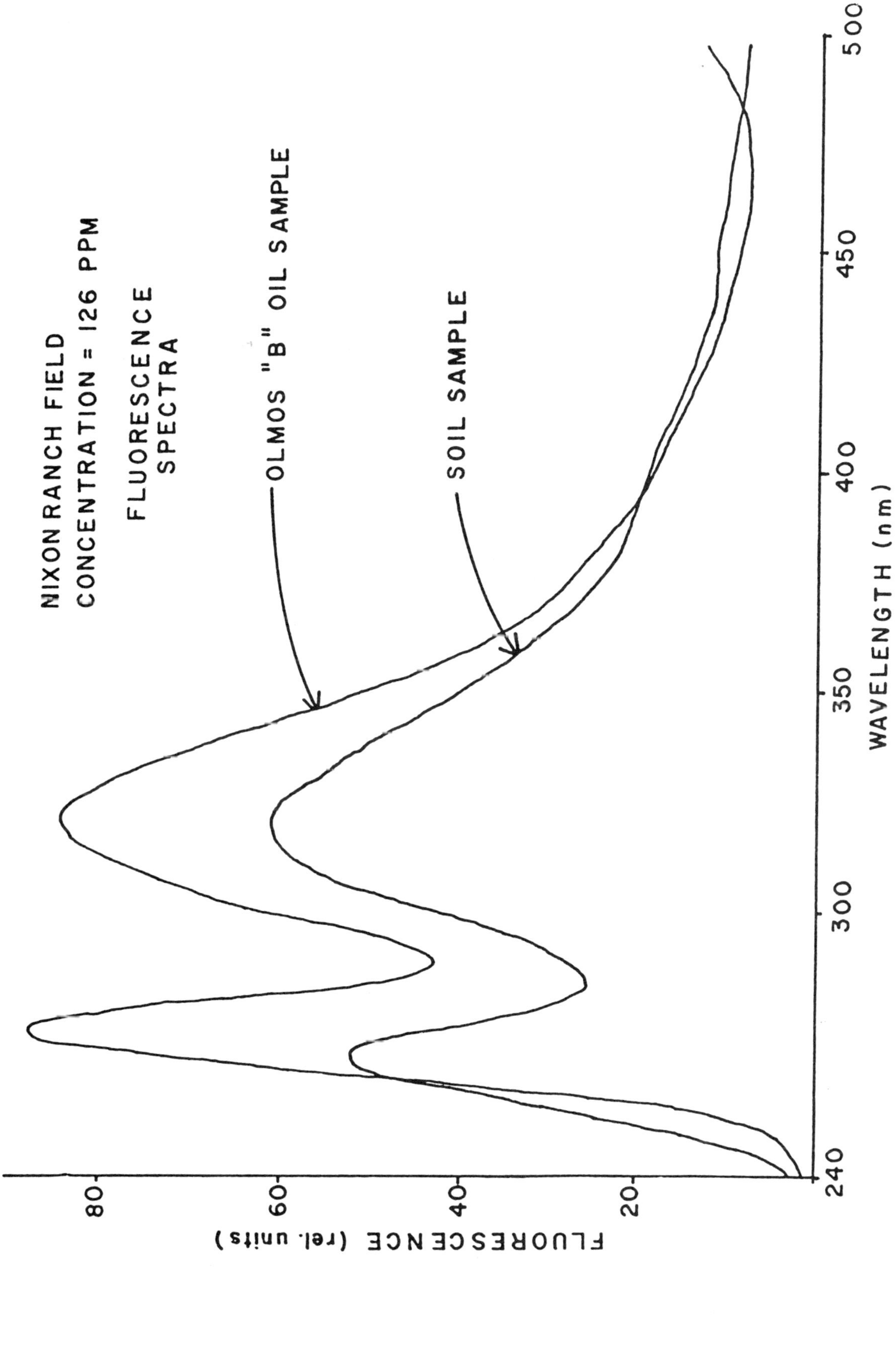

Fig. 6. Comparison of oil spectrum and soil spectrum showing similarity of reservoir hydrocarbons to surface microseepage.

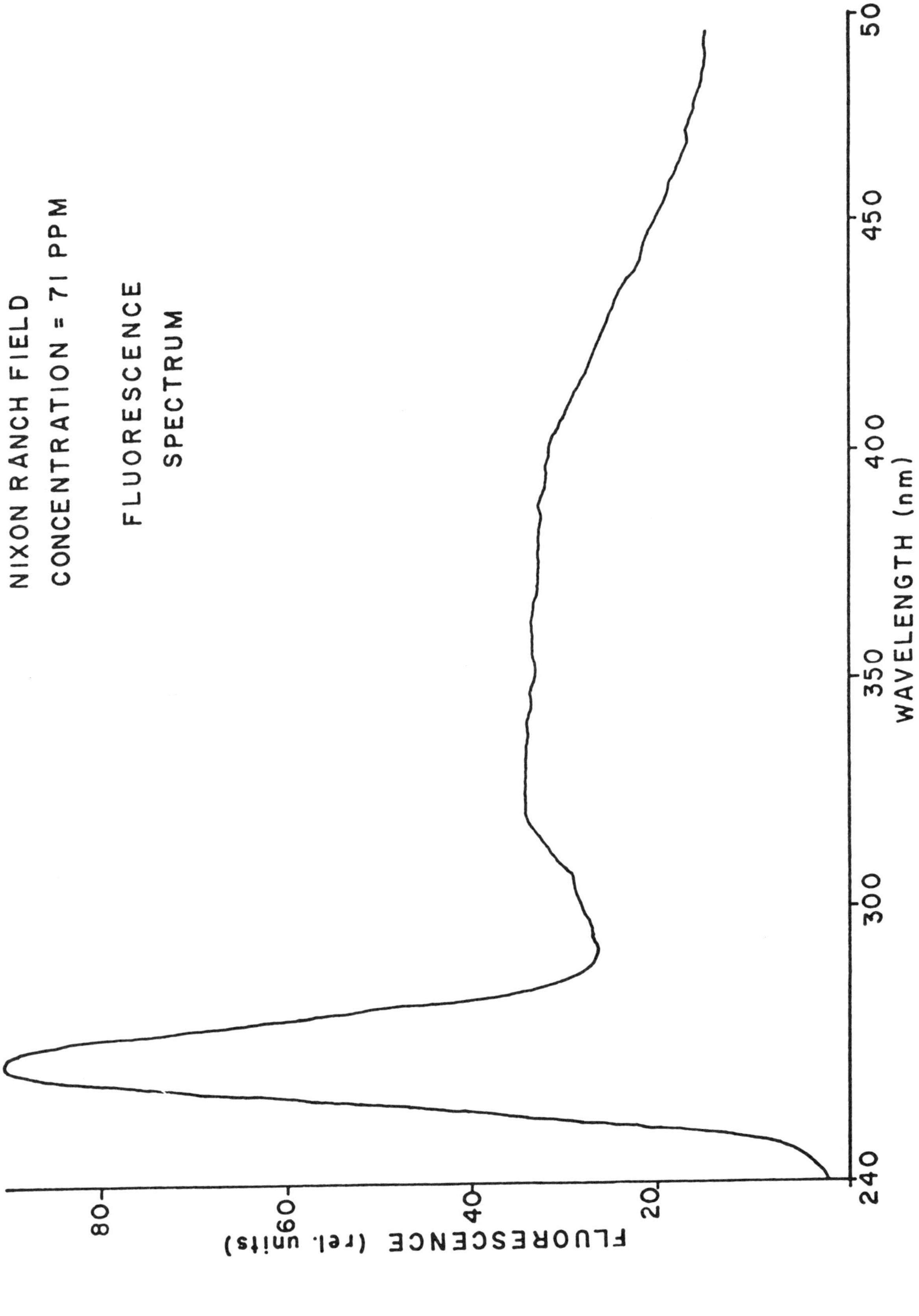

Fig. 7. Soil sample spectrum from an area in Nixon Ranch field having chalk production.

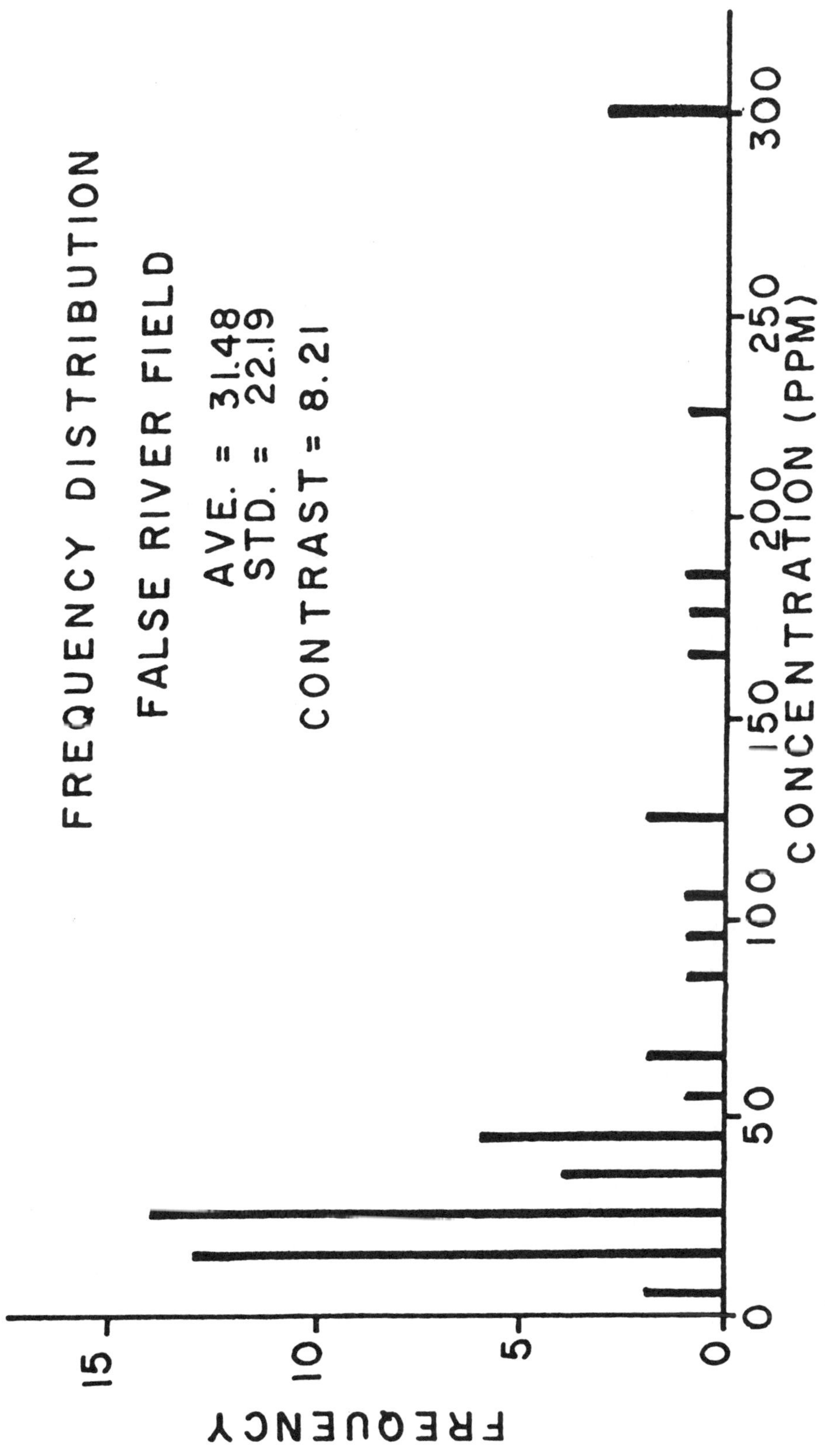

Fig. 8. Statistical analysis of soil samples from the False River field showing high contrast, indicating a subsurface reservoir that is abnormally pressured.

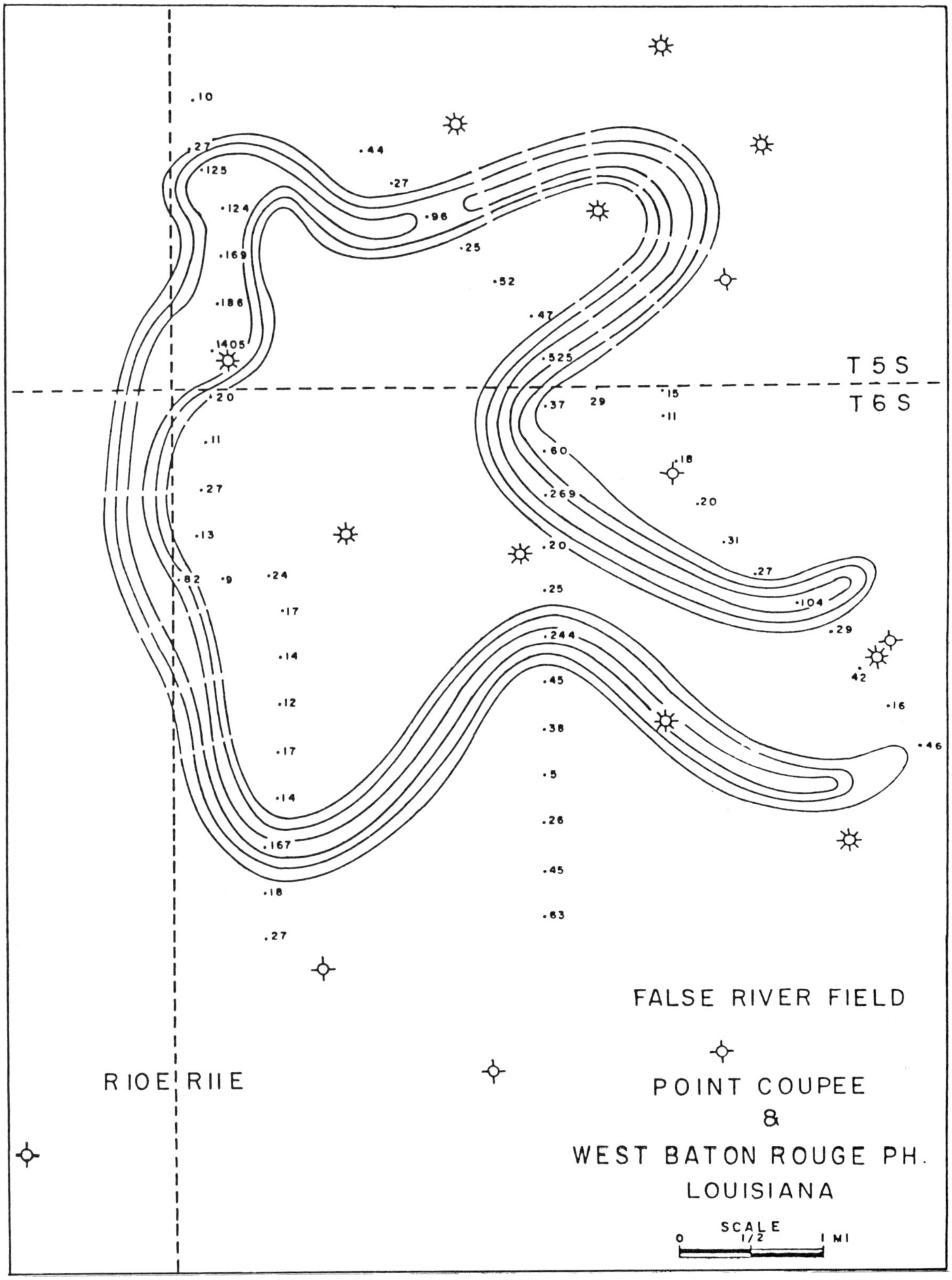

Fig. 9. Map of the geochemical anomaly surrounding the False River field. All drilling shown occurred after the survey.

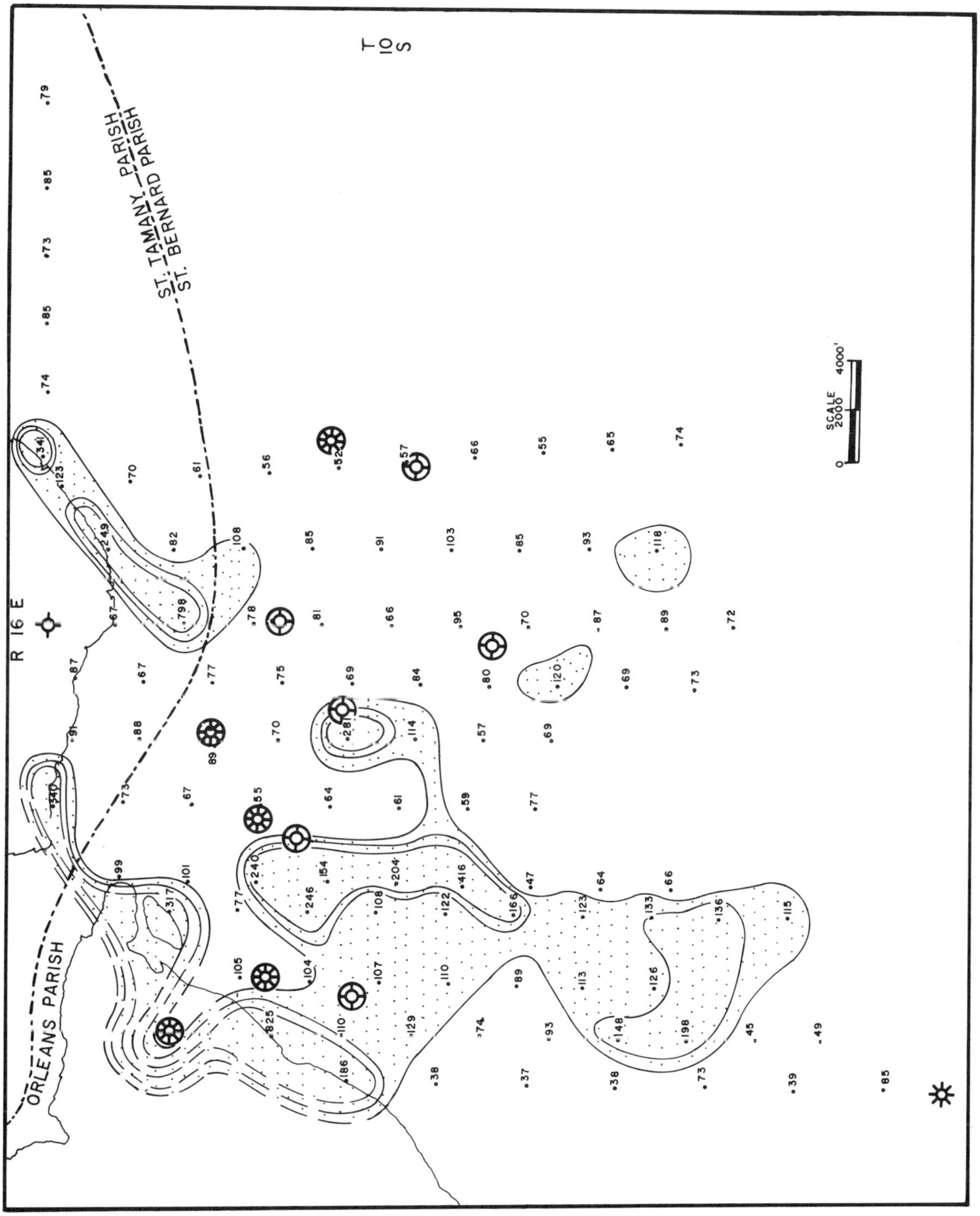

Fig. 10. Little Rigolets field. Geochemical anomaly is shown by the dotted area. Drilling after the survey is indicated by the circled well symbols.

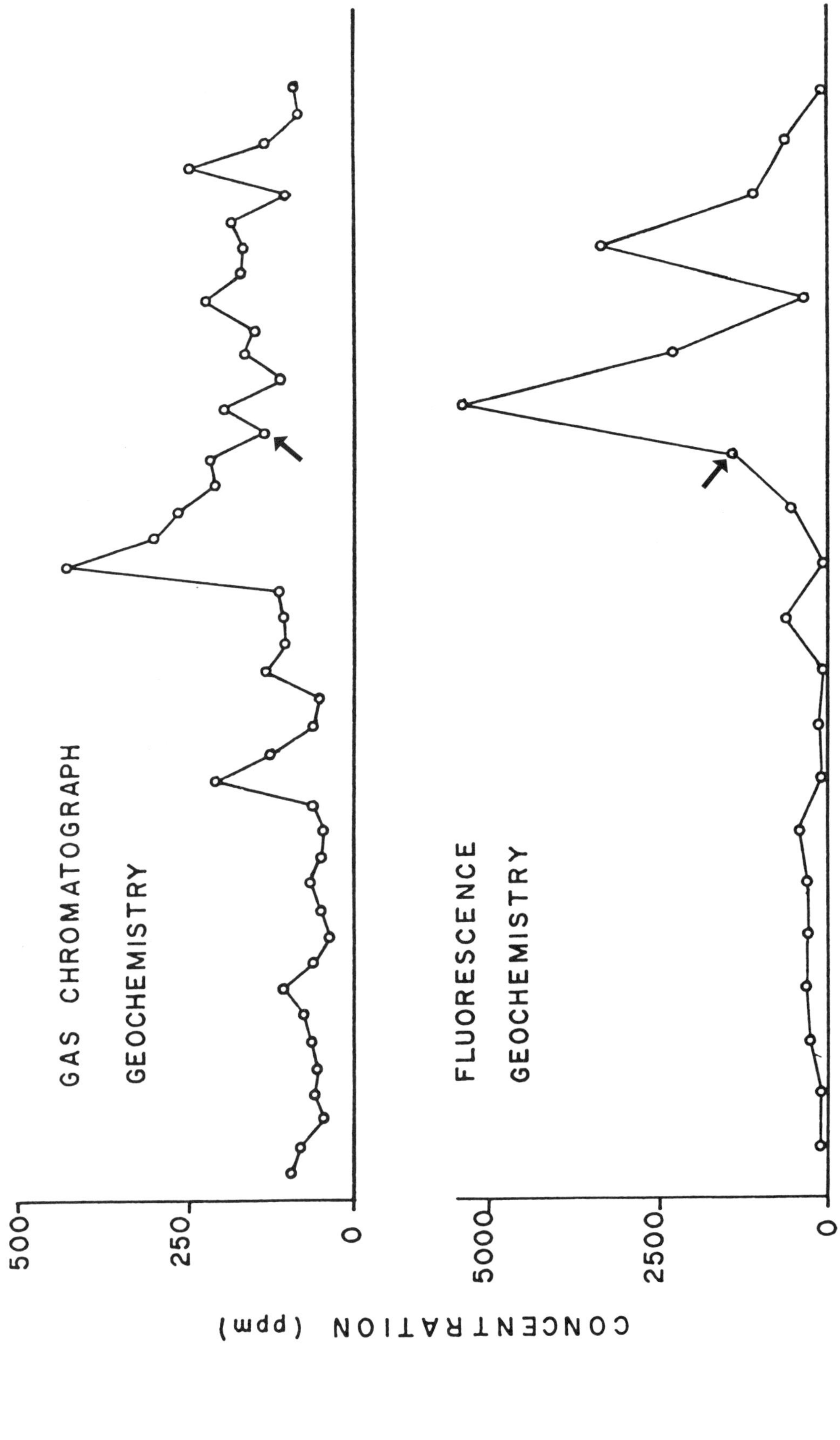

Fig. 11. Comparison of fluorescence and gas geochemistry. Production begins to the left of the graph and dry holes occur to the right of the arrows.

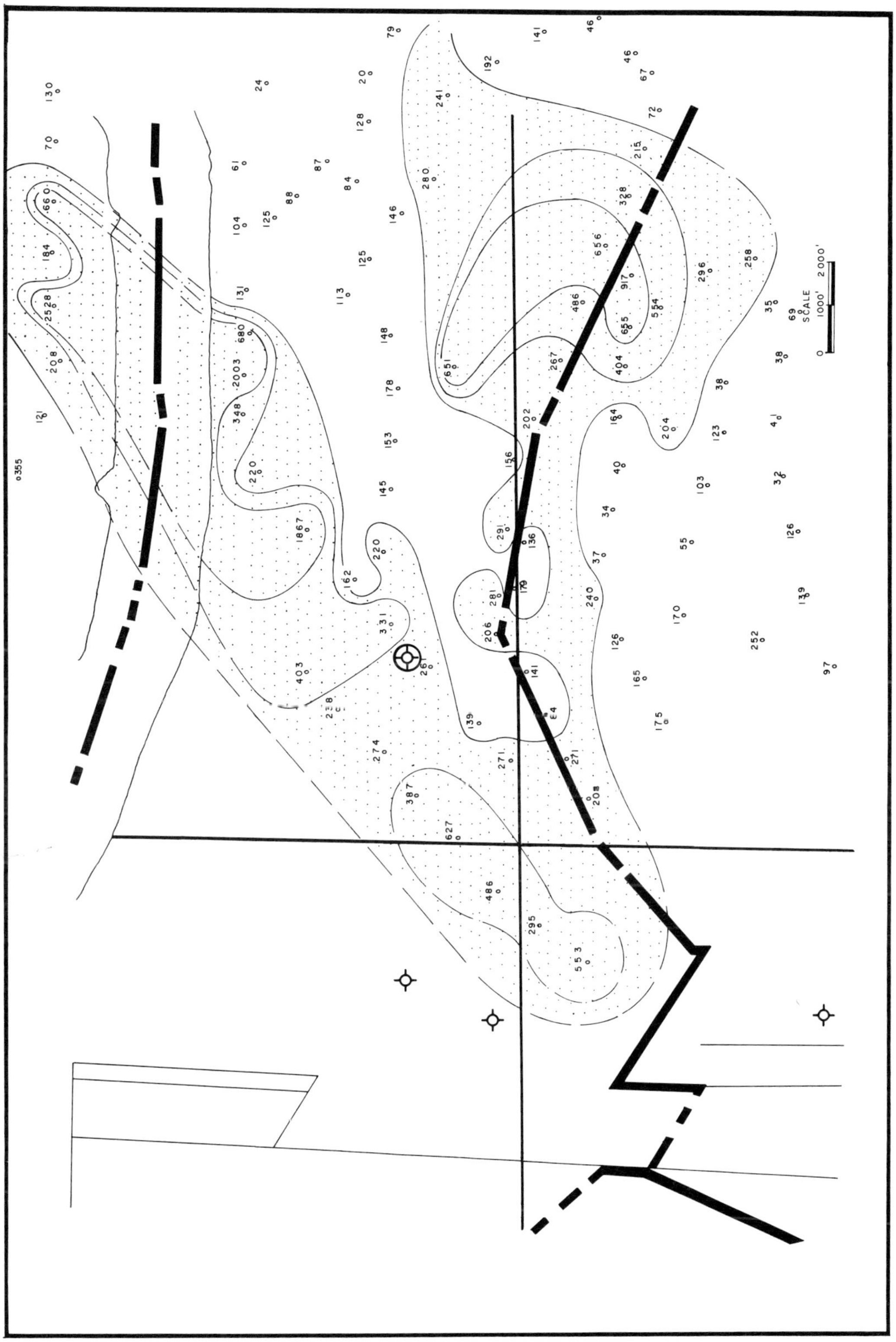

Fig. 12. Algiers Prospect. The geochemical anomaly is indicated by the dotted area on the map. The well symbol that is circled near the center of the map was drilled after the survey.

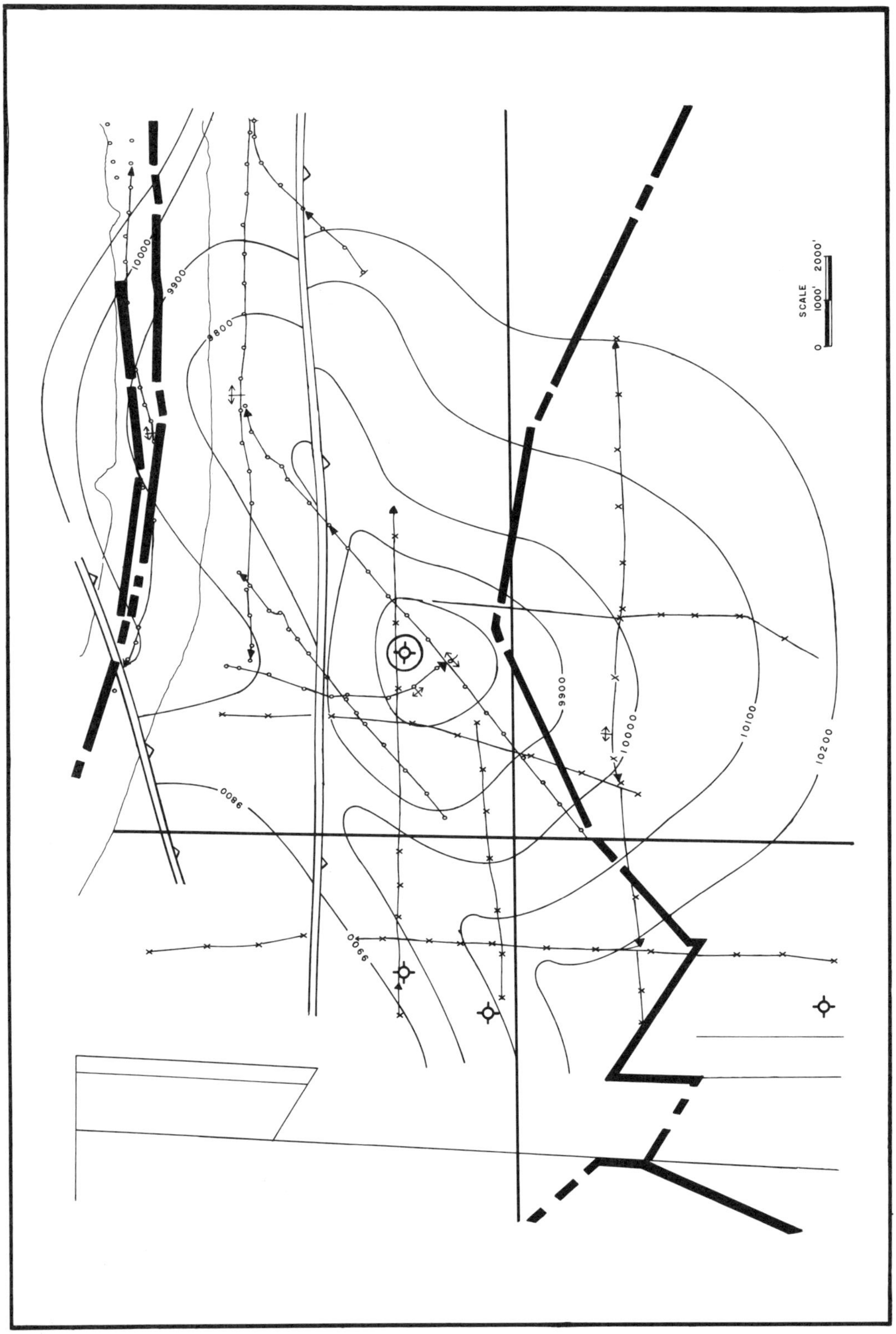

Fig. 13. Algiers Prospect. Interpretation of the seismic survey of the area. Note the location of the well drilled after the survey.

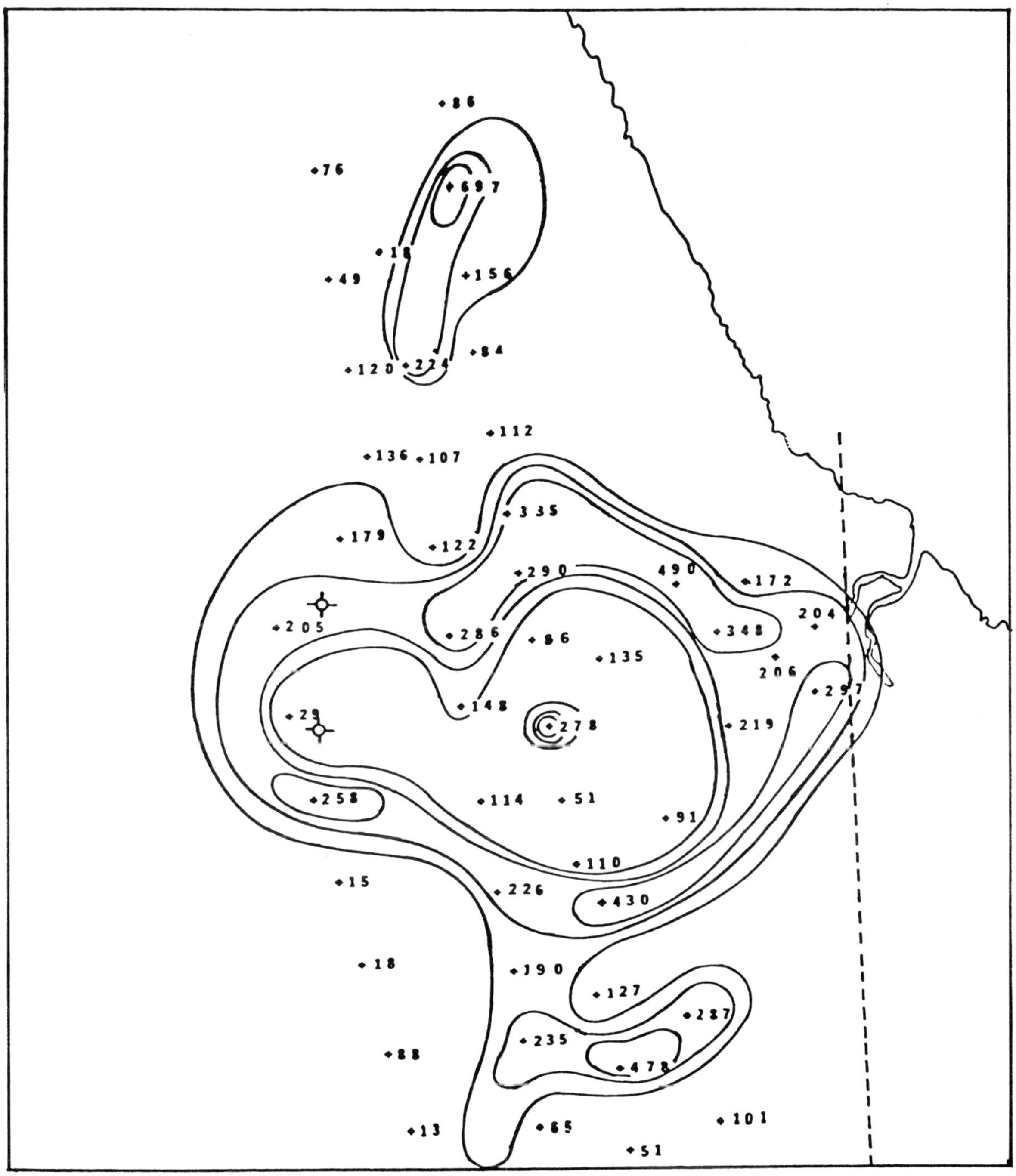

Fig. 14. Irish Bayou Prospect. Geochemical survey shows an anomaly surrounding the prospective area near the center of the map.

Fig. 15. Irish Bayou Prospect. Seismic interpretation showing prospective area similar to the geochemical survey.

CORRELATIONS OF SUSPECTED PETROLEUM-GENERATED BIOGEOCHEMICAL AND AEROMAGNETIC ANOMALIES, BELL CREEK OIL FIELD, MONTANA

Mary C. Dalziel and Terrence J. Donovan

U.S. GEOLOGICAL SURVEY

ABSTRACT

Petroleum microseepage causes chemical changes in surface and near-surface rocks and soils. The reducing environment brought about by the presence of hydrocarbons induces reduction and mobilization of iron. This element, essential for plant growth, is assimilated by vegetation in the reduced divalent state and is found in anomalously high concentrations in pine needles and sagebrush leaves over the Bell Creek Oil Field, Montana.

The reduction of hydrated iron oxides and (or) hematite to form magnetite is also a consequence of petroleum microseepage. High-wave-number magnetic anomalies at Bell Creek are interpreted to reflect increased amounts of diagenetic magnetite in near-surface rocks. Comparison of these preliminary biogeochemical and aeromagnetic data suggests that changes in transition-element chemistry in near-surface rocks and soils owing to petroleum microseepage may be a useful indicator of buried petroleum deposits.

INTRODUCTION

In recent years, the search for subtly hidden hydrocarbon deposits has stimulated the development of unconventional prospecting techniques. Two unconventional approaches that show promise as hydrocarbon exploration tools are biogeochemical surveys and a new concept for aeromagnetic surveys.

These dissimilar techniques share common ground—the detection of changes in iron chemistry brought about by microseepage of hydrocarbons from a buried oil and gas reservoir. The data in this study suggest a correlation between certain biogeochemical parameters and a high resolution aeromagnetic survey over the Bell Creek Oil Field, Powder River County, Montana.

Biogeochemical prospecting for oil and gas has shown encouraging results at the Recluse Oil Field, a stratigraphic trap on the east flank of the Powder River Basin, Campbell County, Wyoming (Dalziel and Donovan, 1980a, 1980b). At Recluse, anomalously large manganese/iron ratios in the needles of pine (*Pinus ponderosa* Laws) and the leaves of sagebrush plants (*Artemisia tridentata* Nutt.) display systematic variations that correlate closely with several kinds of epigenetic anomalies in rocks and soils over the oil field.

The increased concentration of iron and manganese in plant tissue over the Recluse Field is postulated to be a direct result of microseepage of hydrocarbons. Seeping hydrocarbons provide reducing conditions that are sufficient to reduce iron and manganese to the trivalent or divalent state, enabling them to be complexed or adsorbed on solids or humus in the soils. These bound or adsorbed divalent elements are essential for plant metabolism and are readily assimilated through ion exchange at the plant roots. The elements ultimately are concentrated in leaf and needle tissues at levels as much as 26,000 parts per million (ppm, ashweight basis), as determined by atomic absorption spectrophotometry. The overall higher concentrations of manganese relative to iron at the Recluse Field are not understood, but may be ascribed to the reported mutual-element antagonism (Horvath, 1972) that would result in a plant preferentially accumulating iron or manganese, but not both.

Experimental, low-altitude, total-field aeromagnetic surveys over oil fields have previously detected high-wave-number anomalies suggesting the presence of dispersed, shallow magnetic material interpreted to be epigenetic magnetite formed from iron oxides in the presence of hydrocarbons (Donovan et al., 1979; Donovan et al., in press). At the Cement Oil Field, Caddo County, Oklahoma, where airborne magnetic measurements show anomalous high-wave-number peaks suggesting a shallow magnetic source, a systematic survey of magnetite from the uppermost 300 m of section indicates an increase in the amount of magnetite over the crest of the anticline. The interpretation is that the reducing environment provided by seeping hydrocarbons has caused the reduction of hydrated ferric oxides and hematite to form magnetite in the near-surface zone. The sequence may take place as follows:

hydrated ferric oxides → hematite

hematite (Fe_2O_3) → magnetite ($Fe^{+2}Fe_2^{+3}O_4$)

(Donovan et al., 1979)

BELL CREEK

Bell Creek Oil Field, Montana, was the test site for a comparison of anomaly patterns mapped from biogeochemical and low-altitude aeromagnetic surveys. Bell Creek is a stratigraphic trap on the northeast flank of the Powder River Basin, Montana (Fig. 1). It was discovered in 1967 and produces from the Cretaceous Muddy Sandstone (Berg and Davies, 1968). The trap is a barrier bar (Berg and Davies, 1968) and has an areal extent of 18.5 km by 1.5-4.5 km. The average depth to production is 1,372 m with an average reservoir rock thickness of 6.08 m (20 ft). The oil at Bell Creek has been partially degraded by meteoric waters and bacterial activity (Winters and Williams, 1969); consequently, the API gravities range from 32° to 42°. Surface rocks at Bell Creek have been mapped as the Upper Cretaceous Lance Formation, consisting of alternating beds of nonmarine sandstone, sandy shale, and claystone (Robinson et al., 1964), or alternately, as the Upper Cretaceous Hell Creek Formation, a series of gray to greenish gray shales and clays, brown carbonaceous shales, and massive or crossbedded, buff to greenish buff friable sandstone (Bryson and Bass, 1973; Ross et al., 1958; Fig. 2). This sandstone and shale sequence results in a topography characterized by flat-topped buttes, narrow flat divides, and ridges that rise steeply above valley bottoms. Plants studied in the biogeochemical survey, ponderosa pine (*P. ponderosa*), big sagebrush (*A. tridentata*), and silver sagebrush (*A. cana* Pursh), favor different habitats; the pines grow mostly along hilltops and the sagebrush is distributed on uncultivated hills and valley bottoms.

BIOGEOCHEMICAL METHODS

Samples of the pine needles and the leaves of big sagebrush and silver sagebrush were collected over the Bell Creek Oil Field and surrounding area. Samples were washed in an ultrasonic bath using an aqueous, phosphate-free detergent solution and washed twice in distilled water (Mason, 1953). Samples were ground in a coffee grinder, and 4 g of dry material were ashed at 450°C for 6 hours. One hundred mg of ash were digested in 4 ml of 3N hydrochloric acid and gently heated without stirring for 15 min. The supernatant was diluted to 100 ml with distilled water and analyzed for iron by atomic absorption spectrophotometry (Ward et al., 1969).

RESULTS OF BIOGEOCHEMICAL SURVEY

Concentrations of iron in plant tissue for all plant species studied are shown as computer-generated images that were compiled using a program for spacial-array processing and data smoothing adapted from Eliason et al. (in press). Ranges of iron concentrations in plant materials are displayed in shades of gray (Figs. 3, 4, 5). The gray tones were selected using a density-slice program; consequently, the ppm boundary divisions are a reflection of the computer smoothing program and do not represent analytical precision to four or five places.

Iron concentration in pine needles (Fig. 3) shows high values in two areas: one in the north (9,000 ppm), and a second in the south (11,000 ppm). However, paucity of pine sample localities results in data that are too limited for conclusive results. Even though the data are sparse, comparison with areas of high iron concentration in two species of sagebrush indicate that the limited pine data agree with the pattern of iron enrichment displayed by the two sagebrush species.

Big sagebrush is more uniformly distributed than pine, and the pattern of iron concentration in leaves is more clearly defined (Fig. 4). Biogeochemical baseline values for big sagebrush in the northern great plains have a mean of 3,300 ppm iron, with a range of 1,500 to 10,000 ppm iron (Ebens and Shacklette, 1982). The pattern of high iron concentration in big sagebrush over the northern part of the Bell Creek Field extends 1.6-3.2 km west of known petroleum production and reaches concentrations of over 13,000 ppm iron. In the southern part of the field, a lobe of iron enrichment shows a northwesterly trend away from known oil production limits and has concentrations of up to 15,900 ppm iron. The most widely collected species, silver sagebrush (Fig. 5), shows a clear pattern of iron enrichment in leaves. Comparison of point-to-point data among species does not agree well; however, the overall pattern (in this case delineated by values ranging between 4,500 ppm and 5,200 ppm iron) is similar to that of big sage and forms an arcuate pattern over and around the oil field. Previous studies of iron concentration in silver sagebrush from Montana, North Dakota, South Dakota, and Wyoming established a mean value of 1,200 ppm and a range of 520 ppm to 4,400 ppm (Ebens and Shacklette, 1982). These baseline values for sagebrush suggest that anomalous iron concentrations occur in sagebrush found in the locality of the Bell Creek Oil Field.

The iron-enrichment pattern in plants does not precisely correlate with areas of oil production at Bell Creek. Iron enrichment in plant tissues coincides with areas of known oil accumulations in the north central and south central sections of the oil field; the arcuate offset in the north and the off-field extension in the south are not located over known oil and gas reservoirs and are interpreted as halo pattern anomalies surrounding suspected hydrocarbon accumulations.

LOW-ALTITUDE AEROMAGNETIC SURVEY METHODS

Aeromagnetic techniques used in the Bell Creek study were developed for work in northern Alaska and are described in Donovan et al. (in press).

The survey at Bell Creek used a twin-engine aircraft flown at 90 m above terrain at a speed of approximately 130 knots (67 m/sec). Flight lines were east-west, separated by approximately 1.5 km. The aircraft was configured with three proton precession magnetometers, one extended from each wing tip and the third located in a tail stinger.

Three total-field measurements made in the horizontal plane allow transverse and longitudinal vector components of the horizontal gradient to be calculated. The horizontal gradient is particularly sensitive to near-surface magnetic occurrence (Breiner, 1973). The horizontal gradient also provides data for areas between adjacent flight lines, thus allowing for more accurate map production. Additionally, because the gradiometer is a differential magnetometer, it eliminates the need for diurnal corrections because the gradient is computed from simultaneous readings.

Magnetic readings are acquired every 0.7 sec, which is equivalent to a measurement on the ground of approximately every 47 m. Data are recorded in both a digital and analog mode. Ground track is recorded on map strips by the copilot/navigator and is later verified by a videotape system.

The magnetic anomalies in question probably do not result from buried or surface (pipelines, drill pipe, etc.) metal as the wave number, amplitude, and reinspection of ground-track video tape discount the majority of these effects. Magnetic effects of near-surface or surface pipe and casing anomalies have been discussed by Barrett (1931) and Breiner (1973).

Potential sources of error within the system are: electric eddy currents set up in the plane itself as it moves through the Earth's magnetic field; excessive pitch, roll, or yaw of the aircraft; and imperfect magnetic compensation between the airplane and the sensors. Eddy current noise is removed by filtering during data processing. This filtering is possible because the noise has a period less than that of the true signal. Pitch, roll, and yaw are kept at a minimum during survey flights and generally do not exceed 5°, and resulting magnetic variations can be neglected. Magnetic compensation of the aircraft is accomplished by introducing a small, artificial magnetic field between each sensor and the aircraft; this compensation is accurate to ±0.05 gammas.

The horizontal gradient is computed from the three total-field readings and can be transformed to a Transverse Mercator map-projection array with a resolution of 0.08 km per array element. Interpolation routines create continuous coverage, which is displayed as computer-enhanced images (Eliason et al., in press).

RESULTS OF THE AEROMAGNETIC SURVEY

A computer-generated image of the horizontal magnetic gradient (Fig. 6) shows anomalous gradients suggesting a shallow magnetic source over and near petroleum production at Bell Creek. These shallow magnetic disturbances are interpreted as reflecting diagenetic magnetite. The magnetic contrast between typical sedimentary rocks of low magnetic susceptibility and those locally enriched in epigenetic magnetite results in distinctive anomalies of high-wave-number and low amplitude (Donovan et al., 1979). These findings are in agreement with an independent total-field aeromagnetic survey at Bell Creek (Foote, this volume). The computer-generated image of the horizontal gradient at Bell Creek shows a noticeable pattern of high values that closely correlates with biogeochemical patterns (Figs. 3, 4, and 5).

CONCLUSION

Computer-generated images of data collected by two unconventional survey techniques have similar patterns of anomalies. The correlations in patterns of data collected by unrelated survey techniques emphasize the importance of changes in transition-element chemistry in the near-surface environment as a result of the microseepage of petroleum. Both kinds of surveys described here can be completed rapidly and are relatively inexpensive to conduct.

REFERENCES

Barrett, William M., 1931, Magnetic disturbance caused by buried casing: American Association of Petroleum Geologists Bulletin, v. 15, no. 11, p. 1371-1399.

Berg, R.R., and Davies, D.K., 1968, Sandstone at Bell Creek field, Montana: American Association of Petroleum Geologists Bulletin, v. 52, p. 1888-1898.

Breiner, Sheldon, 1973, Applications manual for portable magnetometers: Geometrics, Palo Alto, California, 58 p.

Bryson, R.P., and Bass, N.W., 1973, Geology of Moorehead coal field, Powder River, Big Horn, and Rosebud Counties, Montana: U.S. Geological Survey Bulletin 1338, 116 p.

Dalziel, M.C., and Donovan, T.J., 1980a, Biogeochemical evidence for subsurface hydrocarbon occurrence, Recluse oil field, Wyoming: Preliminary results: U.S. Geological Survey Circular 837, 11 p.

__________, 1980b, Chemical and isotopic data for some plants, soils, and carbonate cements in sandstone outcrops, Recluse oil field, Wyoming: U.S. Geological Survey Open-File Report 80-783, 16 p.

Donovan, T.J., Forgey, R.L., and Roberts, A.A., 1979, Aeromagnetic detection of diagenetic magnetite over oil fields: American Association of Petroleum Geologists Bulletin, v. 63, no. 2, p. 245-248.

Donovan, T.J., Hendricks, J.D., Roberts, A.A., and Eliason, P.T., in press, Experimental low-altitude aeromagnetic reconnaissance for petroleum in the Arctic National Wildlife Refuge, Alaska, using horizontal gradients—a progress report and review, U.S. Geological Survey Professional Paper.

Ebens, R.J., and Shacklette, H.T., 1982, Geochemistry of some rocks, mine spoils, stream sediments, soils, plants, and waters of the western energy region of the conterminous United States: U.S. Geological Survey Professional Paper 1237, 173 p.

Eliason, P.T., Donovan, T.J., and Chavez, P.S., Jr., in press, Integration of geological, geochemical and geophysical data of the Cement oil field, Oklahoma, using spatial array processing: Geophysics, v. 48.

Foote, R.S., this volume, Significance of near-surface magnetic anomalies, *in* Gottlieb, Benjamin M., ed., Unconventional methods in exploration for petroleum and natural gas, III, Dallas, SMU Press.

Horvath, D.J., 1972, Availability of manganese and iron to plants and animals: Geological Society of America Bulletin, v. 83, p. 451-461.

Mason, A.C., 1953, The cleaning of leaves prior to analysis: Maidstone, England, East Malling Research Station, v. 40, p. 104-107.

Robinson, C.S., Mapel, W.J., and Bergendahl, M.H., 1964, Stratigraphy and structure of the northern and western flanks of the Black Hills Uplift, Wyoming, Montana, and South Dakota: U.S. Geological Survey Professional Paper 404, 134 p.

Ross, C.P., Andrews, D.A., and Witkind, I.R., compilers, 1958, Geologic Map of Montana: Montana Bureau of Mines and Geology, scale 1:500,000, 2 sheets.

Ward, F.N., Nakagawa, H.M., Harms, T.F., and VanSickle, G.H., 1969, Atomic-absorption methods of analysis useful in geochemical exploration: U.S. Geological Survey Bulletin 1289, 45 p.

Winters, J.C., and Williams, J.A., 1969, Microbiological alteration of crude oil in the reservoir: Paper presented at Symposium on Petroleum Transformation in Geologic Environments, American Chemical Society, Division of Petroleum Chemistry, Sept. 7-12, New York City, Paper PETR 86, p. E22-E31.

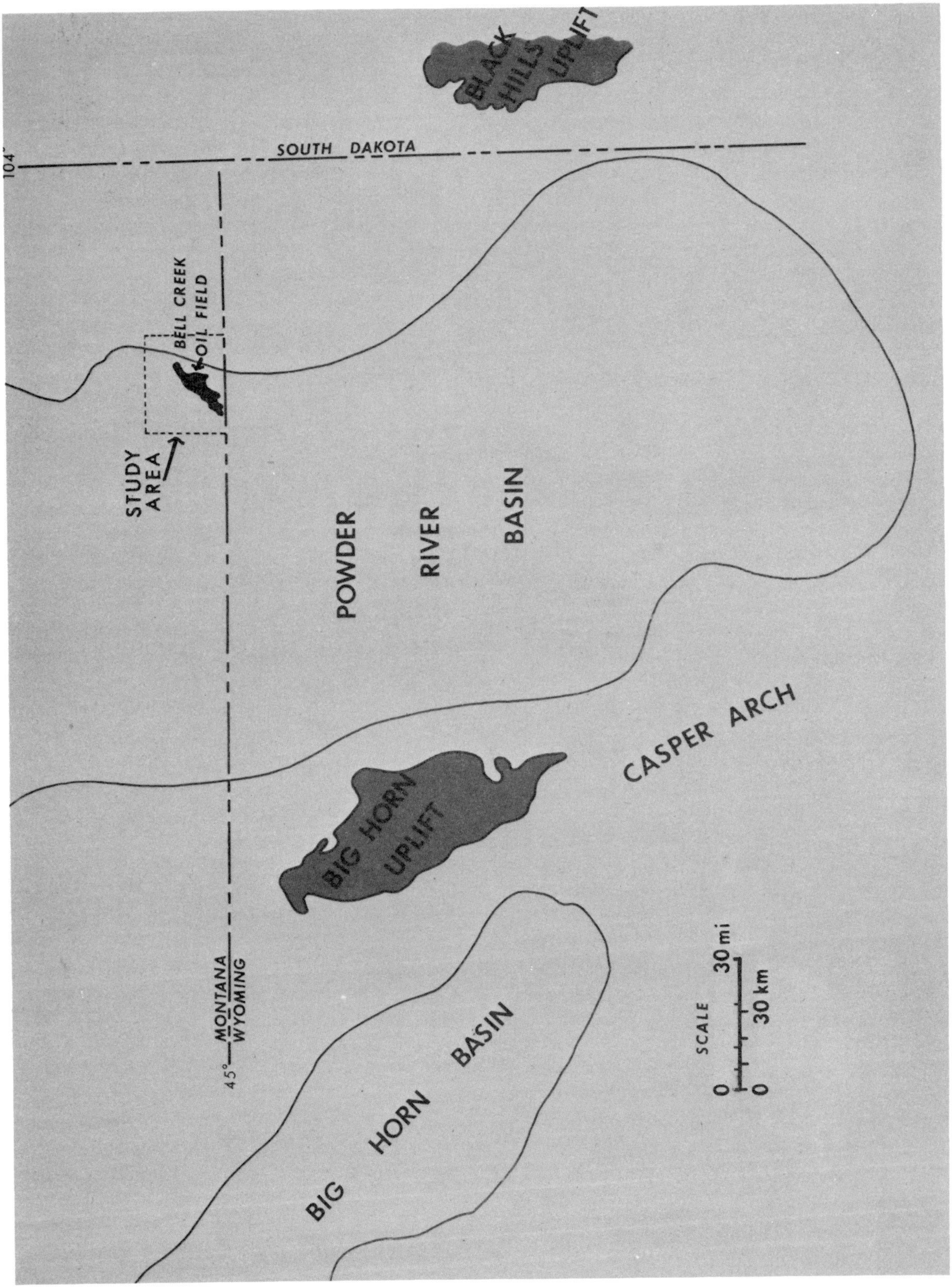

Fig. 1. Location of Bell Creek Oil Field, Montana.

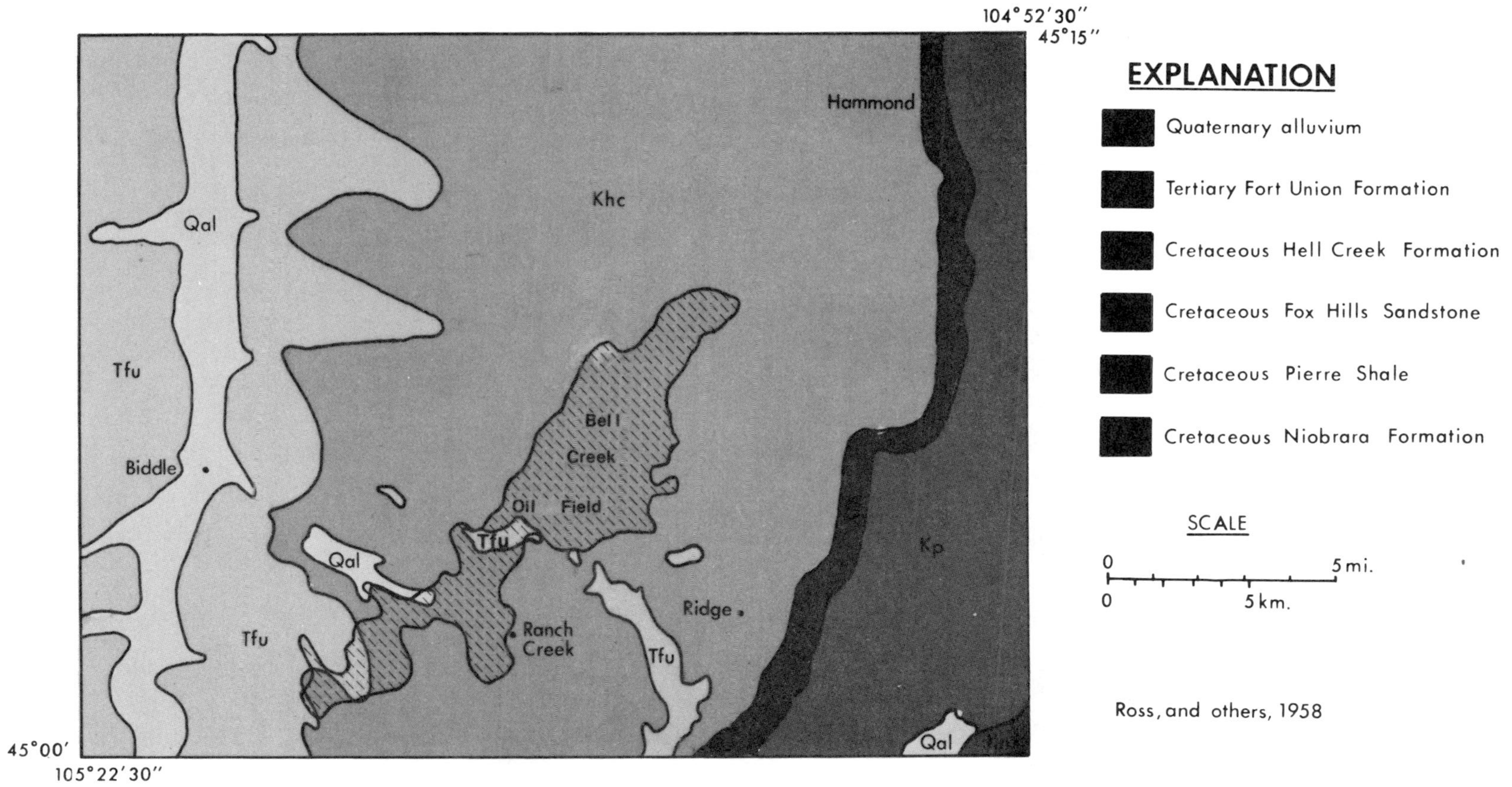

Fig. 2. Surface geology of the Bell Creek Oil Field area, Montana.

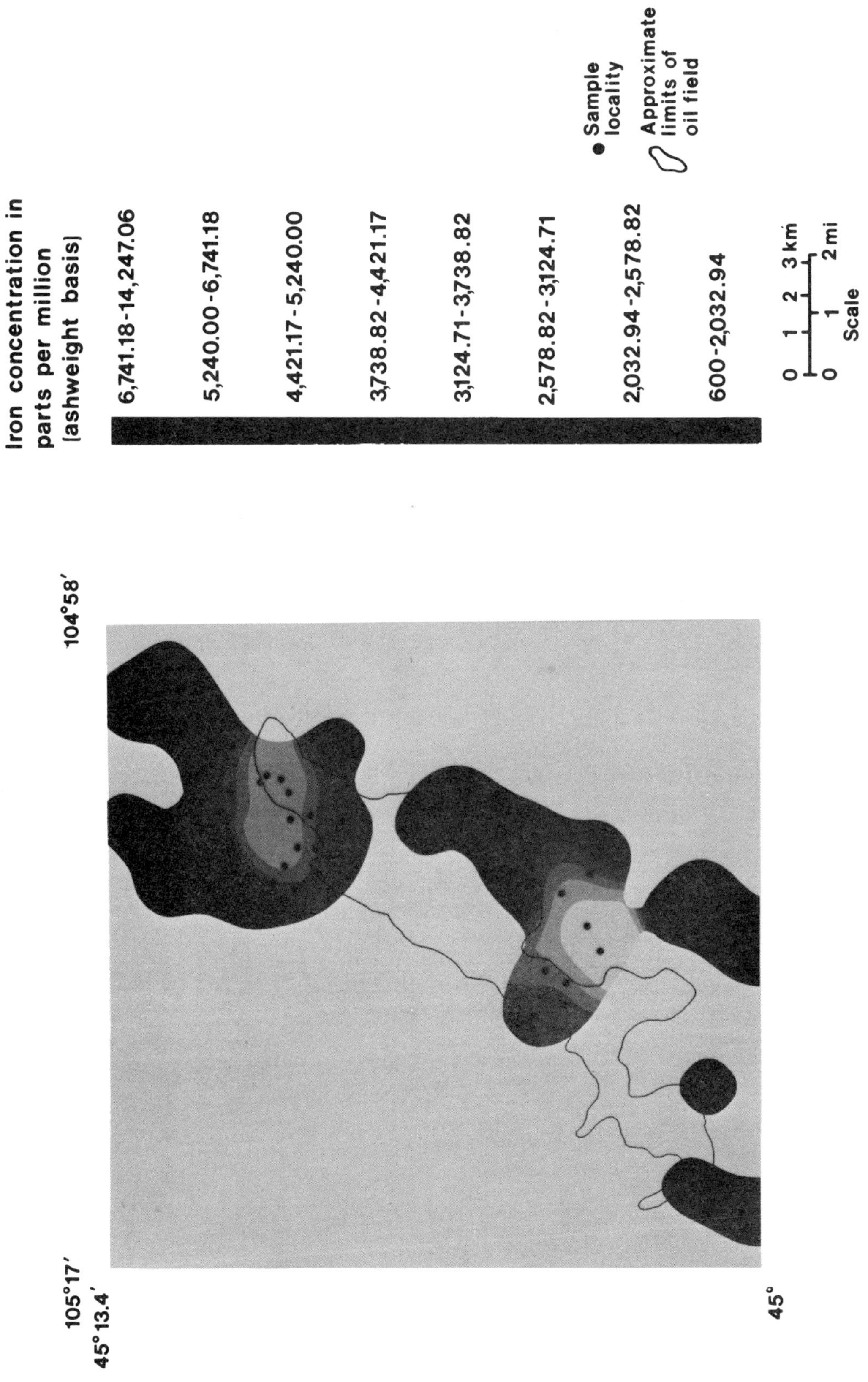

Fig. 3. Computer-generated image of iron concentration in pine needles (*Pinus ponderosa* Laws) at Bell Creek Oil Field, Montana.

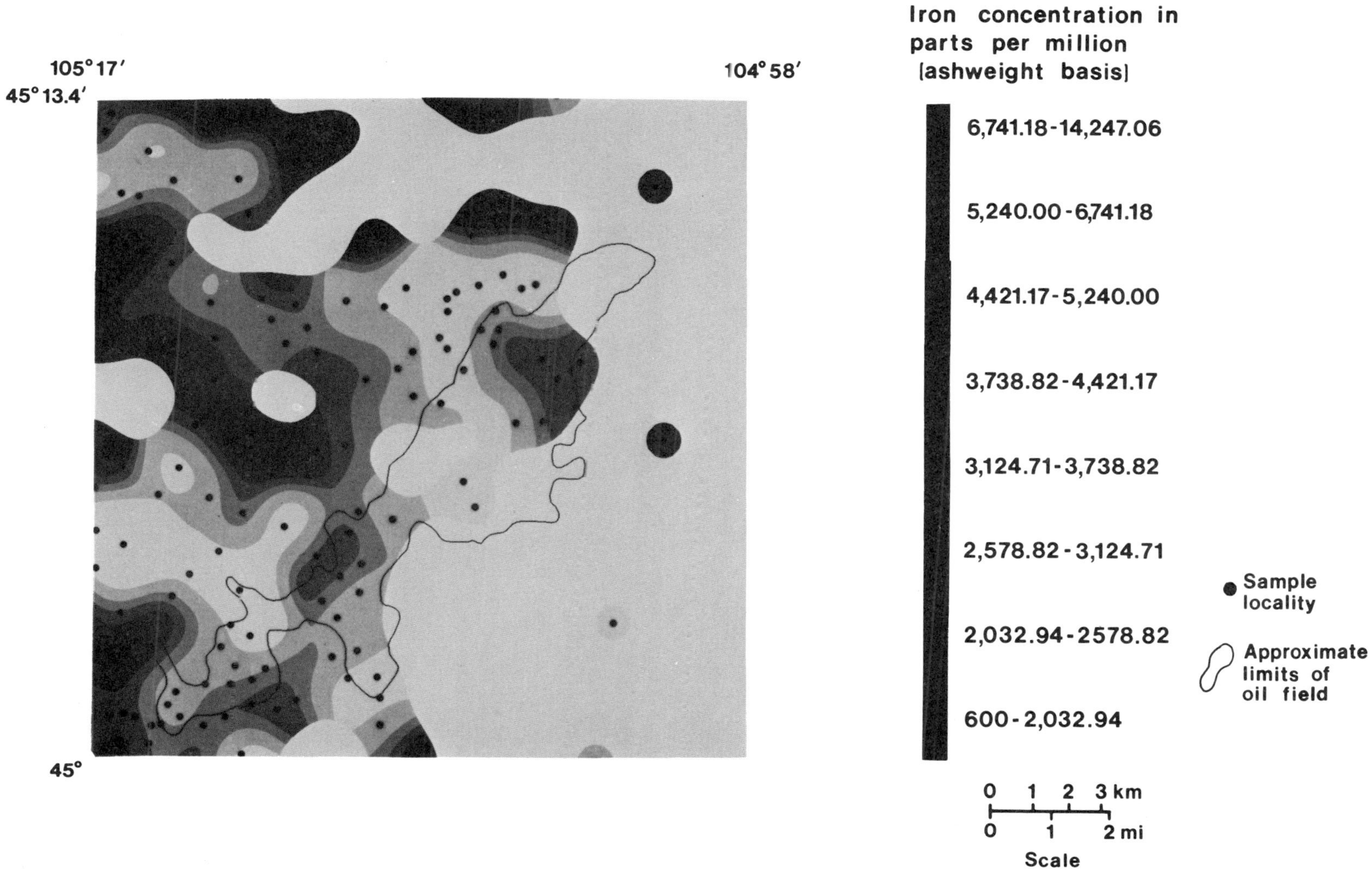

Fig. 4. Computer-generated image of iron concentration in big sagebrush leaves (*Artemisia tridentata* Nutt.), Bell Creek Oil Field, Montana.

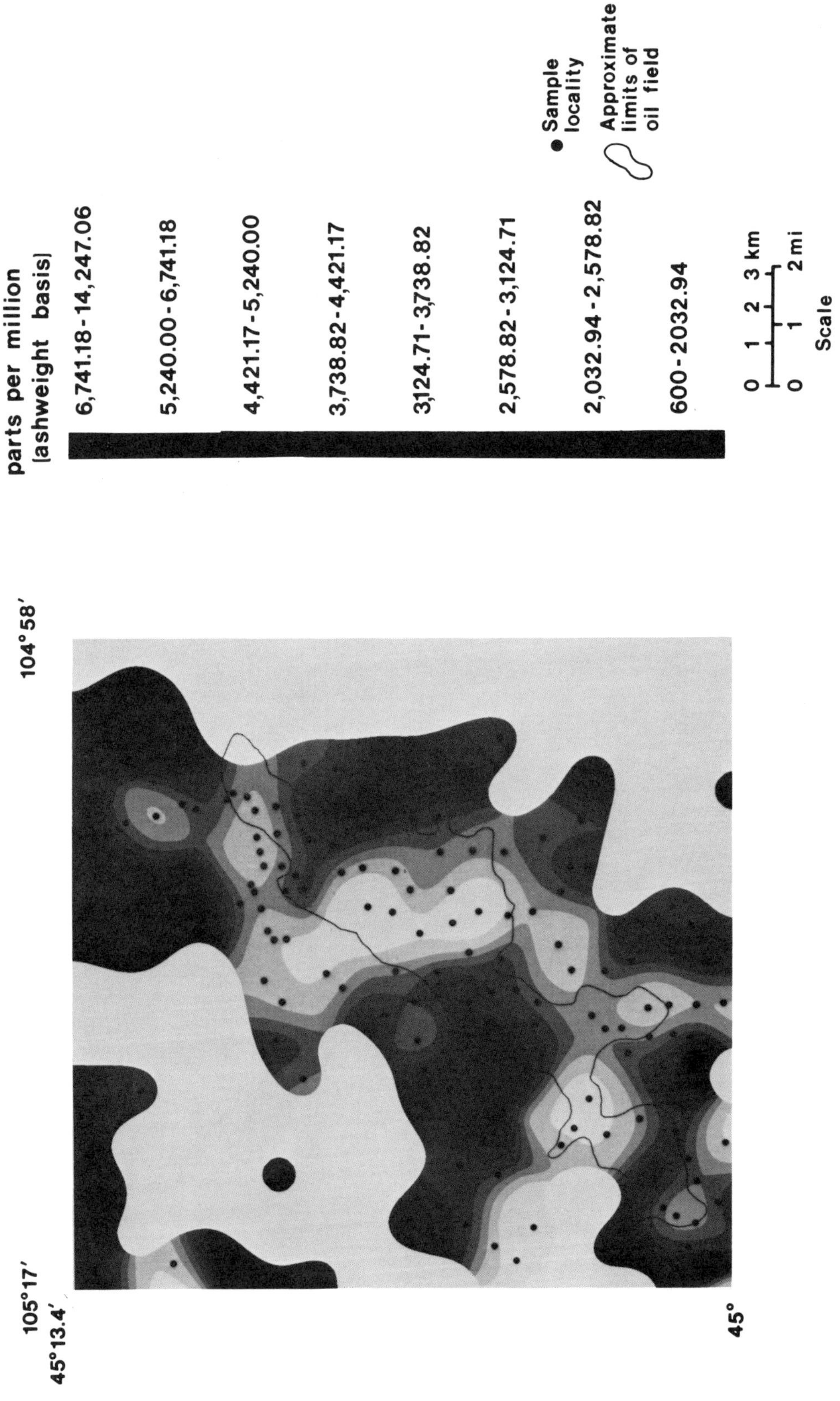

Fig. 5. Computer-generated image of iron concentration in silver sagebrush leaves (*Artemisia cana* Pursh), Bell Creek Oil Field, Montana.

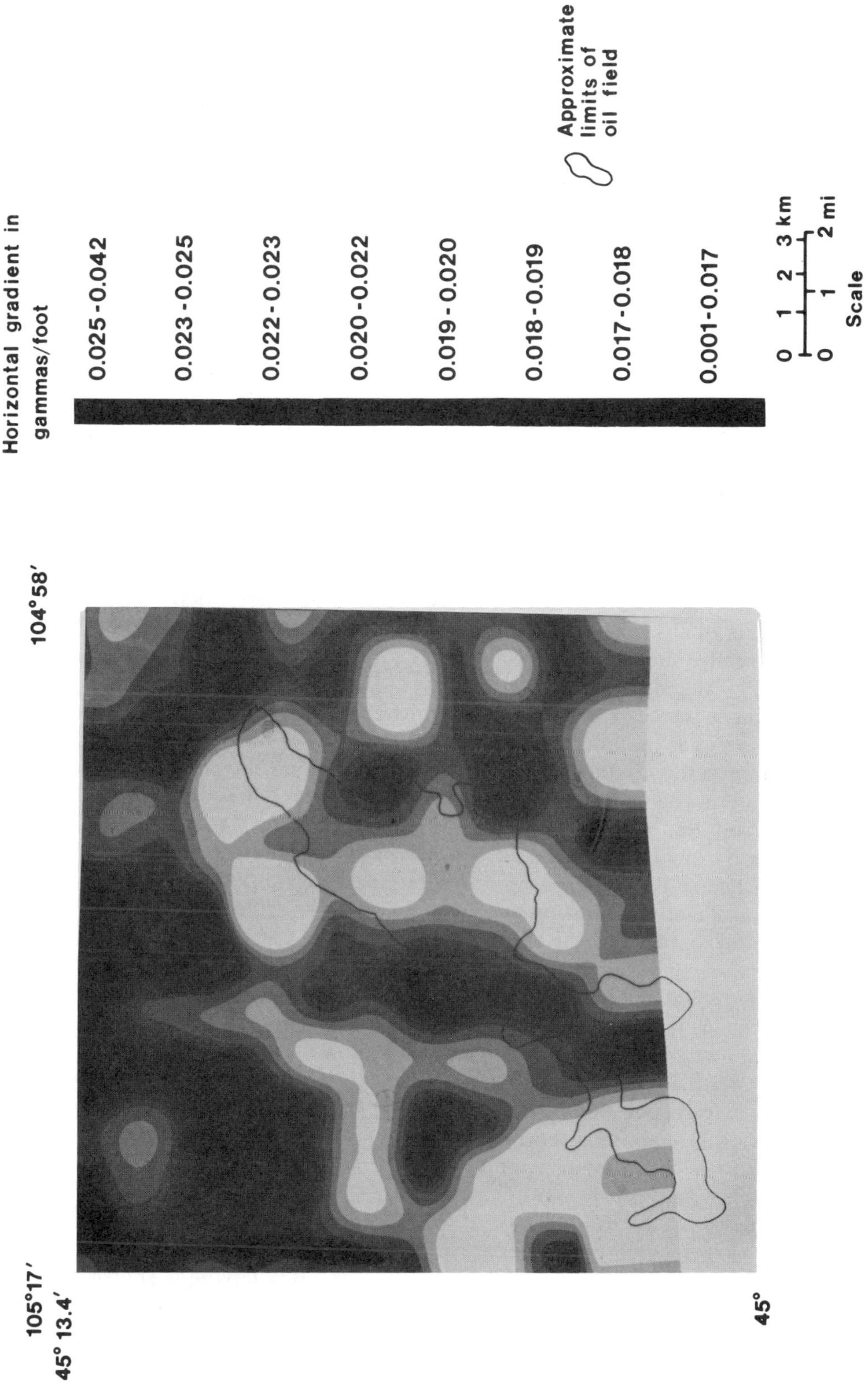

Fig. 6. Computer-generated image of aeromagnetic horizontal gradient at Bell Creek Oil Field, Montana.

THE ACHIEVEMENT OF AIRBORNE GRAVITY EXPLORATION

Sigmund Hammer

UNIVERSITY OF WISCONSIN-MADISON

ABSTRACT

The achievement of successful airborne gravity surveying represents the culmination of twenty years of intensive and expensive research. Commercial operations have now been conducted for nearly three years. In this author's opinion, based on more than fifty years' employment in gravity exploration, the precision and anomaly resolution now available are adequate for rapid initial reconnaissance of prospective areas. Industrial acceptance of airborne gravity (with simultaneous magnetics and topography) should double the effectiveness in both time and costs of geophysical exploration for petroleum.

INTRODUCTION

I especially appreciate this opportunity to discuss "The Achievement of Airborne Gravity Exploration" for a personal reason. Thirty-three years ago I gave a paper on "Modern Developments in Gravity Exploration" (Hammer, 1950) at the 75th Anniversary Celebration of the Colorado School of Mines. This event took place shortly after the achievement of airborne magnetic surveying and, quite naturally, the topic of airborne gravity came up in the discussion that followed. In response, I made a categorical statement that airborne gravity was an impossible dream. Fortunately, this dream did not fade away, and now several decades later it gives me great personal satisfaction to proclaim that I was wrong.

The reason for my negative opinion thirty-three years ago is illustrated in Figure 1. When gravity measurements are made on a platform moving in simple harmonic up-and-down motion of one foot (30 cm) with a period of ten seconds, an apparent change in gravity of nearly 12,000 milligals occurs every five seconds. This change is nearly a million times greater than that allowed for the required precision. At that time, long before the development of advanced electronic techniques, this problem seemed insuperable. But happily, airborne gravity exploration has now been in commercial operation for three years.

My purpose in this discussion is twofold: (1) to show some airborne gravity results that demonstrate the precision of the data, and (2) to discuss the usefulness and future role of airborne gravity in a more effective strategy of petroleum exploration. A few operational aspects that pertain directly to precision will be discussed in general terms to provide necessary background.

AIRBORNE GRAVITY DATA

The first gravity measurements in a moving vehicle were made sixty years ago in a submerged submarine (Heiskanen and Vening-Meinesz, 1958). The spectacular gravity anomalies found by Vening-Meinesz on the Sumatra-Java Island Arc were a precursor of the modern concept of plate subduction. The need for gravity measurements from a moving vehicle has been in the minds of geophysicists ever since.

Modern airborne helicopter gravity surveying was brought to fruition by Carson Helicopters, Inc., of Perkasie, Pennsylvania, three years ago (Navagio and Gamert, 1981; Hammer, 1982a) after twenty years of intensive and expensive research. A report on the technique was published in the *Oil and Gas Journal* of January 11, 1982. The procedure that succeeded involves the observation of a more or less square grid of intersecting lines. Flying is done only at night in a large, stable helicopter, Sikorski S61, usually in two shifts: 10 P.M. to 4 A.M., and 4 A.M. to 10 A.M. By ten o'clock in the morning, solar heat begins to stir up the turbulent atmosphere. Flight speeds range from 30 to 60 knots (55 to 100 km/hr), and the data are digitally recorded every second for rapid computer processing. At a speed of 60 knots, the data interval represents forward motion of 100 ft (30 m). Normal output of line data is 6,000 mi (10,000 km) per month.

Besides gravity and navigation data, total field magnetics and terrain clearance are also recorded every second. This record provides a very important bonus at little cost.

PRECISION OF AIRBORNE GRAVITY

The flight pattern of a recent helicopter gravity survey for an oil company is shown in Figure 2. The line spacing is 1 km in both azimuths. The area covered is about 1,000 km² (400 mi²). Flight altitude was 1,000 ft (310 m), and ground speed was 30 knots (55 km/hr). The area is a swampy jungle of extremely difficult access. The airborne survey was completed in two weeks of field work.

This survey provided a total of 2,200 km (1,400 mi) of line data, and there are more than 1,000 line crossings. At each line intersection the two independently observed values should be equal. The actual intersection difference, therefore, is a direct measure of the error. A histogram of the gravity differences (Δg) at all of the intersection points is shown in Figure 3. Eighty-five percent of the points are within the range ± 1 mgal. Each Δg is the statistical combination of the errors in two independent readings. The error in a single observation is smaller by $\sqrt{2}$. The expanded scale for this factor, drawn across the center of the diagram, shows that 85% of the points are within the range ± 0.70 mgal. This gives the probable error of a single observation as ± 0.37 mgal. It is important to point out that this procedure provides observations twice as precise as those made from current ship-borne gravity data, which, as we know, are widely used in offshore geophysical exploration.

The Δg histogram from another commercial airborne gravity survey is shown in Figure 4. This survey was made over mountainous terrain in which the topographic relief is about 5,000 ft (1,500 m). There are nearly 1,400 line intersection points in this histogram, and the indicated probable error of a single observation is 0.39 mgal, which confirms the previous determination.

A critical check of accuracy is repeat observations. A line that was reflown, not because the gravity data were unsatisfactory but because the line location deviated in part from specification, is shown in Figure 5. The flight locations are shown in the upper part of the figure. The two observed gravity profiles, observed on different days, are plotted below. The maximum difference in the Bouguer anomaly values in the two profiles is less than 1/2 mgal.

COMPARISON WITH GROUND DATA

A special airborne survey was made by Carson in August, 1981, to permit direct comparison with conventional ground data. The flight pattern (Fig. 6) includes two well-known Gulf Coast salt domes: High Island and Big Hill. Both domes have thick, shallow caprock, and their typical strong gravity minima have sharp central gravity maxima. The line spacing of the survey is 1/2 mi. Flight altitude was 1,000 ft (310 m), and flight speed was 45 knots (83 km/hr). Total length of line data is 436 mi (702 km) and was acquired in 8½ hours of flying time. The entire project was completed in six days of field work. There are 75 line intersections in the survey, and the indicated probable error is 0.38 mgal, in perfect agreement with previous determinations above.

Conventional ground gravity data were made available (but not until after print-out of the airborne gravity map) by oil companies that were interested in seeing the results of the test. A comparison of the airborne and ground data is exhibited in Figure 7. The profile crosses the centers of both salt domes, shown in cross section at the bottom of the figure. Note the great vertical exaggeration of the cross section. The caprock is very shallow.

The gravity profile at ground level is shown in the center of the figure. The airborne profile is the solid line in the upper figure. The general agreement of the two profiles is excellent except that both sharp caprock anomalies on the ground profile are greatly attenuated in the airborne data. The result of three-dimensional upward continuation of the ground data to the flight altitude is shown by the dashed line in the upper figure. The caprock attenuation in the airborne data is confirmed by the continuation. Aside from the areas of the central caprock, the two sets of data agree to better than one-half mgal. A more convincing test hardly seems necessary.

ANOMALY ATTENUATION

A simulated calculation of the caprock attenuation problem at Big Hill and High Island domes is shown in Figure 8. The anomalous mass that represents the caprock in the simulation is a thin circular disk of appropriate diameter and depth. Curve 1 is the calculated gravity anomaly on the ground surface. Curves 2 to 6 are the anomalies at increasing flight elevations indicated by depths to the corresponding dashed masses. For a flight elevation equal to the depth of burial of the mass, Curve 2 shows an anomaly attenuation of 15%. Curve 6 shows that attenuation of the anomaly at a flight elevation ten times the depth of the mass is 81%. This simulation confirms the weak observed airborne anomalies of the Big Hill and High Island caprock.

These calculations simulate three-dimensional atenuation. For a two-dimensional mass extending to infinity in both directions from the plane of the figure, the anomaly at ground level is essentially identical to Curve 1, but becomes Curve 7 at flight elevation corresponding to Curve 6. The anomaly attenuation on Curve 7 is 58%. This demonstrates weaker attenuation than would have been indicated if the ground data had been continued upward from the profile data alone.

A theoretical analysis of the gravity attenuation problem is illustrated in Figure 9. Attenuation affects both magnitude and width. Note that when the radius of the circular disk becomes large, the attenuation reduces to zero. The rapid attenuation for small anomalies and the negligible attenuation for large anomalies is characteristic of anomaly resolution in airborne potential field surveys and must not be ignored in planning survey specifications, such as flight elevation and speed.

OPERATIONS

A full discussion of the many operational problems in airborne gravity surveying (Navazio and Gumert, 1981) is beyond the scope of this paper. There are, however, three factors that have been of special interest to many geophysicists. These are (1) the Eötvös correction, (2) elevation and navigation control, and (3) terrain corrections for irregular topography. Brief explanations of these three operations follow.

The Eötvös Correction

The Eötvös correction removes the effect of centripetal acceleration in a vehicle that moves along the curved surface of the earth. The mathematical theory gives the following equations:

$$E = 7.5028V \cos\phi \sin\alpha + 0.004154V^2$$

Mgals for V in knots

$$dE = (\delta E/\delta\phi)d\phi + (\delta E/\delta\alpha)d\alpha + (\delta E/\delta V)dV$$
$$= 0.13095V(\cos\phi \cos\alpha\, d\alpha - \sin\phi \sin\alpha\, d\phi) + (7.5028 \cos\phi \sin\alpha + 0.008308V)dV$$

V = ground speed in knots
α = heading clockwise from North
ϕ = latitude
$d\alpha$ and $d\phi$ in degrees

The Eötvös correction (E) is maximum for east-west headings. But the sensitivity to error is maximum in north-south headings. To avoid the latter problem, the flight grid is skewed 10° from the north-south orientation, as shown in Figure 2.

A typical example of the attained precision of the Eötvös correction is indicated in Figure 10. The Eötvös correction in Carson's procedure is determined by electronic tracking from a pair of fixed navigation-control stations. One track-pair will determine the correction, but with four control stations a maximum of six pairs is possible. The curves on the figure show the variations in two 2-track and one 4-track determinations for the line in Figure 5. Regular field practice is to discard and re-fly lines for which the average inconsistency of the tracking pairs exceeds 0.2 mgal.

Elevation Control

Elevation control for Carson's airborne gravity is obtained by a combination of three sets of data: pressure altimeters, radar terrain clearance over known points, and multiple electronic navigation tracking. A locked-in vertical channel of the autopilot flies on a given barometric pressure surface. A human pilot cannot sustain the necessary flight accuracy for more than a few minutes. The parking spot for the helicopter is surveyed with satellite positioning equipment, and precise X, Y, and Z are determined. The elevation of the flight with respect to sea level is established by radar over the parking spot at the beginning of a flight and is checked over the same spot at the termination of the flight four to six hours later. Relative barometric elevations along a flight line are essentially constant to within a few centimeters. Differences in actual elevation of intersecting flight lines due to barometric changes are adjusted to the same average altitude for final data reductions. Flights over water are useful to confirm the procedure and also to calculate vertical accelerations in the helicopter.

Terrain Corrections

Terrain corrections are evaluated from the topographic contour map that is derived from the continuous terrain clearance data along the intersecting flight lines. It is not generally appreciated that terrain corrections are much less of a problem in airborne gravity measurements at some distance above the topography than in ground surveys (Hammer, 1982b). The order-of-magnitude attenuation with altitude is illustrated in Figure 11. A gravity survey in an area of rugged topography with relief up to 3,000 ft (900 m) had terrain corrections for the airborne data of less than one mgal. For ground stations, the terrain corrections were an order of magnitude larger.

In addition, for unmapped areas the topographic map provided by the airborne survey is useful for geological and engineering purposes, a valuable bonus.

FUTURE ROLE OF AIRBORNE GRAVITY

To examine the future role of airborne gravity in geophysical exploration for petroleum, let us begin by taking a look at our present strategy. There are essentially two steps in any effective strategy. The first step is to make a rapid, low-cost reconnaissance to find and localize any interesting prospects that may exist in an area. The second step is to make a detailed examination of the indicated prospect to guide wildcat drilling. The methods to be used for such a multitrack procedure are, in principle, quite different.

In the early days of geophysical exploration, initial reconnaissance was done by gravity and magnetics. The seismograph followed to detail the localized prospects. This multitrack exploration strategy was spectacularly successful and prospered right through the great depression of the 1930s.

But times have changed. The seismograph now dominates both steps. Trade journals are filled with advertisements for seismic reconnaissance surveys with line spacing of 10 to 20 mi and more. This method of surveying is very slow and expensive. Seismic work costs about $5,000 per mi, and the average rate of progress is 40 mi per month. Nonseismic exploration has dwindled. Gravity and magnetics, the best reconnaissance tools in our kit, now represent less than one percent of our geophysical effort. The ratio of the costs is more than 100 to 1. How much of the work should be classified as reconnaissance is not recorded. At sea it is undoubtedly almost all of it. But on land, which is by far our major interest, I believe that probably 50% of the seismic work is reconnaissance—say about $500,000,000 in 1981 on land in the United States. The relative ratio of the two methods for reconnaissance is about 50 to 1. The rate of progress is no more than 1 in 20. Including both factors, the overall ratio of seismic reconnaissance to gravity is 1,000 to 1. Is this a proper, or even sensible, balance of our geophysical strategy?

A closer comparision of the seismic and gravity effort is tabulated below.

Geophysical Exploration
For Petroleum
1981
U.S. Dollars in Millions

	World		United States	
	Land	Total	Land	Total
Seismic	$2,371.0	$2,968.0	$1,139.0	$1,344.0
Gravity	14.2	15.2	8.9	9.4
Ratio	168	195	128	143

Another important element in an effective exploration strategy is adequacy of the data. Figure 13 illustrates a useful analysis of the probability of success as a function of the spacing with a square grid of data (Agocs, 1955). Success is defined as having at least one line of data to intersect a prospect. As an example, assume that the line spacing is six miles in a square grid. A circular oil field covering 10,000 acres (diameter 4½ mi) has a probability of discovery of 75%; in other words, only three out of four prospects have a reasonable chance of discovery. If the data consisted of parallel lines, without cross lines (which would be typical for seismic surveys), the odds are reduced by one-half and the probability of success becomes 50-50. For elongated anticlines and faults, the odds are 34% less. This is not adequate reconnaissance. With airborne gravity, the needed closer line spacing, entirely independent of accessibility conditions on the ground, is feasible and economic.

Advantages of Airborne Gravity

It is hard to overestimate the advantages of airborne gravity in our present strategy of exploration. A partial list of these advantages follows:

1. Virtually unlimited accessibility
2. Economic close-line spacing
3. Very rapid coverage and reports
4. Precise interpolation between widely spaced, reconnaissance seismic surveys

5. Double overall effectiveness of geophysical exploration for petroleum in both time and cost
6. Regional tectonic studies of faults and trend alignments
7. Geodetic, tectonic, and military applications
8. Simultaneous total-field magnetics and topography

I have no doubt that there are other advantages.

Is there a need for continued geophysical reconnaissance in petroleum exploration? Opinions regarding this question are not unanimous. Halbouty and some others are emphatic that there is urgent need. Halbouty's (1977) world map (Fig. 14) shows that there are 400 prospective sedimentary basins on earth that have been explored inadequately or not at all.

CONCLUSION

To conclude, let us look at a spectacular case of "what might have been." Figure 15 is a gravity map of a portion of the Texas Gulf Coast that I made in 1937, immediately after the advent of the modern gravimeter. We are deeply indebted to the Gulf Oil Corporation for granting publication of this interesting old map. The cost of the gravity survey was $50,000 in 1937 dollars. The coincidence on the map of gravity anomalies and oil fields (areas outlined in black) is outstanding. If this map had been available in 1901, the year the Spindletop Oil Field was discovered, nearly all the oil in the area could have been discovered at a geophysical cost of 100 barrels per penny.

Is it unrealistic to think that in the 400 unexplored basins on Halbouty's map (Fig. 14) there is not one or several such examples waiting to be found? Airborne gravity and magnetics are surely the way to lead the search, particularly because seismic interests keep calling for more and more seismic work at enormous costs. What the petroleum industry really needs is fast, adequate land reconnaissance at 100 times the rate of coverage and 1/10 the cost. This capability is now available in airborne gravity and magnetics.

In my opinion, the achievement of airborne gravity is the major advance in exploration geophysics in this decade. I predict that we are on the verge of a new era.

REFERENCES

Agocs, W.B., 1955, Line spacing effect and determination of optimum spacing illustrated by Marmora, Ontario, magnetic anomaly: Geophysics 40, p. 871-885.

Halbouty, M.T., 1977, Future trends in exploration to discover new hydrogen carbon reserves, Organization of Petroleum Exporting Countries, Proceedings, October 10-12, Vienna, Austria.

Hammer, Sigmund, 1950, Recent developments in gravity prospecting: Colorado School of Mines Quarterly, v. 45(4a), p. 87-103.

_________ , 1982a, Airborne gravity is here!: Accepted for publication in The Leading Edge, Society of Exploration Geophysicists, Tulsa, Oklahoma.

_________ , 1982b, A note on airborne gravity terrain corrections: Accepted for publication in Geophysics.

Heiskanen, W.A., and Vening-Meinesz, F.A., 1958, The earth and its gravity field; McGraw-Hill, New York, p. 115-118.

Navazio, Fred, and Gumert, Wm. R., 1981, Airborne gravity—1980's. Presented at Annual Meeting and Exposition of the Society of Exploration Geophysicists, 51st, Oct. 11-15, 1981, Los Angeles. See also Oil and Gas Journal, v. 8, no. 2, p. 123-126, January 11, 1982.

Senti, R.J., 1982, Personal communication, July 1982.

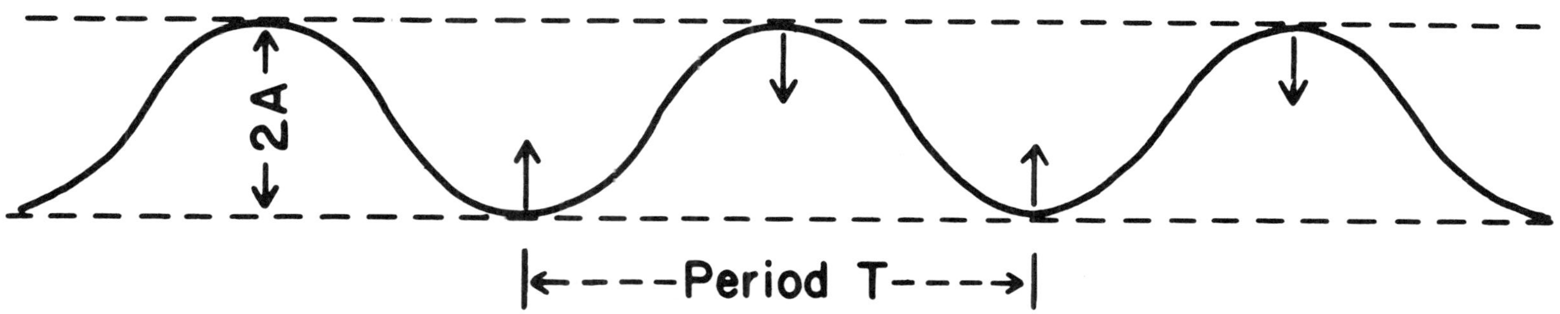

$$\pm a_{max} = 4\pi^2 A / T^2 = 592{,}176 / T^2 \text{ mgals}$$
$$A = 15 \text{ cms}$$

Period T	1"	5"	10"	30"	1'	1 hr	12 hrs
2a mgals	1,184,353	47,374	11,844	1,316	329	0.091	0.0006

Fig. 1. Acceleration in simple harmonic motion.

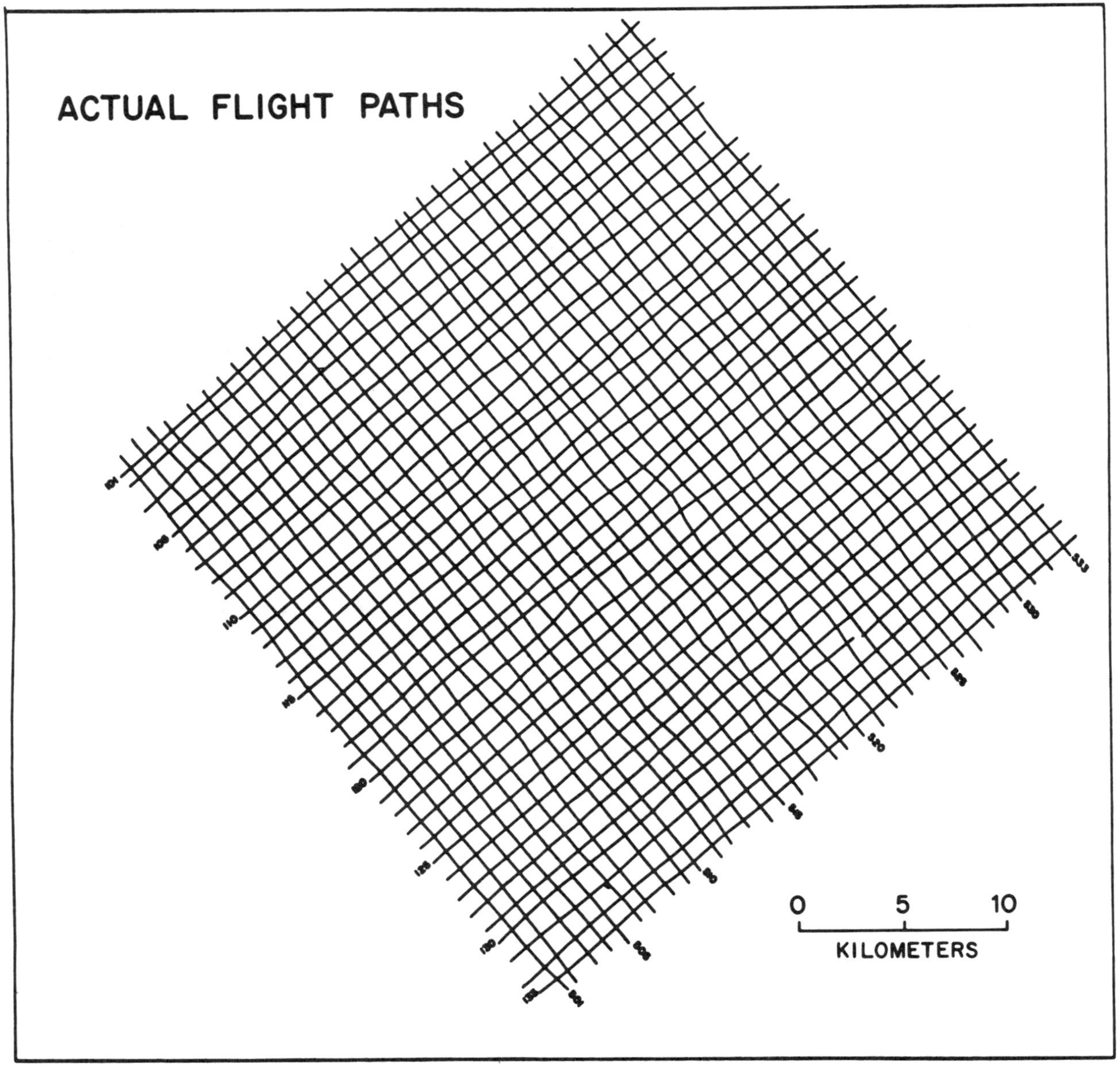

Fig. 2. Flight pattern of airborne gravity survey for an oil company. The line spacing is one km in both azimuths; the area covered is about 1,000 km² (400 mi²).

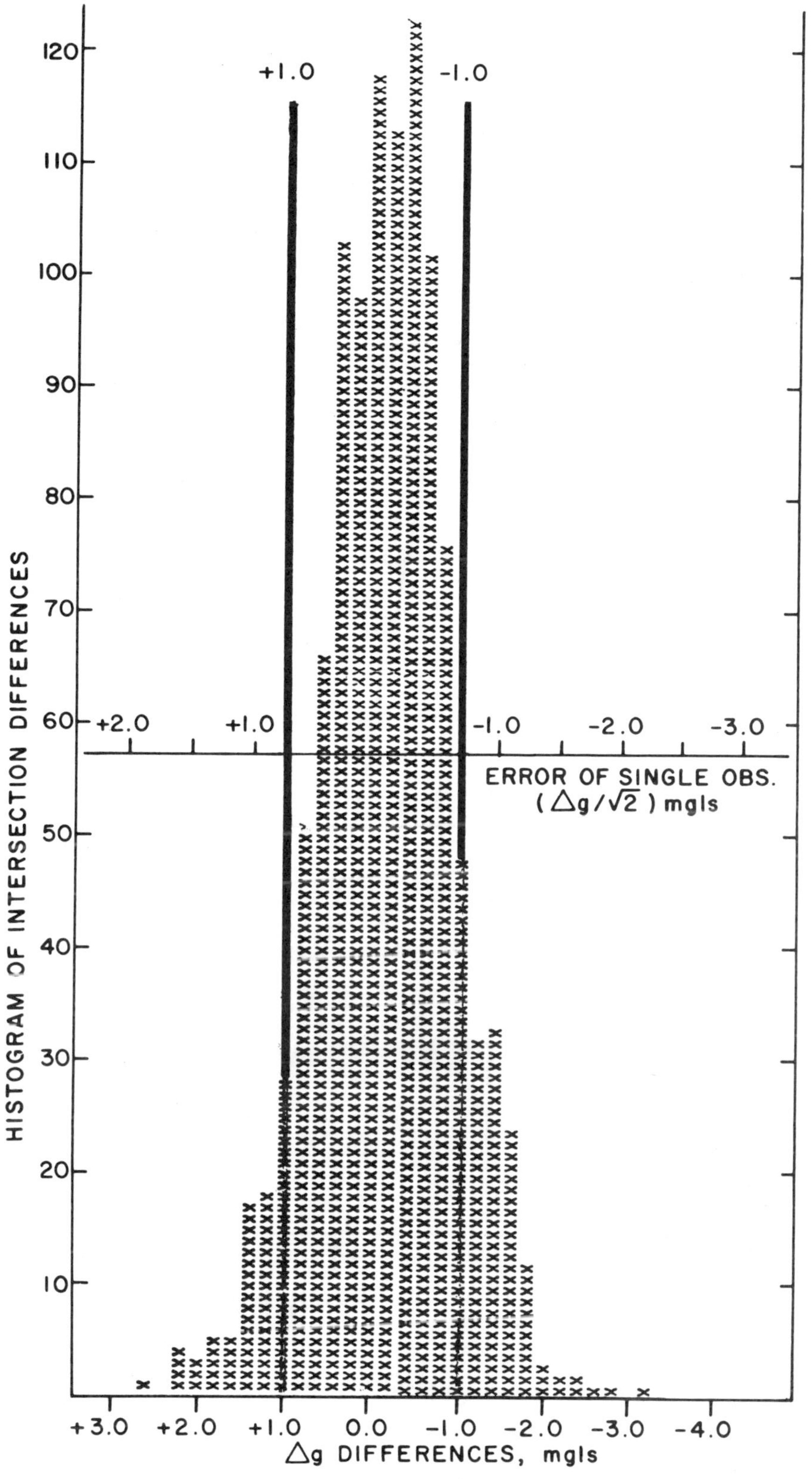

Fig. 3. Histogram of gravity differences (Δg) at line intersections in Figure 2.

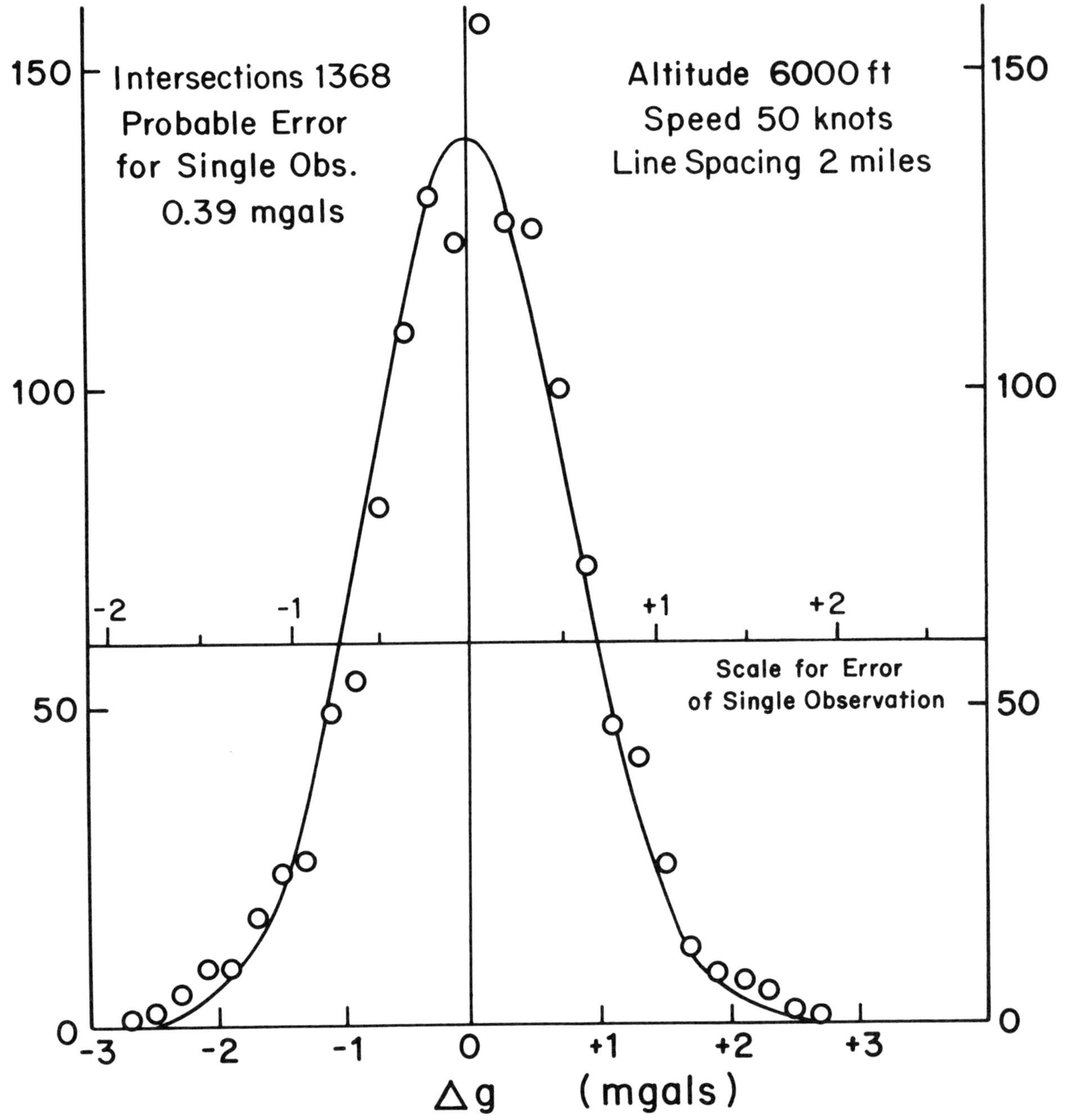

Fig. 4. Histogram of gravity differences (Δg) at line intersections in another survey.

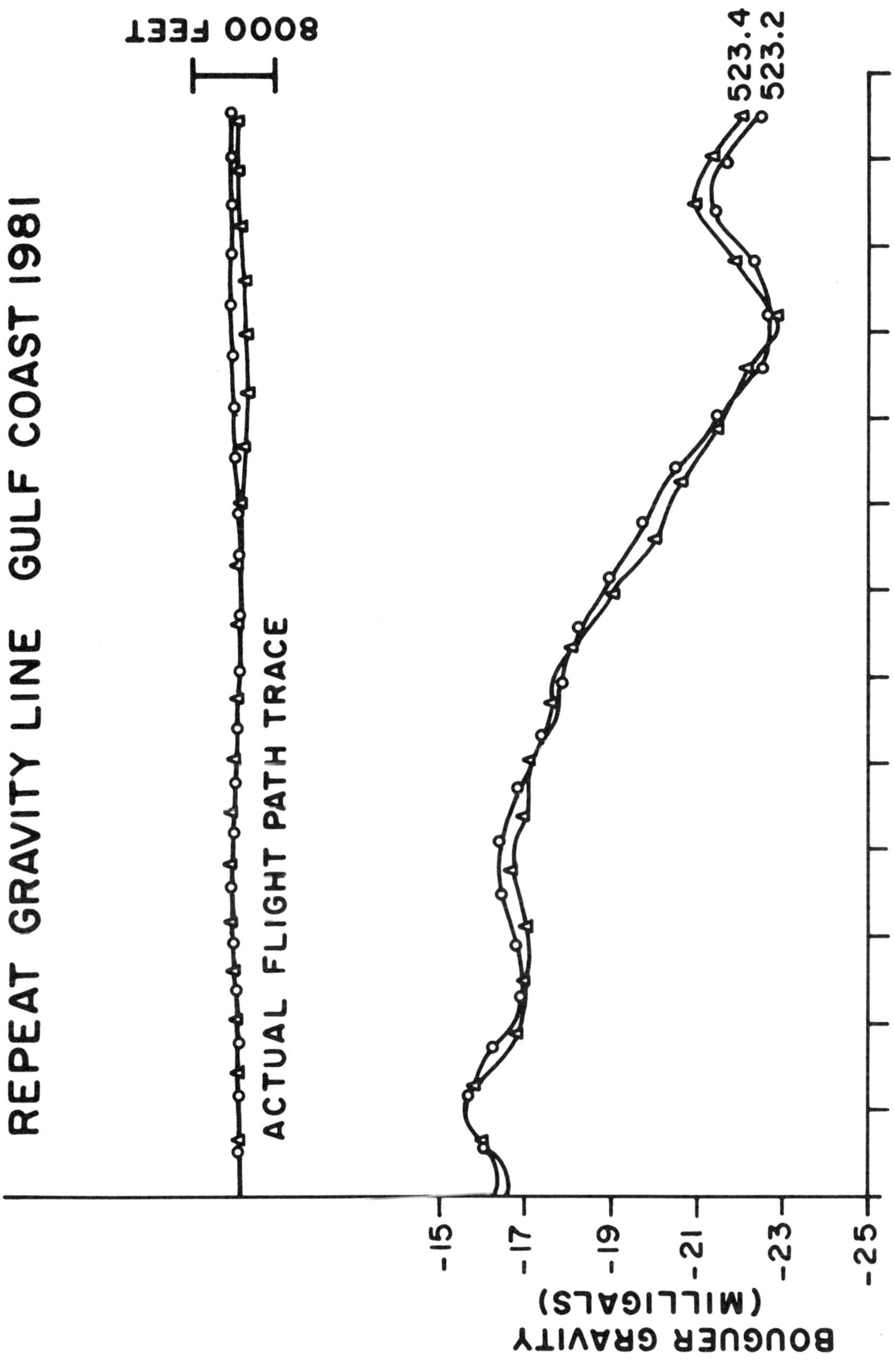

Fig. 5. Gravity profiles and location recovery in a line observed twice.

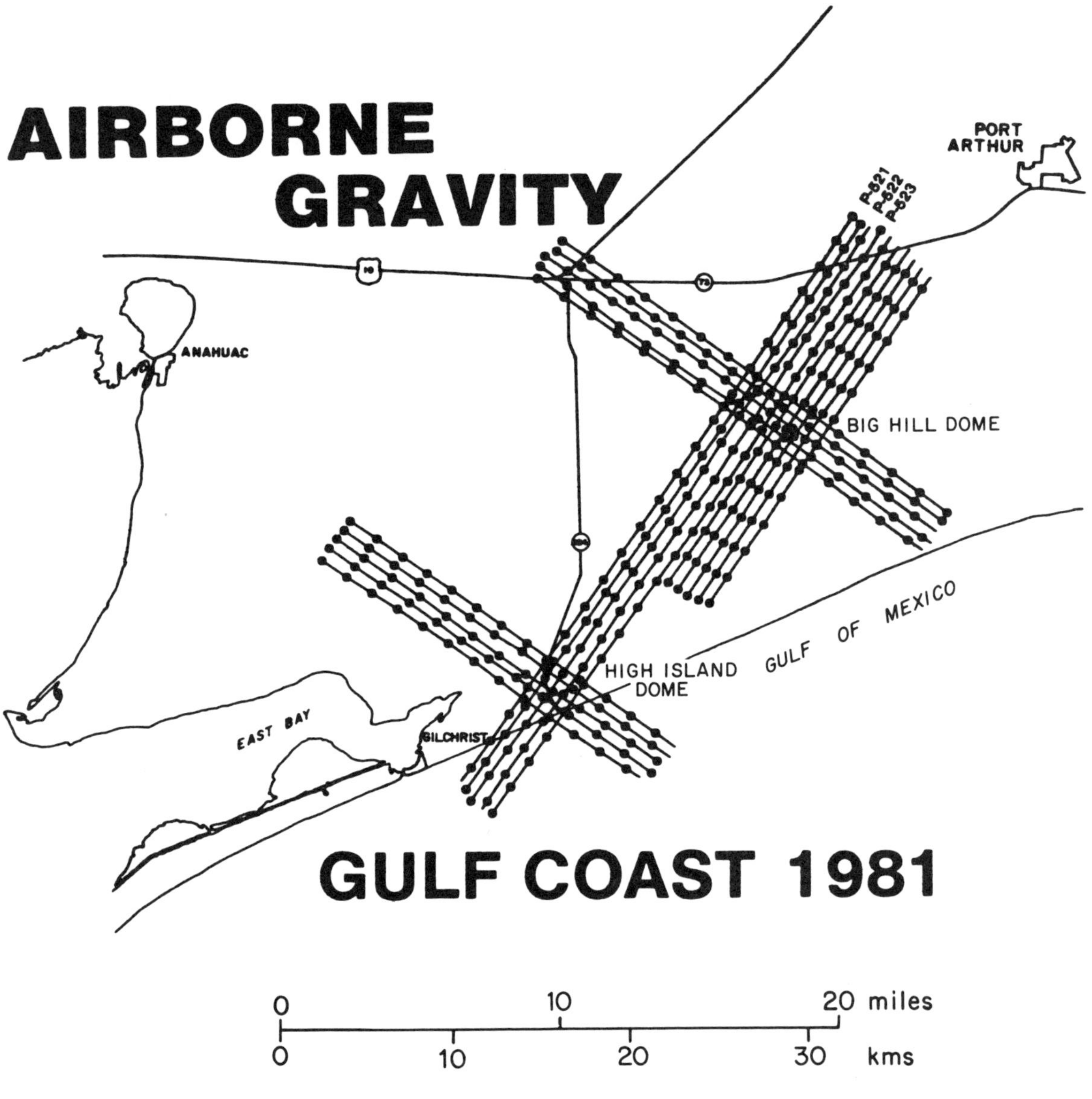

Fig. 6. Flight pattern of a test survey to compare airborne and conventional ground data.

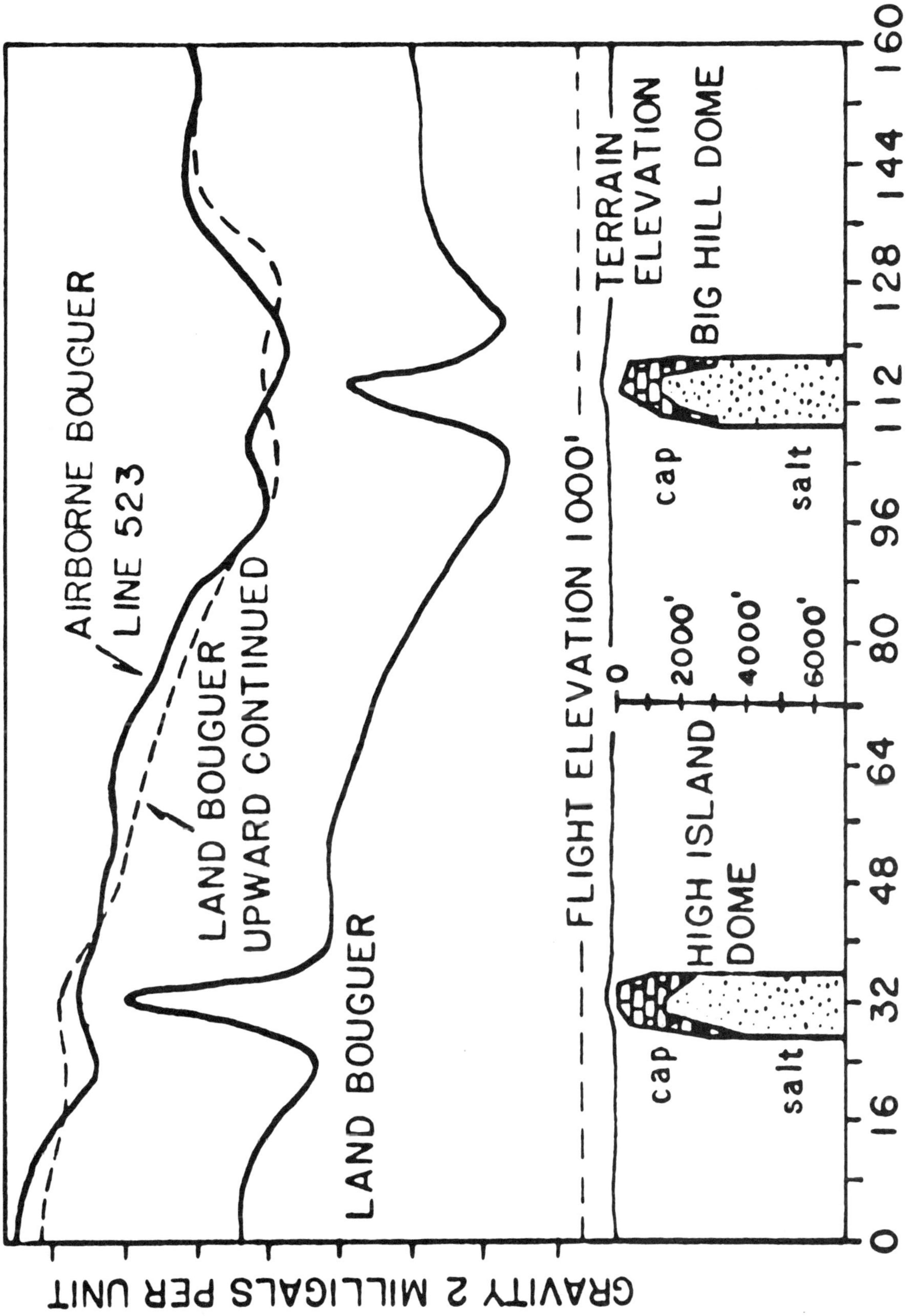

Fig. 7. The achievement of Airborne Gravity.

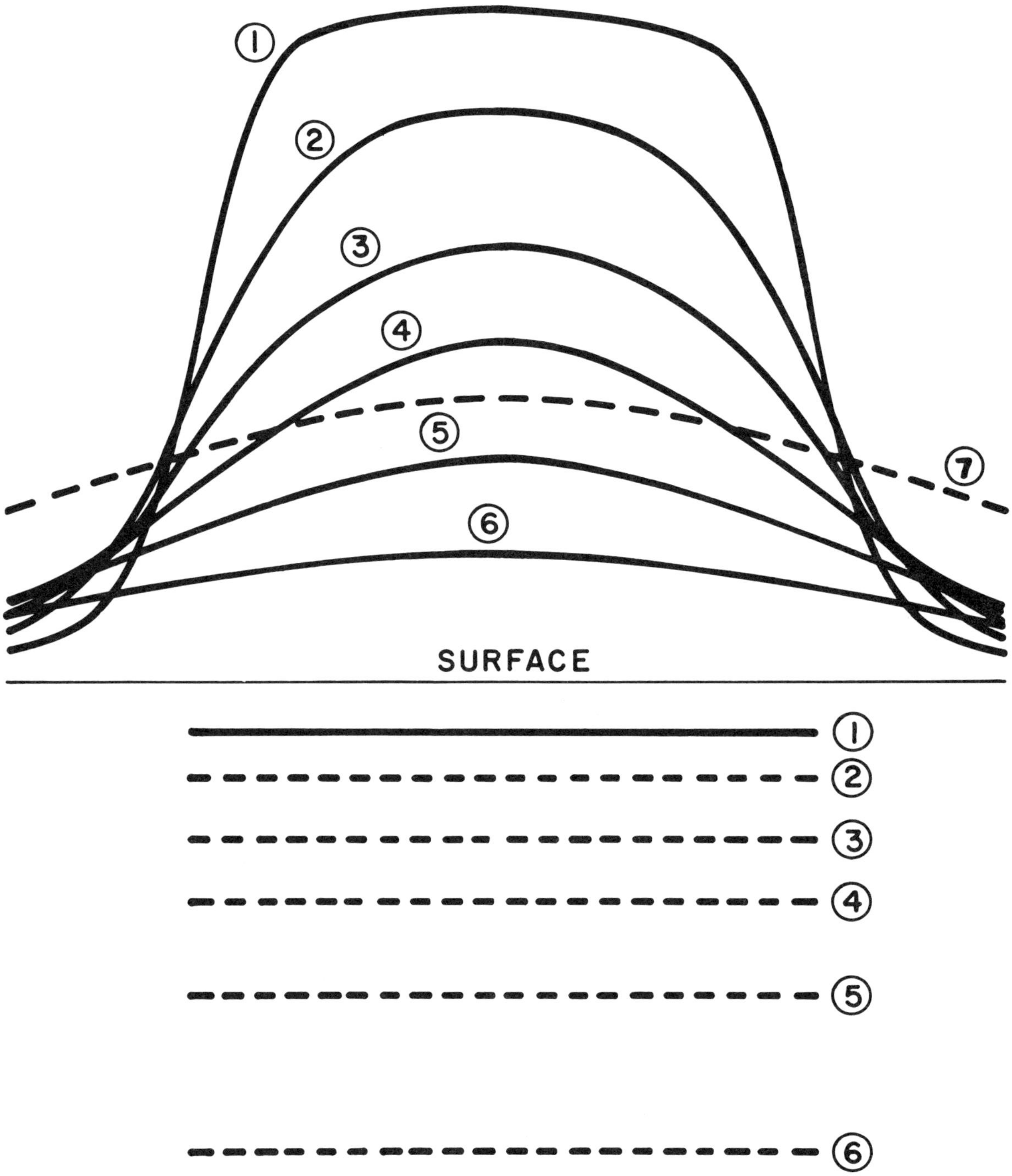

Fig. 8. Simulation of attenuation with altitude of gravity anomaly over a shallow local mass.

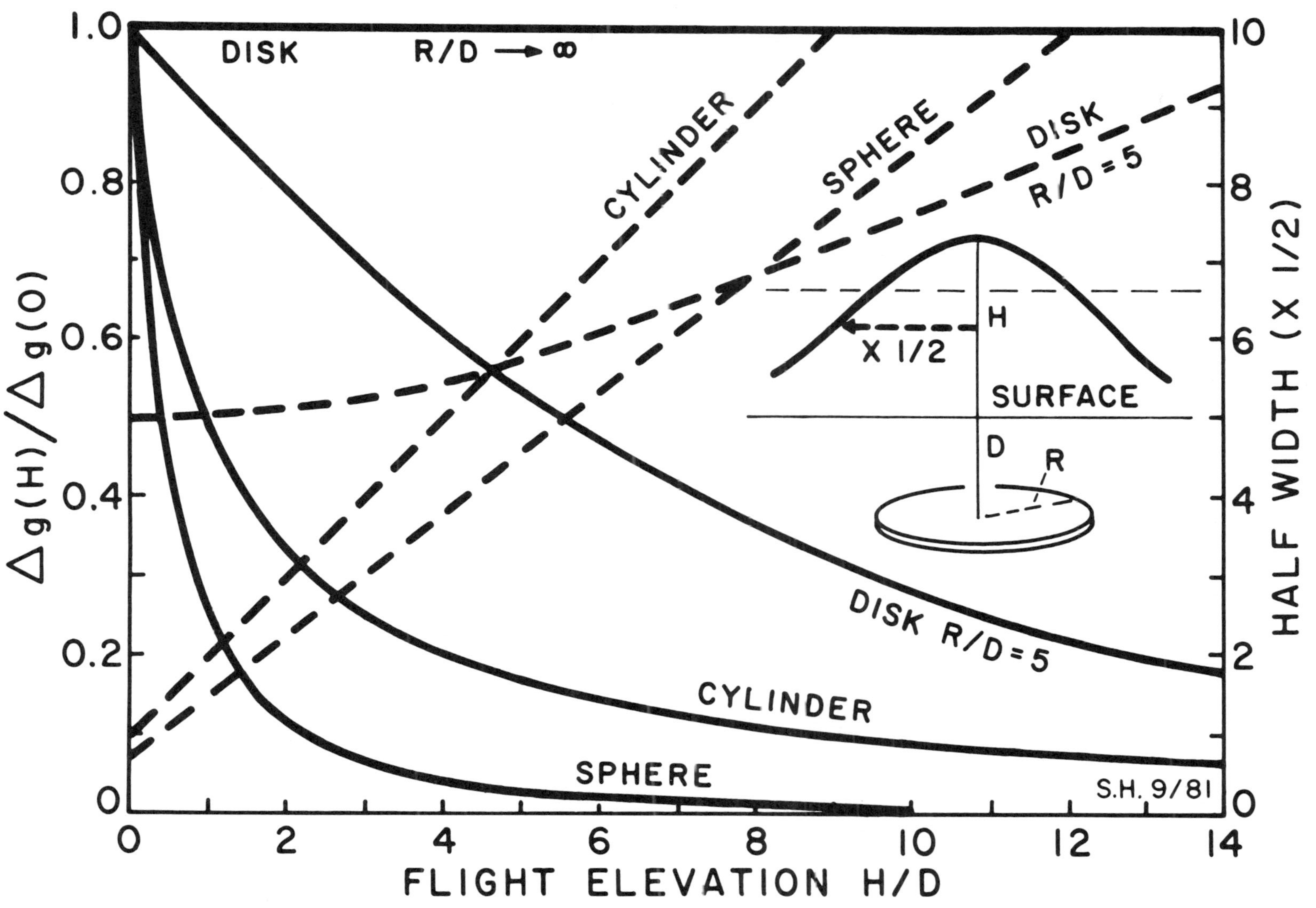

Fig. 9. Theoretical attenuation in airborne gravity of anomalies for simple masses. Solid curves anc left scale show anomaly magnitude; dashed curves and right scale show anomaly broadening $X_{1/2}$.

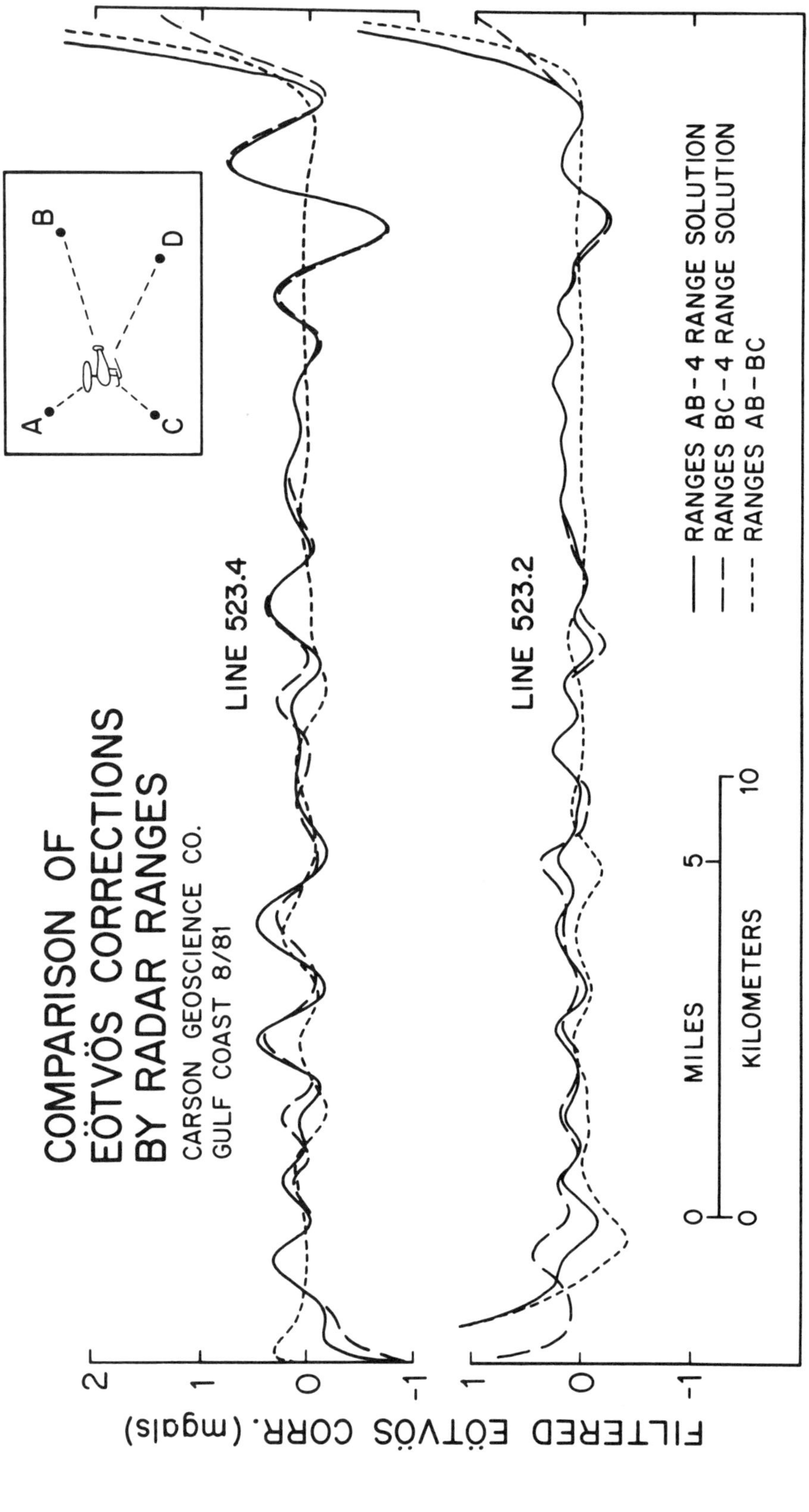

Fig. 10. Precision of the Eötvös correction in airborne gravity by the Carson procedure.

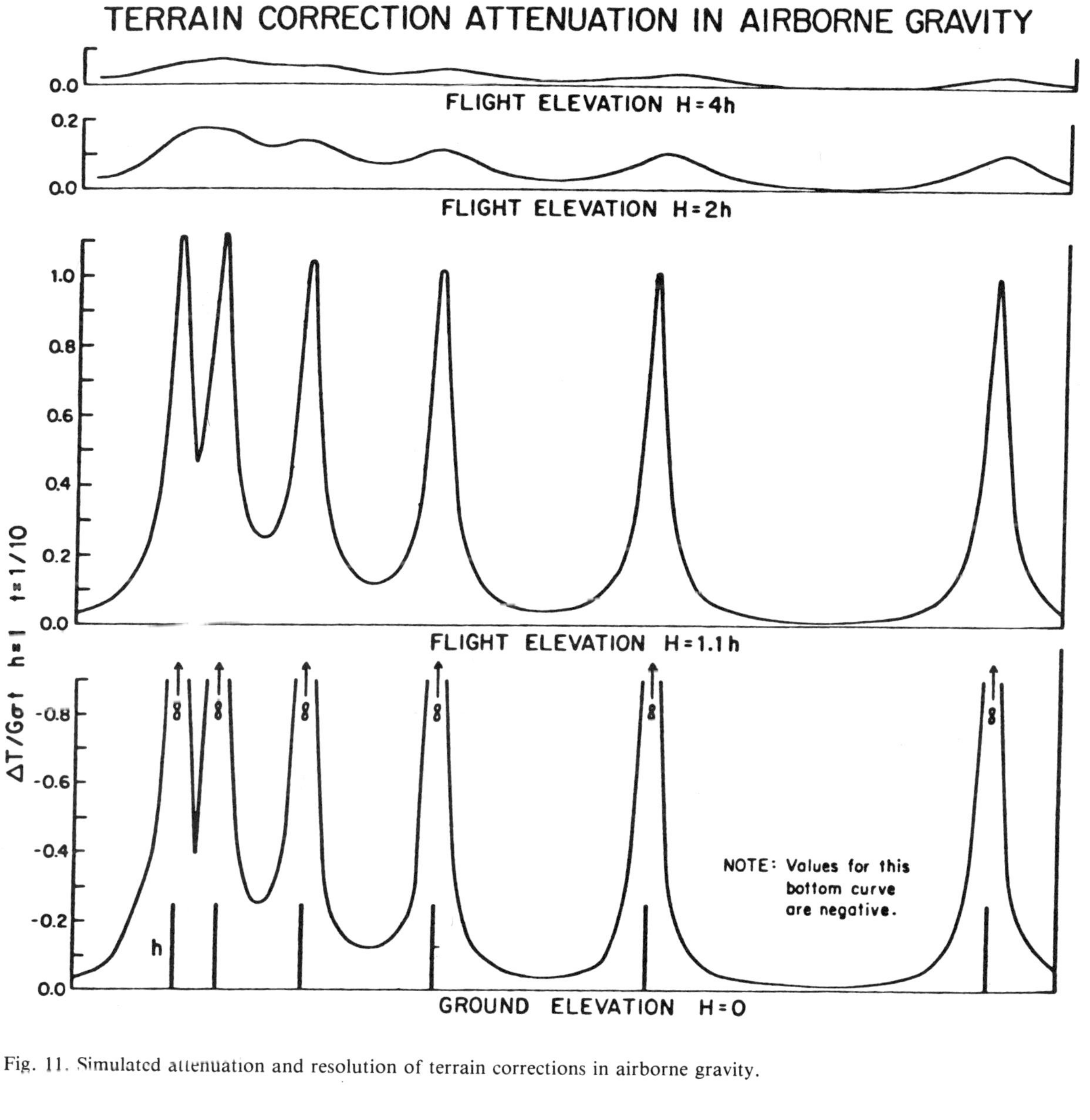

Fig. 11. Simulated attenuation and resolution of terrain corrections in airborne gravity.

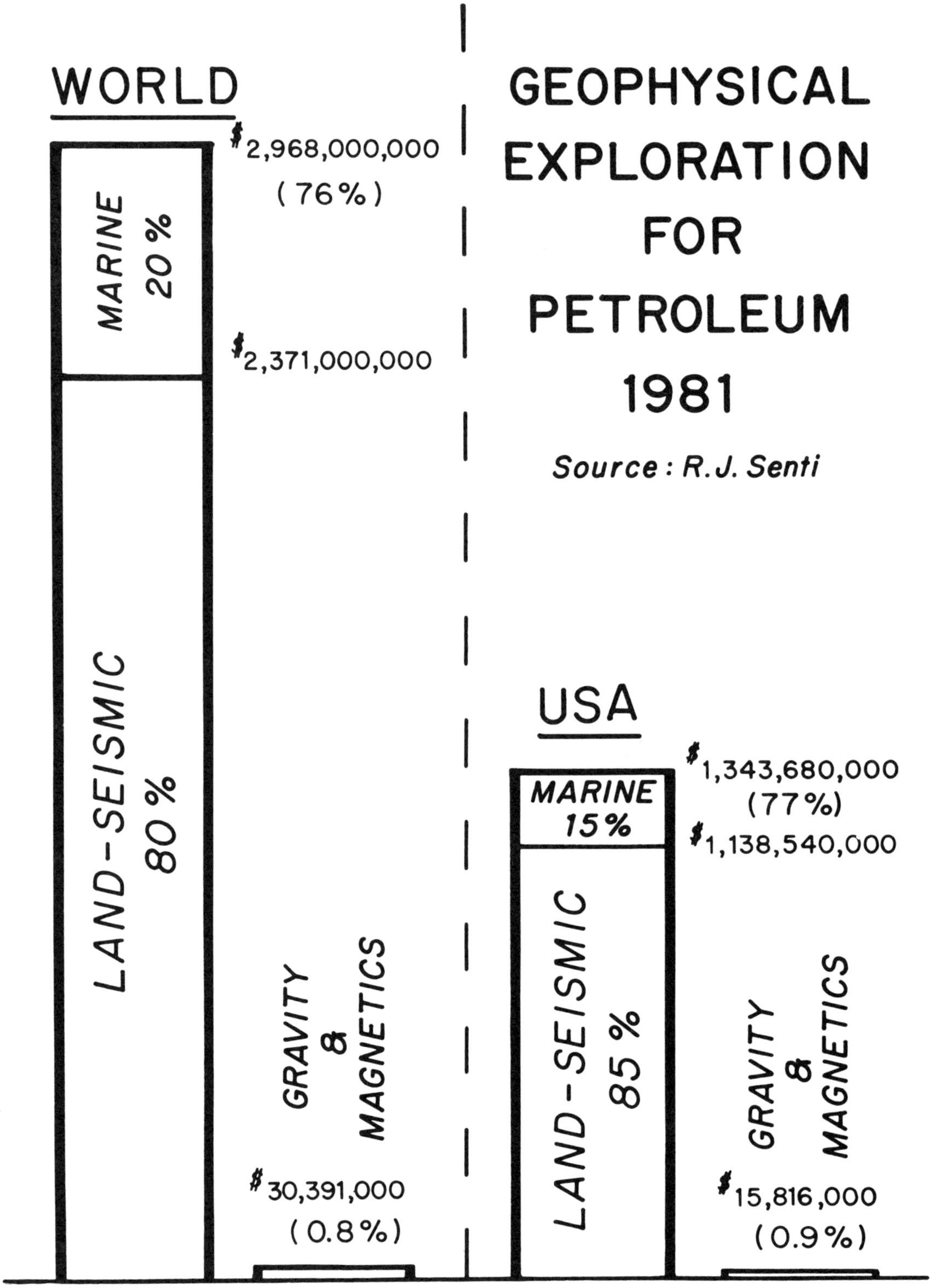

Fig. 12. Geophysical costs for petroleum exploration in 1981. By special permission of R.J. Senti, Chairman of the SEG Geophysical Activities Committee.

CIRCULAR

$P = D/S$

$(D<S)$

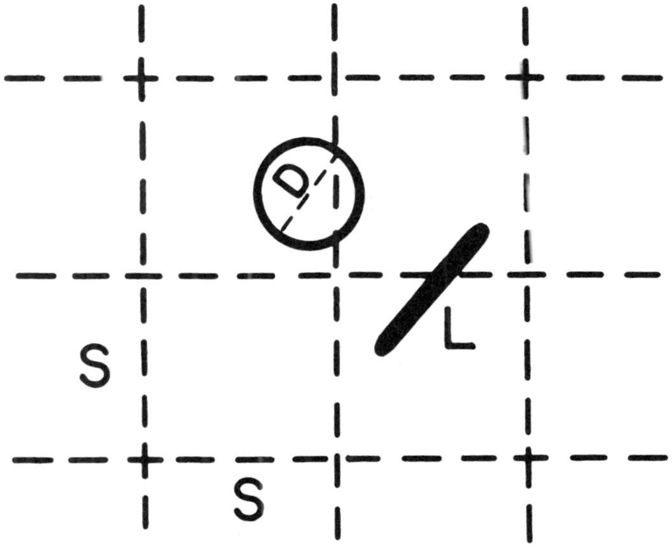

LINEAR

$P = \frac{2}{\pi} \cdot \frac{L}{S}$

$(L<S)$

CIRCULAR D/S	1.00	0.64	0.50	0.25	0.10
LINEAR L/S	—	1.00	0.9	0.39	0.16
PROBABILITY P	1.00	0.64	0.50	0.25	0.10

Fig. 13. Probability of success versus spacing of data in exploration reconnaissance.

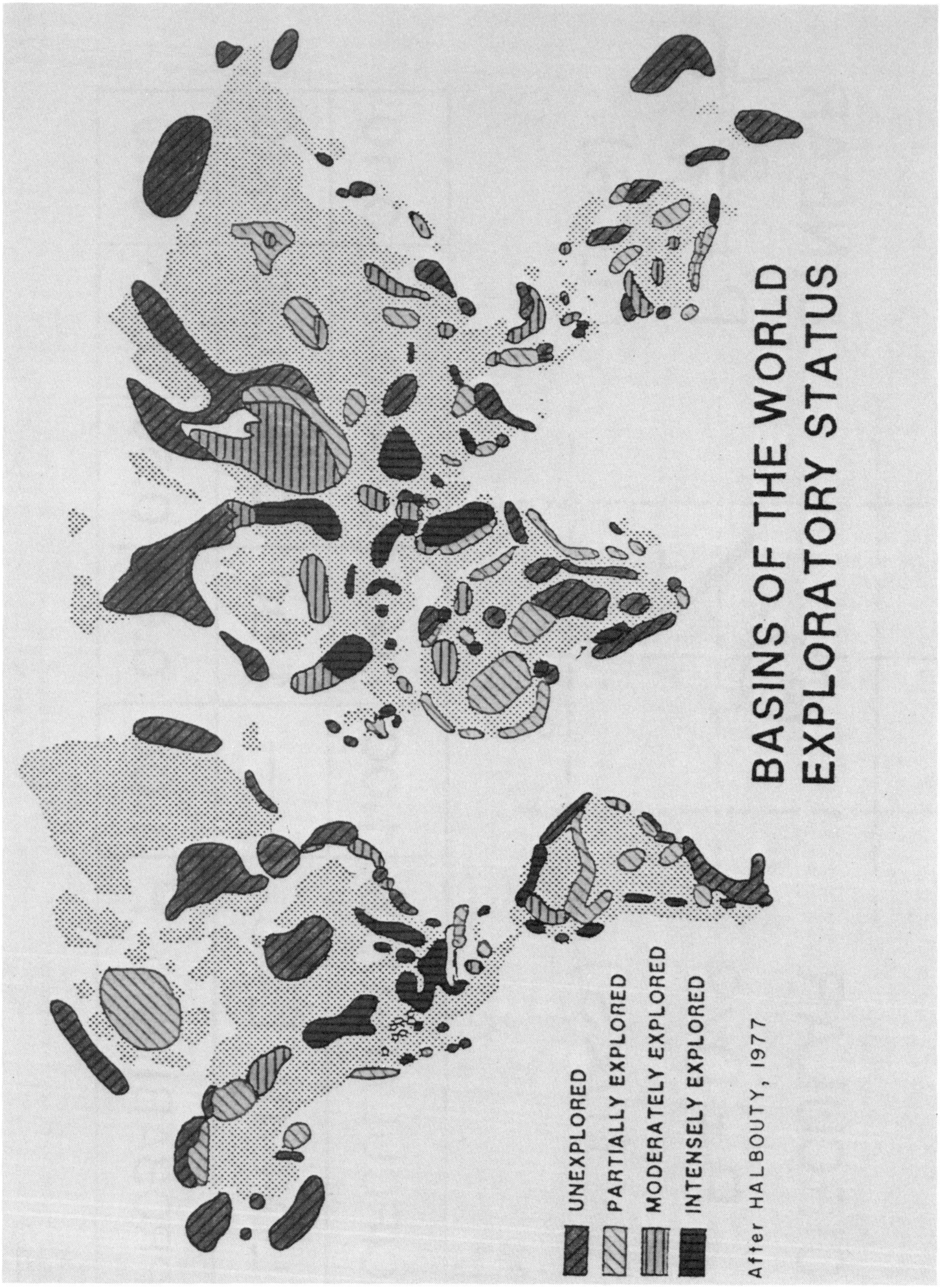

Fig. 14. Halbouty's world map (1977) showing status of exploration in prospective sedimentary basins. By special permission of M.T. Halbouty.

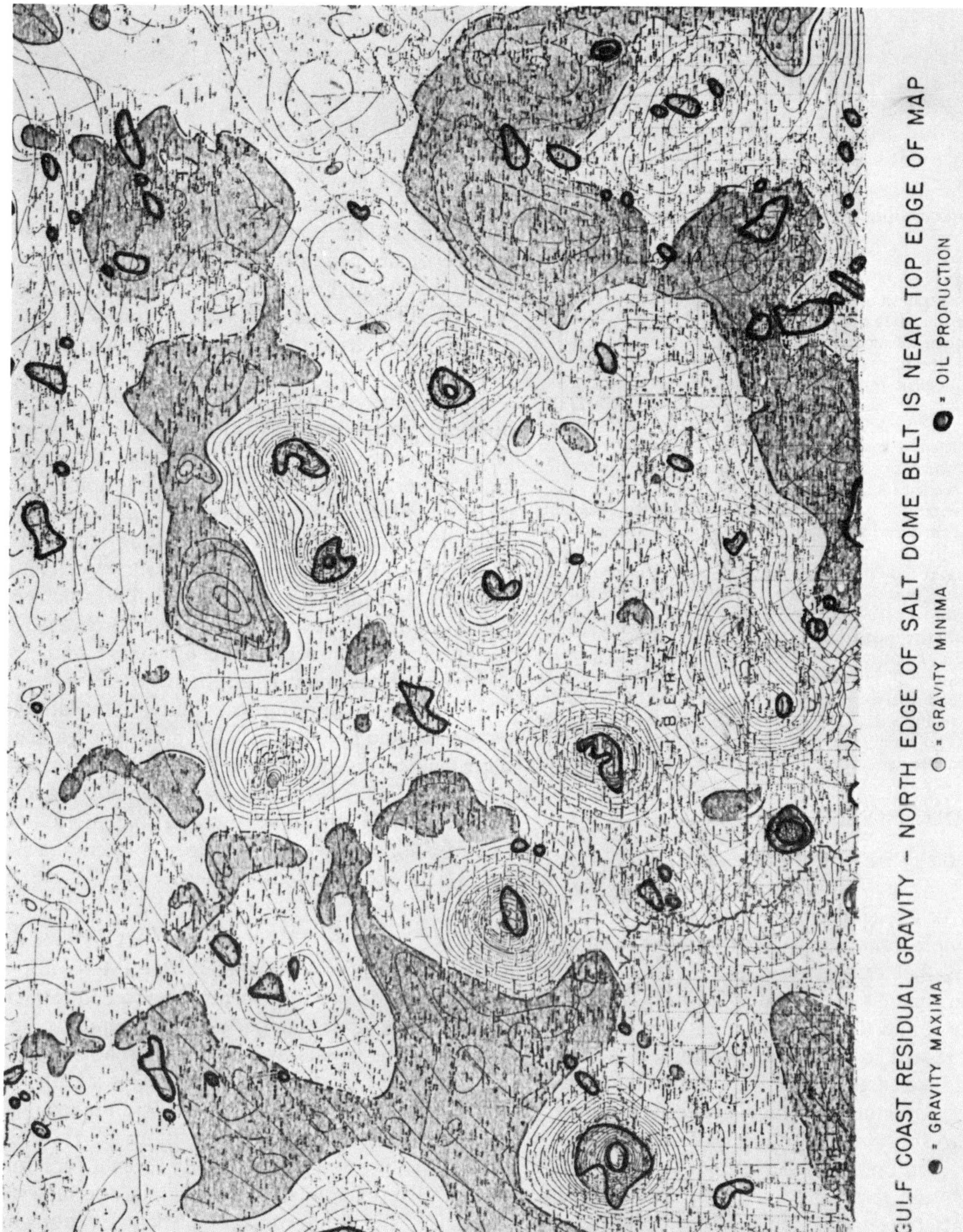

Fig. 15. A gravity map of a portion of the Texas Gulf Coast made in 1937 to illustrate "what might have been." Note the coincidence of gravity anomalies and oil fields (outlined in black). By special permission of the Gulf Oil Corp.

BOREHOLE GRAVITY USED TOGETHER WITH SURFACE GRAVITY

Noel F. Rasmussen

BOREHOLE EXPLORATION CORP.

ABSTRACT

Structural interpretation is one of the uses of borehole gravity surveys. The effectiveness of structural interpretation is enhanced by including high resolution gravity and airborne magnetic surveys. All types of data that can be used to improve the interpretation should be included. While the use of subsurface and seismic data is obvious, the use of surface gravity and airborne magnetic surveys is not.

Prior to conducting and interpreting the borehole gravity survey, geological and geophysical data are used to create a detailed computer model. This model is used to calculate the predicted structural effects curve for the borehole gravity survey. This curve can be used as part of a feasibility study to determine whether or not a borehole gravity survey can resolve the detail needed for interpretation.

Once the feasibility is demonstrated and the survey is conducted, the predicted structural effects curve is compared with the actual structural effects curve. Significant differences between the two indicate the existence of structural configurations not previously interpreted. The resulting changes in the interpretation are incorporated as changes in the computer model used. The theoretical effects calculated from the model must agree with both the surface and the borehole gravity data.

The theoretical applications include the interpretation of structures and reservoirs below the total depth of the borehole. Distances to basement, the presence of deep faults, and the presence of potential reservoirs are some of the significant factors that can be determined by this method.

INTRODUCTION

Borehole gravity surveys are an established part of well logging. Although they are not as widely used as other geophysical logs, there is a very distinct need for these surveys to solve particular problems that cannot otherwise be solved. Uses can be grouped into two categories: (1) formation evaluation, and (2) geophysical interpretation.

With regard to formation evaluation, some of the published cases demonstrate results that can be considered spectacular. These concern turning dry holes into substantial producers. For example, as a result of a borehole gravity survey in one dry hole, it was possible to locate a porous zone, which, after the casing had been perforated, had a calculated open flow (CAOF) of 52.5 MMCF gas and 500 bbls of condensate per day (Bradley, 1974). In another case, a dry hole was reentered and completed for an initial production (IP) of 360 bbl/day oil with 450 MCF gas (Rasmussen, 1975). These completions were the results of borehole gravity surveys that detected porous pay zones missed by conventional logs. These pay zones were not penetrated by the borehole because they were situated beyond the radius of investigation of conventional logs. The reservoirs were reached by hydraulic fracturing from the borehole.

Subsequently, uses and potential uses of borehole gravity surveys in formation evaluation have been published by several writers. The principal writers are noted in the bibliography. The purpose of this paper, however, is to call attention to potential geophysical interpretation uses for borehole gravity meter (BHGM) surveys. These include (1) the interpretation of possible structures and potential reservoirs below the borehole, (2) the presence of nearby deep faults, and (3) distances to basement from the borehole.

To do these things successfully, all available geological and geophysical data must be used. The use of subsurface geological data and seismic data is obvious. However, airborne magnetic surveys and surface gravity surveys must also be used, and one of the purposes of this paper is to emphasize this very important requirement.

BASIC PRINCIPLES

The borehole gravity meter (BHGM) is not a continuously recording logging device, inasmuch as the readings are made when the sonde is stationary. The points within the borehole where readings are made are called "gravity stations." Pairs of stations are used to calculate the average bulk density of the intervening interval. The formula (Fig. 1) shows bulk density to be a function of the difference in gravity readings, the interval thickness, and some constants. The derivations of this and other formulae were discussed by McCulloh et al. (1968).

Interpretation methods are based on density changes away from the borehole. A frame of reference must be used; in this case it is the sidewall density measured by the gamma-gamma type density logs. The correct sidewall density values are sampled and averaged through the intervals between gravity stations. The averaged sidewall densities are subtracted from the corresponding BHGM densities. These values are density differences. The sidewall bulk density, BHGM bulk density, and density difference curves are plotted by depth. This is the basic log used by the log analyst and geophysicist. An example is shown in Figure 2.

A method of interpretation using Bouguer gravity values rather than bulk densities was presented by Snyder (1976). The writer prefers, however, to use density differences.

The Bouguer layer is the basic model for the density calculation. A Bouguer layer is a homogeneous, horizontal bed of finite thickness and density extending to infinity. In practice, detectable variations from this model are interpretable. These variations are geological changes from the Bouguer layer that produce detectable gravity gradients. These geological changes are called "perturbing bodies." Since the BHGM does not emit a signal, it does not have a radius of investigation. Instead, each perturbing body has a radius of effect.

Some geological variations close to the borehole do not produce a significant gravity gradient. For example, the borehole, casing, and most washouts can be disregarded because the gravity gradients produced by them are quite small. Some perturbing bodies are at a great distance. For example, a tidal correction must be made for the apparent motion of the sun and the moon. Usually, several perturbing bodies are present within any given interval. Distant perturbations produce density differences that affect large intervals. Perturbations close to the borehole produce density differences through short intervals. The long interval effects are separated by drawing a curve. Since this curve includes the effects of large structures, it is called by some workers "the structural effects" curve. Most formation evaluation techniques are based on this approach.

An example interpretation is shown on Figures 3 and 4. Figure 3 shows the neutron and density porosity curves for the Amoco 1-14 Prizy Schiff well (Bradley, 1974). This well was drilled into a tall pinnacle reef in northern Michigan. The conventional logs showed no porosity, and the well was scheduled to be plugged and abandoned. However, a borehole gravity survey was run, a density difference (anomaly) was noted, and the well was a spectacular success.

The density curves are shown in Figure 4. The thin interval density differences are caused by the presence nearby of an irregular porous reef reservoir not intersected by the borehole. The broad form of the structural effects curve is produced by the presence of salt beds lying outside the reef. Figure 5 illustrates a simple reef model that shows the source of the structural effects.

INTERPRETATION RULES

There are some basic "rules of thumb" that are helpful in making borehole gravity interpretations from the density and density difference curves. These concern the shape of the perturbing gravity field as penetrated by a borehole. Figure 6 shows a perturbing body of irregular shape located on one side of the borehole. The gravity field and density difference curve are also shown. The density contrast of the perturbing body is positive in sign. As the BHGM descends, it records an increasing gravity field. The sign of the gravity field is the same as that of the body. The downwardly increasing gravity gradient produces a negatively signed density difference. At a point somewhat above the perturbing body, the gravity field forms a maximum value and is lobe-shaped. A similar minus-signed lobe occurs at some point below the body. Between the lobes, the gravity field decreases downward, and the density difference is positive and relatively large. Below the minus-signed lobe, the gravity field increases downward and produces a minus-signed density difference.

The rules of thumb are as follows:

1. Between the lobes, the density difference curve is the same sign as the density contrast of the perturbing body.

2. Above and below the lobes, the density difference curve has a sign which is the opposite of the density contrast of the perturbing body.

In the case of the pinnacle reef discussed above, the porous zones are side-located and produce minus-signed density differences. The sidewall material is nonporous and dense. The reservoir is much less dense. The density contrast thus produced by the perturbing body is estimated to be about − 0.20 gm/cc. The A1 and A2 salt beds also are side-located. The rule of thumb states that their perturbations produce a minus-signed density difference within the reef. From above, the reef would be a high density perturbation and would produce a minus density difference component. Some additional rules of thumb follow:

3. If the borehole penetrates the perturbing body, the lobes tend to be at the surfaces of the penetrated body.

4. If the borehole penetrates the perturbing body, the sidewall reference density changes to that of the perturbing body.

Thus, the density difference curve is minus within and above the reef because of the change in the sidewall reference.

Another handy rule concerns elongate bodies such as the borehole, casing strings, and vertical fractures:

5. For vertically elongate bodies, the density difference anomalies occur only at the upper and lower edges.

PROCEDURES

The structural effects curve of the borehole gravity survey can be used to interpret relatively distant structural phenomena. The success of the application depends on proper preparation. This preparation includes detailed gravity modeling and proper station selection.

The first use of gravity modeling is in making a feasibility study. The purpose of the feasibility study is to determine whether the purpose of the survey can be achieved. The structural configurations and the density contrasts are interpreted from all available data. The interpreted structure is then closely approximated by a computer model. The theoretical gravity curve is used to calculate a theoretical density difference curve. Several borehole locations and structural configurations can be modeled. The geophysicist then determines whether the geological parameters under study can be resolved. The answer will be "yes," "no," or "maybe." Even if the answer is "maybe," the survey should proceed.

The selections of gravity stations should be made differently for structural interpretation than for formation evaluation. Pairs of stations should be somewhat farther apart. This increases the precision of density resolution by the BHGM survey. It is not necessary to have continuous density and density difference curves. Instead, the structural effects curve should be sampled by pairs of stations that include intervals of nonanomalous densities. The best lithologies are wet sandstones. There is less likelihood of lateral density variations in wet sandstones. The sidewall densities should be the same as the BHGM densities, except for structural effects. Carbonates will require many interval measurements for the interpretation of structural effects. This is illustrated in Figure 4. A sidewall density log also will be needed for the entire range of the BHGM survey.

Some lithologies, such as salt, have electron densities that yield erroneous density values on gamma-gamma type logs. Density differences from these intervals should not be used. Shale intervals may or may not produce reliable sidewall densities. Some shales yield erroneous, very low gamma-gamma log densities. Density differences calculated from these shale zones should not be used.

The extent of the BHGM survey and its cost can be reduced by focusing on those depths through which the density difference curve shape changes are significant. These zones will be shown by the feasibility study.

CONSTRUCTING THE MODEL

The configurations of the structures are defined by seismic and subsurface data. These must be augmented by surface gravity and airborne magnetic information.

Airborne magnetics is used to detect basement faults and compositional changes within the basement. Intrusives and extrusives also may be present and would be additional magnetic anomaly sources. Compositional changes within the basement also may produce density contrasts that should be included in the model. Basement faults are especially important and are most readily revealed by magnetics.

Airborne magnetics is preferred over ground and marine magnetics. The precision and resolution of marine and ground magnetic surveys are usually one order of magnitude less than those of high resolution airborne magnetic surveys. Figure 7 shows a basement fault interpretation made from a high resolution airborne magnetic survey. The basement to sediment density contrast and the fault configurations are included in the model.

Finally, the theoretical surface gravity field is calculated from the model. The field is compared with the Bouguer gravity map. The model is further adjusted until the two are essentially in agreement. The theoretical borehole curves are then calculated.

Examples are shown in Figures 8 and 9. Figure 8 shows the geological configuration of a drape fold over a large basement fault block. A potential sandstone reservoir lies just above granitic basement. A thick carbonate sequence overlies the sandstone, which is overlain in turn by terrigenous clastic formations in which densities decrease upward.

Figure 9 shows theoretical density difference curves for wells located at positions A, B, and C. The overall negative deflections of the density difference curves A and B are caused as per rule of thumb 2. The downward increase in densities of the overall structure causes a negative density difference curve in anticlines and a positive one in synclines. The reversal of density differences toward the positive indicates that lower densities exist below the carbonates. The sharp deflections near the bed boundaries can be used to predict the density of the bed below the bed boundary.

The borehole at location C has missed the structure. By comparing the density difference curve from an actual survey with one of the three theoretical curves, the geophysicist could interpret the relative structural position of the borehole.

CONCLUSIONS

The BHGM survey can be a powerful tool in evaluating phenomena related to basement, the presence of faults, structures beneath the total depth of a well, and the presence of possible reservoirs. Although the use of the BHGM for detecting possible reservoirs is coming to be appreciated, its use in the geophysical interpretation of structure seems not to be well known. Experience and examples shown here demonstrate that the BHGM can be a powerful exploration tool.

REFERENCES

Beyer, L. A., 1971, The vertical gradient of gravity in vertical and near vertical boreholes, Ph.D. thesis, Department of Geophysics, Stanford.

Bradley, J. W., 1974, The commercial application and interpretation of the borehole gravimeter: Symposium of Houston Geophysical Society.

Gourney, L. S., and Maute, R. E., 1982, Detection of bypassed gas using borehole gravimeter and pulsed neutron capture logs: Log Analyst, v. 23, n. 3, p. 27-32.

Godell, R. R., and Fay, C. H., 1964. Borehole gravity meter and its application: Geophysics, v. 29, p. 774-782.

Hammer, Sigmund, 1950, Density determinations by underground gravity measurements: Geophysics, v. 15, p. 637-652.

_________ , 1965, Density determinations by underground gravity measurements—Sequel: Geophysics, v. 30, p. 1133-1134.

Howell, L. G., Heintz, K. O., and Barry, A., 1966, The development and use of a high precision downhole gravity meter: Geophysics, v. 31, p. 764-772.

Jageler, A. H., 1975, Improved hydrocarbon reservoir evaluation through the use of borehole gravimeter data: Society of Petroleum Engineers of the American Institute of Mining Engineers, Dallas.

Jones, B. R., 1972, The use of downhole gravity data in formation evaluation, Society of Professional Well Log Analysts Annual Logging Symposium, 13th.

McCulloh, T. H., 1965, A confirmation by gravity measurements of an underground density profile based on core densities: Geophysics, v. 30, p. 1108-1132.

_________ , 1966a, Gravimetric effects of petroleum accumulations — a preliminary summary: United States Geological Survey Circular 530.

_________ , 1966b, The promise of precise borehole gravimetry in petroleum exploration and exploitation: United States Geological Survey Circular 531.

_________ , 1967, Mass properties of sedimentary rocks and gravimetric effects of petroleum and natural gas reservoirs: United States Geological Survey Professional Paper 528-A, p. A1-A-50.

_________ , La Coste, L. J. B., Schoellhamer, J. E., and Pampeyan, E. H., 1967, The United States Geological Survey — La Coste and Romberg precise borehole gravimeter system — instrumentation and support equipment, *in* Geological survey research, 1967: United States Geological Survey Professional Paper 575-D, p. D92-D100.

_________ , Schoellhamer, J. E., Pampeyan, E. H., and Parks, H. B., 1967, The United States Geological Survey — La Coste and Romberg precise borehole gravimeter system — test results, *in* Geological survey research 1967: United States Geological Survey Professional Paper 575-D, p. D101-D112.

__________, Kandle, J. R., and Schoelhamer, J. E., 1968, Application of gravity measurements in wells to problems in reservoir evaluation: Society of Professional Well Log Analysts Annual Logging Symposium, 9th.

Nettleton, L. L., 1940, Geophysical prospecting for oil, 1st ed: New York, McGraw-Hill.

Rasmussen, N. F., 1973, Borehole gravity survey planning and operations: Society of Professional Well Log Analysts Annual Well Logging Symposium, 14th.

__________, 1975, The successful use of the borehole gravity meter in northern Michigan: Canadian Well Logging Society Formation Evaluation Symposium, 5th.

__________, 1975, Borehole gravimeter finds bypassed oil, gas: Oil and Gas Journal, Sept. 29.

__________, 1977, Gravity logs promise new applications: Petroleum Engineer, July.

Snyder, D. D., 1976, The borehole Bouguer gravity anomaly: application to interpreting borehole gravity surveys: Society of Professional Well Log Analysts Annual Logging Symposium, 17th.

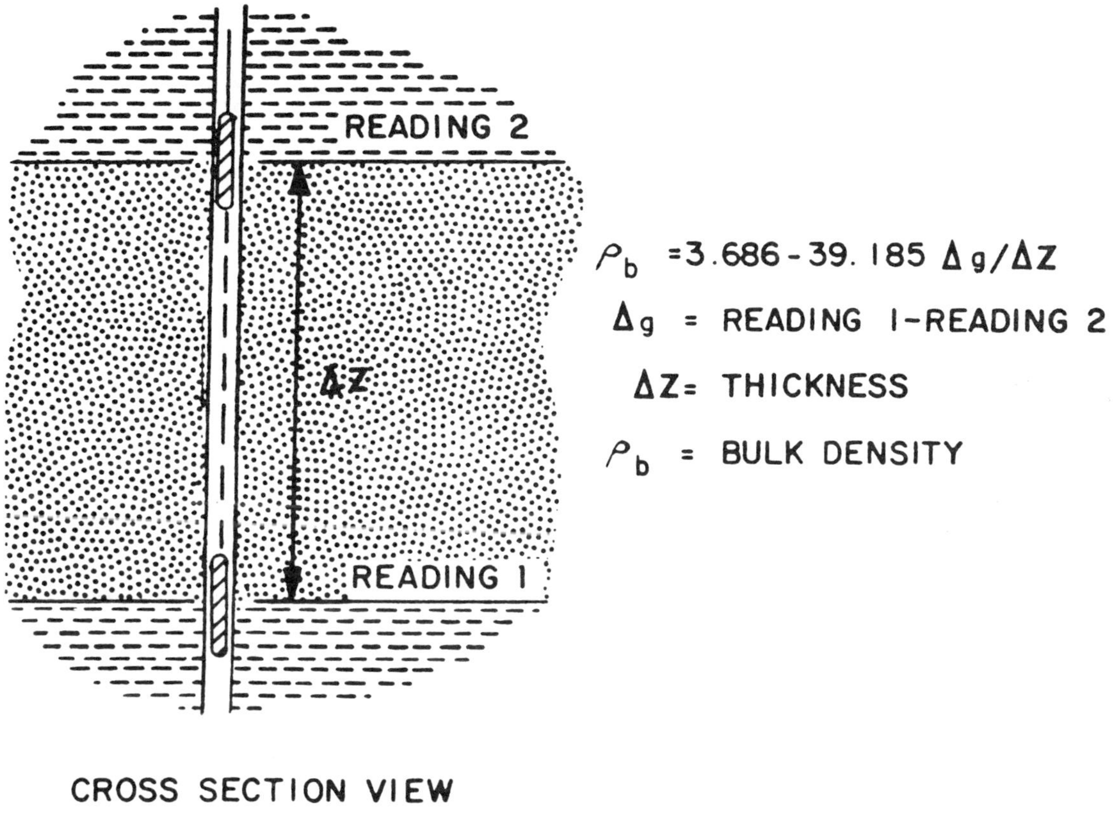

Fig. 1. Basics illustration.

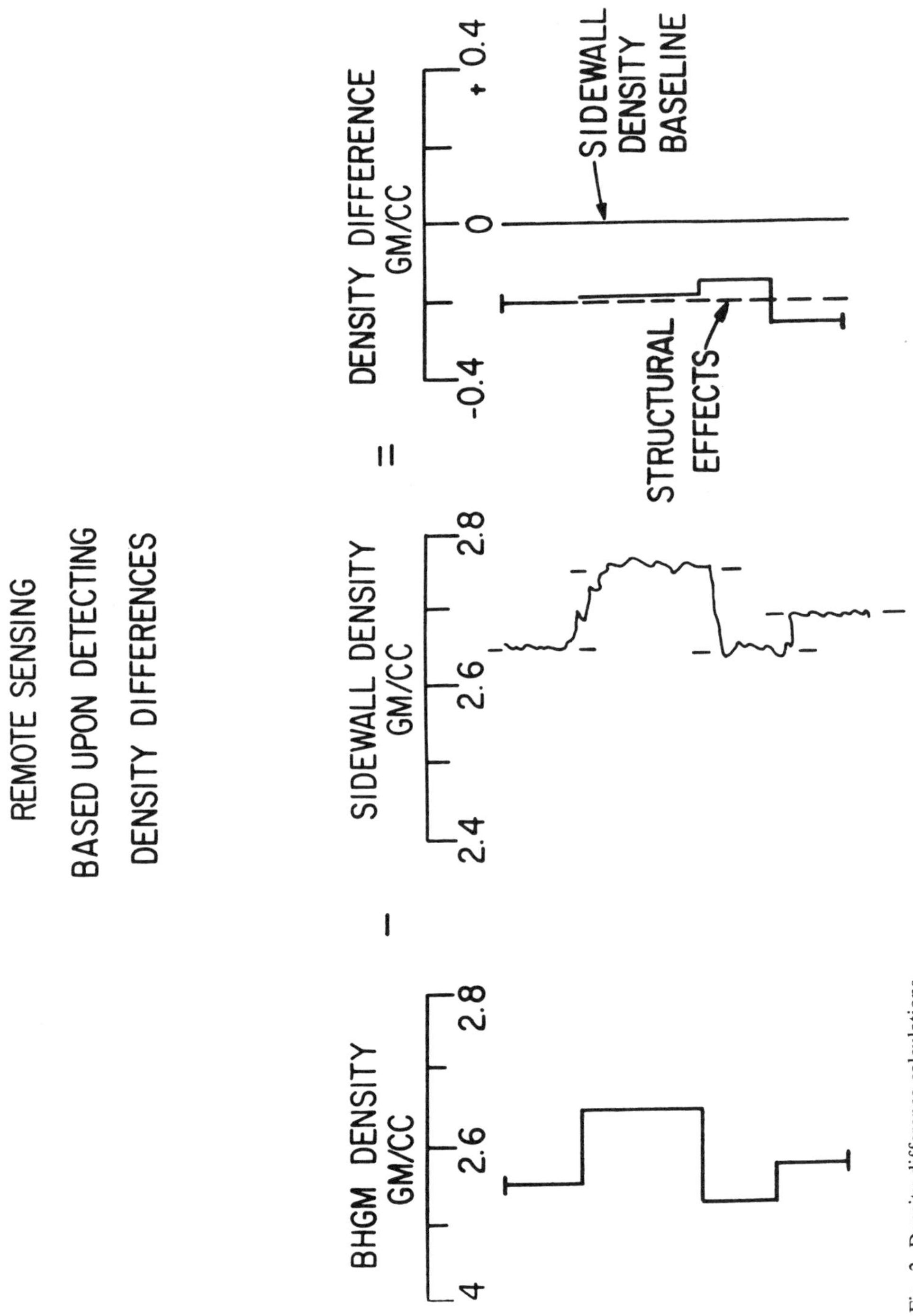

Fig. 2. Density difference calculations.

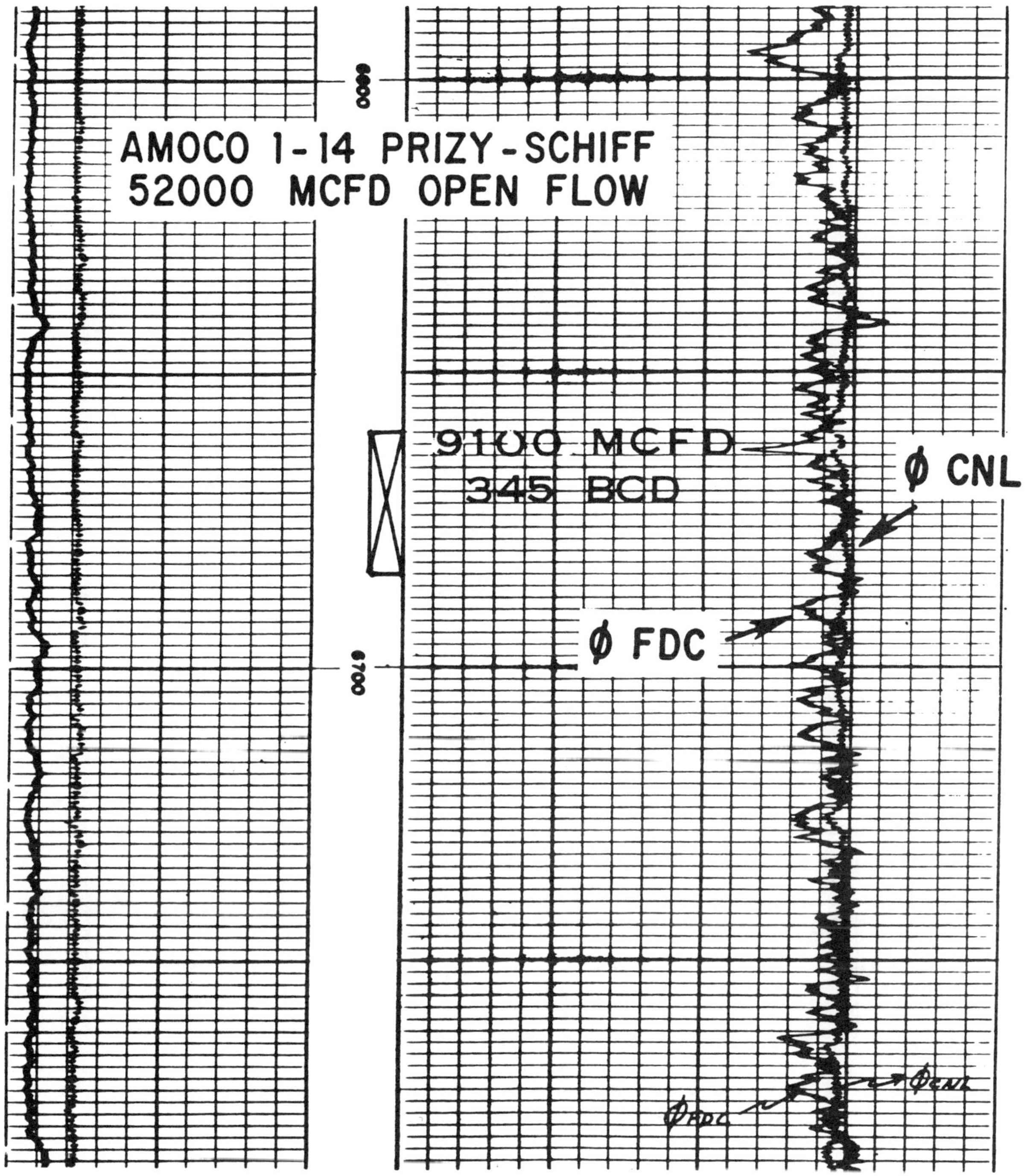

Fig. 3. Amoco 1-14 Prizy Schiff CNL-FDC.

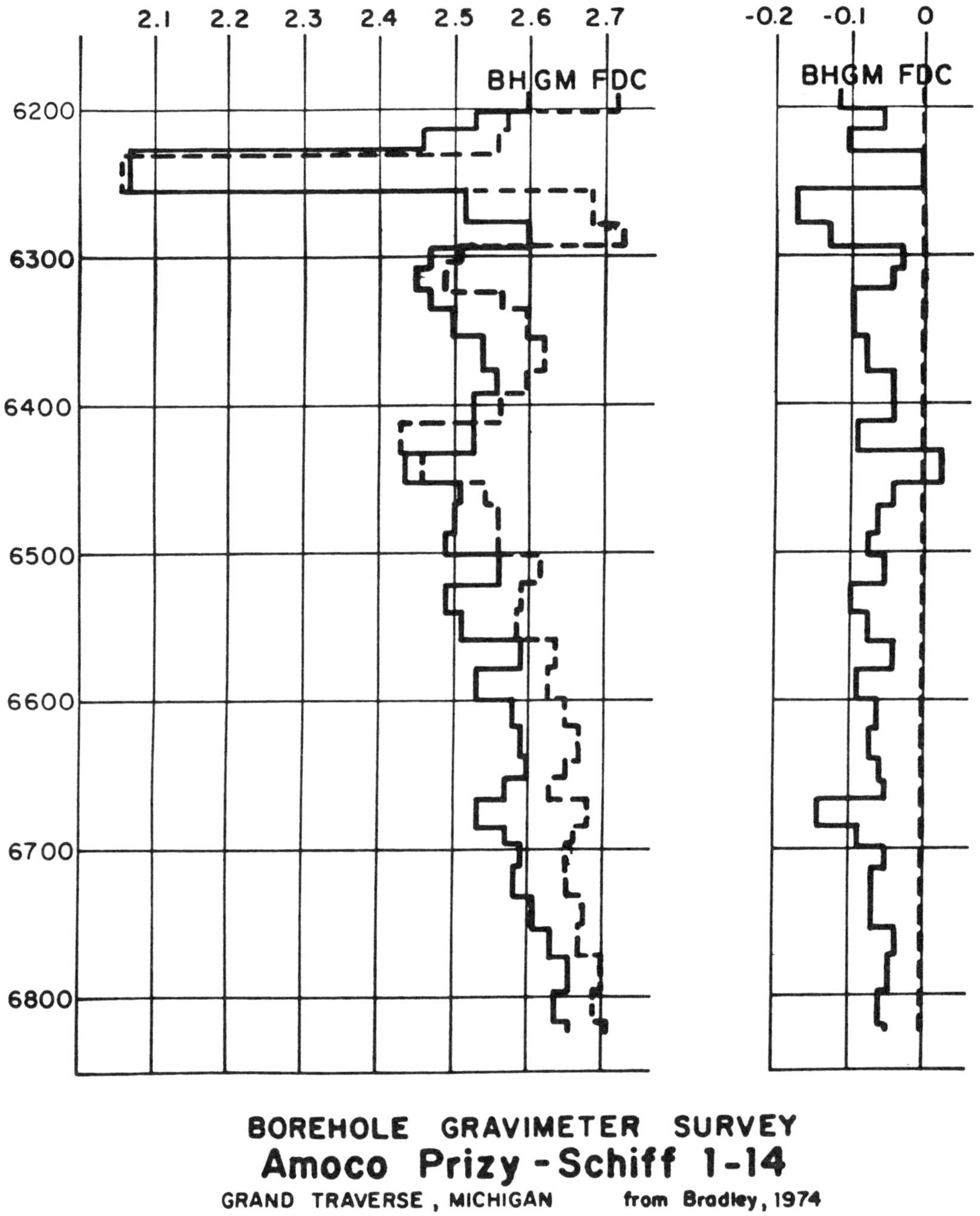

Fig. 4. Density plots with DD.

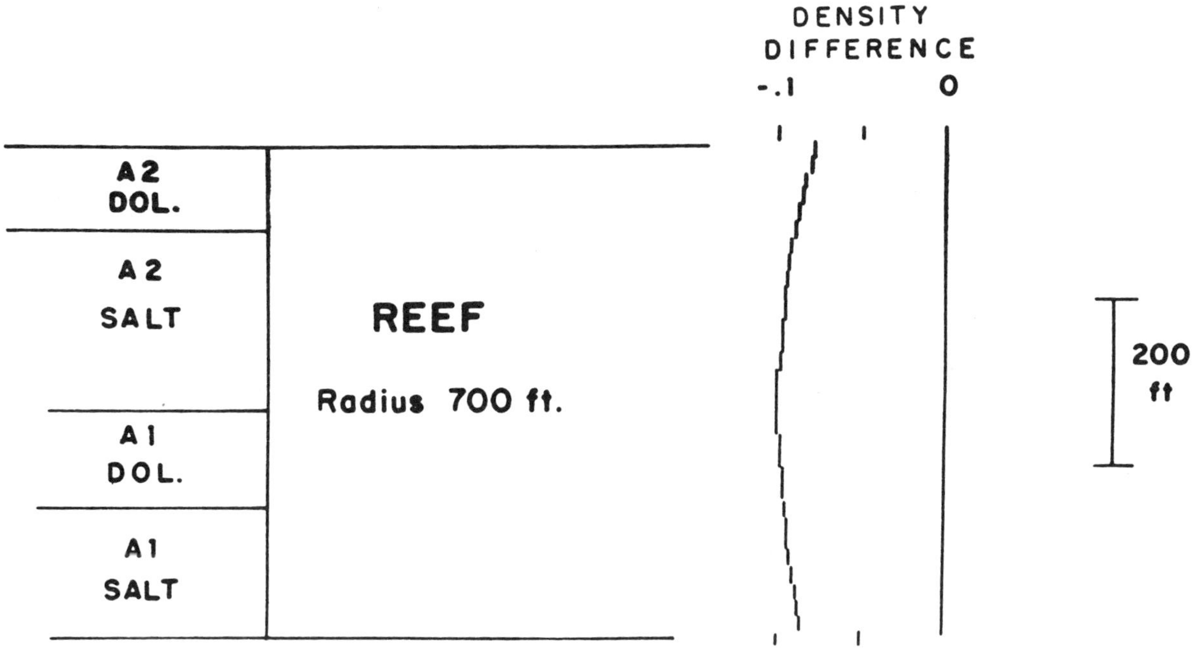

Fig. 5. Simple reef model.

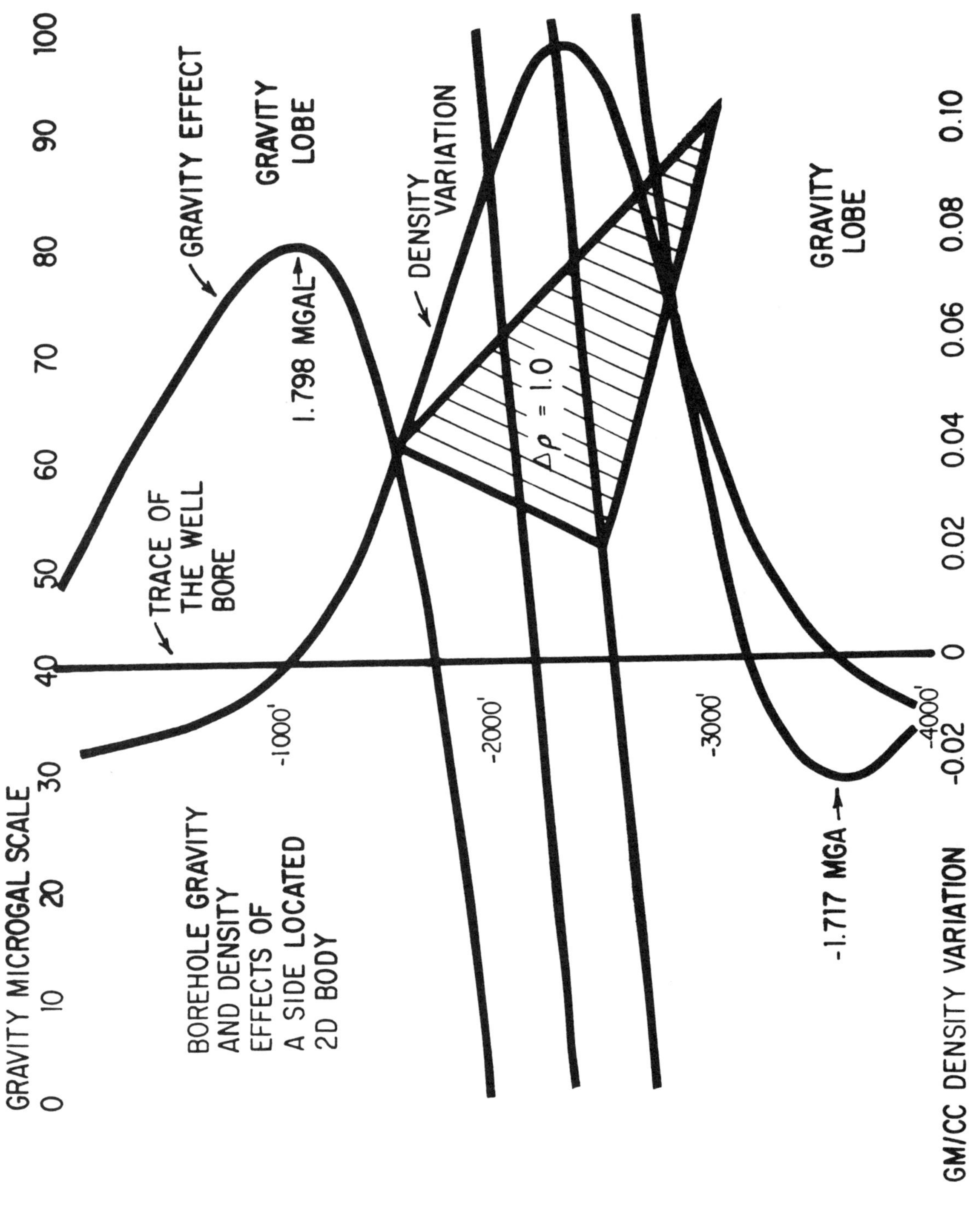

Fig. 6. Side located body with G field and density difference.

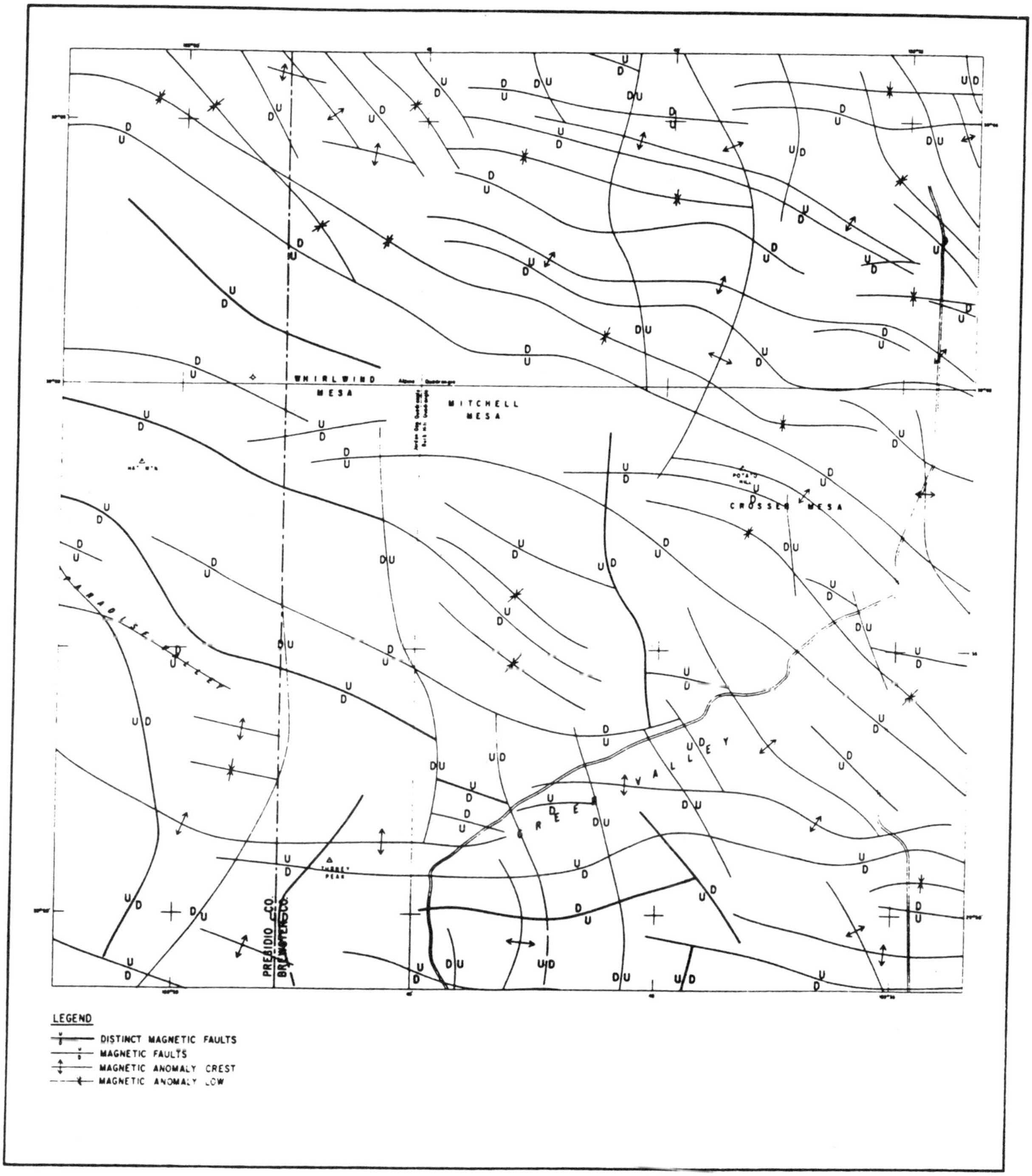

Fig. 7. Basement fault interpretation, high resolution, made from airborne survey.

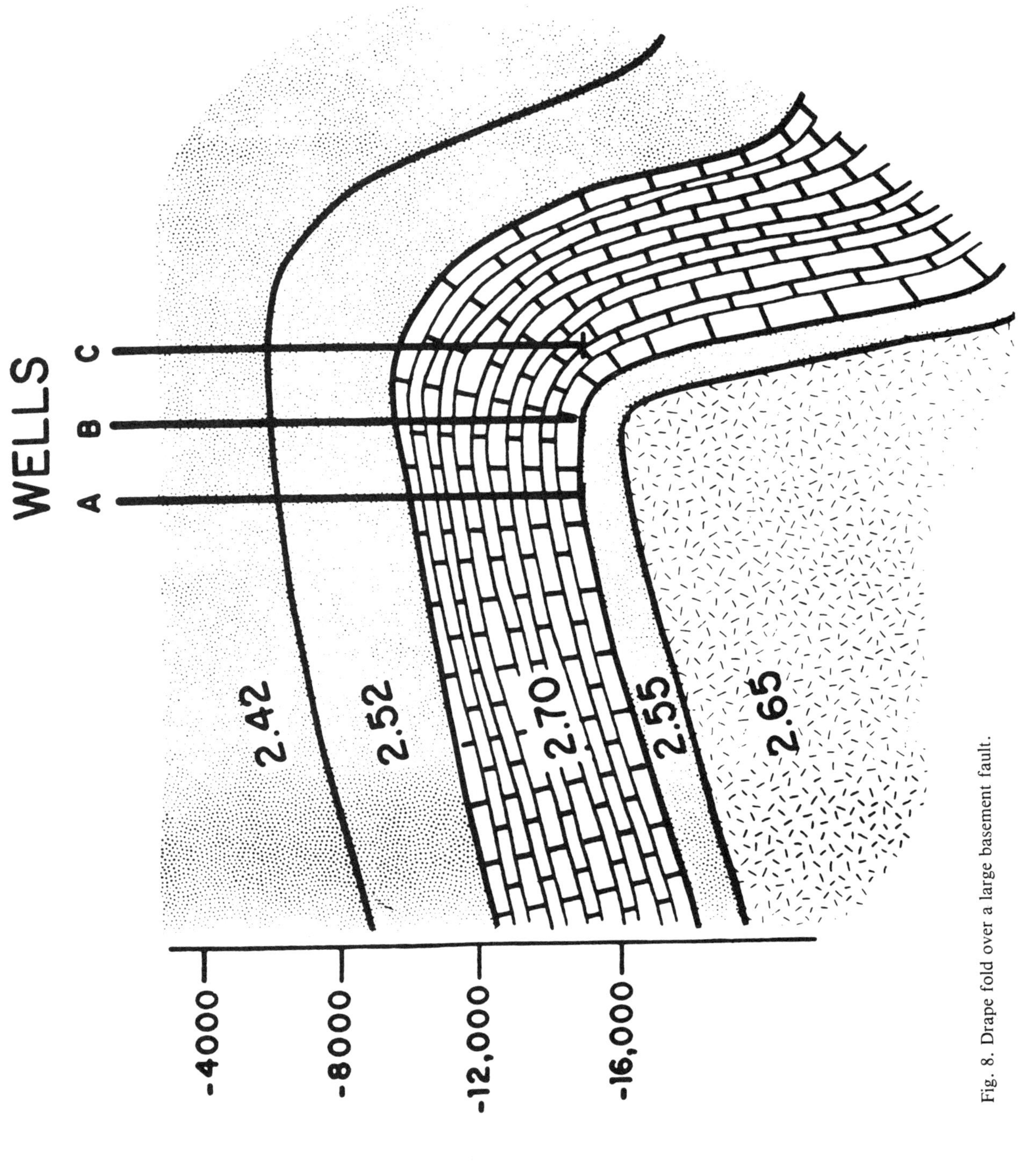

Fig. 8. Drape fold over a large basement fault.

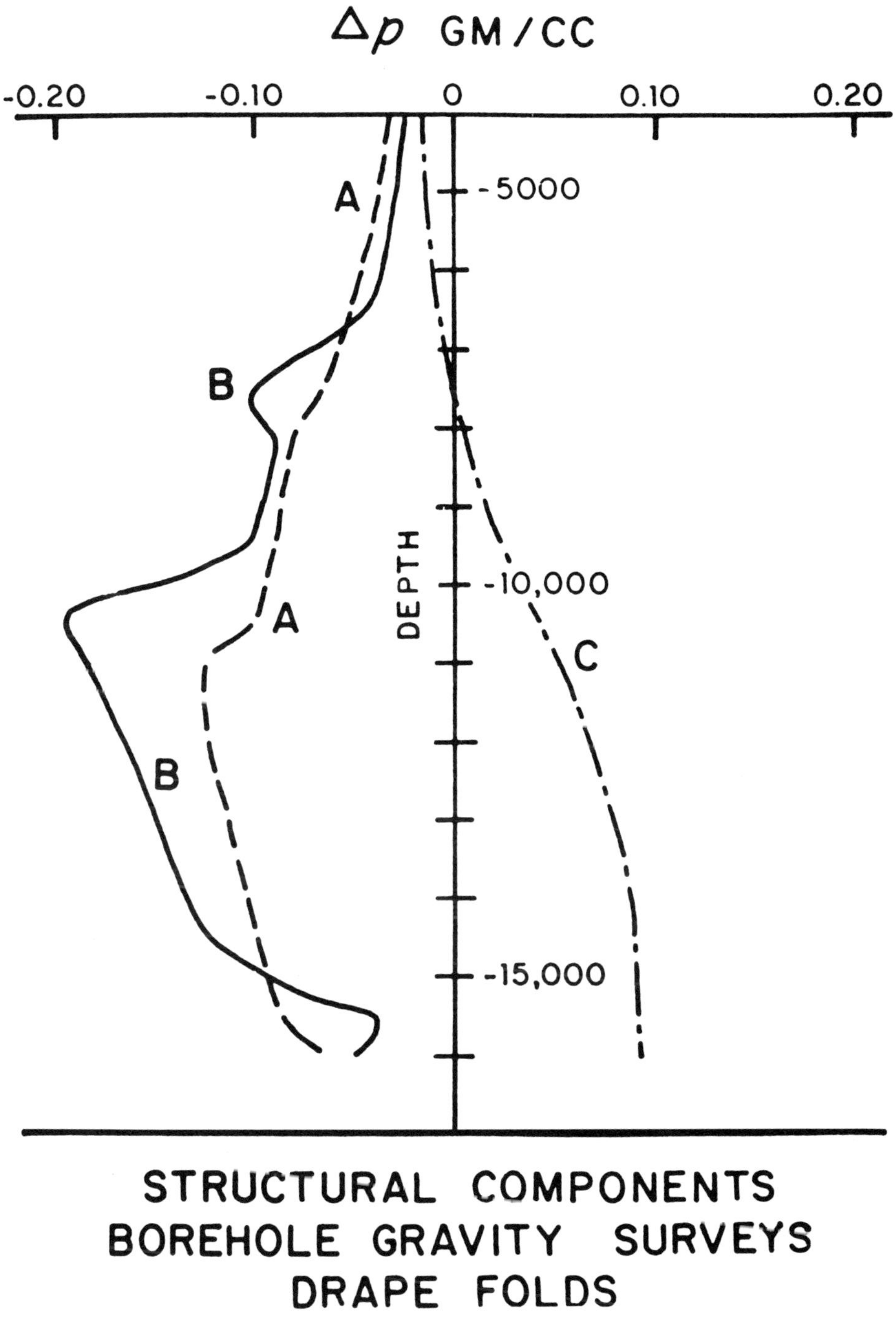

Fig. 9. Wells at three positions: A, B, and C.

STATE OF THE ART IN ELECTROMAGNETIC SOUNDING

Howard Renick, Jr., and James I. Pritchard

INDESCO, INC.

ABSTRACT

State of the art electrical methods include both passive source and controlled source techniques. This paper centers on the far zone (early time) plane wave active approach to electrical prospecting. A case study of reef exploration on the eastern shelf of the Midland Basin is presented. The data from this controlled source electromagnetic (EM) survey are used to simulate the response of a passive source magnetotelluric (MT) survey and of a shallow controlled source integrated transient survey.

The controlled source approach to electrical surveying achieves an improved signal to noise ratio as compared to passive source methods. Such improved data allow greater resolution of the earth's electrical structure. By using controlled sources in combination with "state of the art" multichannel recording techniques, data collection with dense spatial sampling can be achieved. This dense sampling affords a more detailed and a more credible interpretation.

To be fully effective, electrical prospecting needs to incorporate as many methods as possible into the final interpretation. The need for further development of these methods is discussed.

The magnetotelluric method (MT), along with audio magnetotelluric techniques (AMT) given by others at this symposium, comprises an important part in the state of the art of electrical prospecting. Consideration of these methods forms a basis for starting this discussion of controlled source electrical prospecting techniques.

The need for electrical techniques in addition to MT methods results from certain difficulties involved with the MT approach:

1. There is a weak energy frequency range in the earth's natural EM spectrum from 7.65 Hz to 0.1 Hz, which corresponds in average to sedimentary section depths of penetration from 2,000 ft to 20,000 ft (Fig.1). This frequency range can be filled only by recording over an extended time and by using statistical techniques to enhance signal relative to noise. The time required to fill this frequency range impairs the data acquisition time efficiency of the MT method.

2. At this time, it is not possible to record MT data economically at a high station density. Yet to be properly evaluated, many prospects require that the subsurface be sampled at closely spaced intervals, on the order of 220 ft or less.

3. The MT method is relatively insensitive to resistive layers encased in more conductive sediments. It is important to define accurately all rock sequences if electrical anomalies are to be accurately mapped in depth.

These shortcomings of the MT technique can be overcome by using controlled source EM techniques. The controlled source techniques also have limitations for deep mapping. So MT methods and controlled source EM methods, rather than being mutually exclusive, are complementary for deep section mapping.

Field application of the controlled source method can take a variety of source-receiver forms. Generally, sources and receivers are of two types (Fig. 2).

1. Grounded dipoles. For sources, conductive electrodes are placed in the ground and current is introduced into the ground at the two source electrodes; for receivers, the change in voltage is measured between the two grounded receiver electrodes.

2. Loop dipoles. Sources induce electromagnetic fields into the ground, and multiturn loop receivers measure the time rate of charge of the electromagnetic fields within the earth. The source current input used is normally a time domain "boxcar" transient generated by a 50 to 100 amp source. The received transient is analyzed after source current turnoff. Grounded electrodes are the most effective source and receiver types, because they provide conduction field measurements as well as induction field measurements for analysis. They are most efficient in areas of heavy ground cover because they may be laid along a line. Grounded sources do not work well in areas with resistive soil such as that found in deserts and in volcanic and dense limestone overburden. The loop source is used in these areas.

The geometric configurations of source and receiver in field operations are variable. Figure 3 depicts some of the configurations for grounded dipole receivers and sources. The configuration chosen for an area is selected on the basis of the geologic problem to be resolved, including the depth of penetration, the terrain, and cost considerations. For example, the modified Schlumberger array provides the best signal-to-noise ratio and reliable DC data of any of the arrays shown. This array is most often used for high quality DC data which can be acquired at a rapid rate. The broadside array gives the optimum yield of production in EM measurements.

The grounded source broadside array with parallel receiver dipoles was chosen for the example to be shown from the eastern shelf of the Permian Basin. This array was chosen because it easily achieves the depth of penetration needed for this problem. In limestone-covered country such as this, a trial and error effort must be expended to find efficient locations for current electrodes. In extreme cases, water wells may be used to provide a current conduit to the water table.

The source current electrode pairs were placed three miles to the north, three miles to the south, and three miles to the west of the project area. The lines in the middle of the project were recorded separately from each source to establish data repeatability and intertie EM field strengths from the three sources.

The electrical survey performed consisted of eleven east-west lines separated by approximately 660 ft and one 2 mi north-south tie line. Receivers along the profiles are at intervals of 220 ft. The entire area was, then, sampled on a grid of 220 by 660 ft. Two profiles were omitted because of access problems (Fig. 5). Such a fine grid allows thorough evaluation of the area and provides confidence in the repeatability of the measurements.

The geologic problem to be solved was to identify Strawn Reef buildups along the eastern shelf of the Midland Basin (Fig. 4). There are a number of such buildups in the area. Additional control in the immediate area of interest included three wellbores and a seismic line. The seismic line has been processed with the Seislog technique. The seismic data and well data suggest that carbonate buildups in the Strawn could exist in the area surveyed. Increases in the resistivity above normal indicate areas of Strawn buildup, just as velocity increases above normal indicate areas of the Strawn Reef development.

The ultimate aim of the controlled source resistivity profiling is to re-create an induction log from the surface. Figure 6 shows a six-layer Bostick type interpretation of one of the traverses. The solid bars show the layered modeling sounding distributed over the traverse. The hatched anomalies show resistivity increase above the average sounding resistivities. The data appear to tie grossly to an induction log in the control well.

The sounding models for each layer show a general resistivity increase in each layer from left to right. This increase could indicate a geologic section change from offshore to near-shore geology during these geologic times. The residual resistivity anomalies (resistivity above the normal sounding resistivity) do not align vertically, but change from level to level (i.e., different geologic times) as we should expect for reef buildup. It also assures that the data are not locked on upper section geologic noise. The residuals for the fifth layer (Strawn age) project on trend to the seismic velocity highs. The dry hole is in the area of the less resistive portion of the line.

It is instructive to use these data to simulate the results that would be expected by the magnetotelluric and integrated transient approaches on a common survey cost basis. In Figure 7, the simulation of MT data is displayed. It is seen that only three stations of MT data would be recorded as opposed to the thirty-three recorded with the controlled source approach for the same cost. Also, MT data will probably only permit a three-layer interpretation keying on the most conductive layer around 6,500 ft.

The integrated transient response is shown in Figure 8. For a cost comparable to the data recorded above, the typical transient surveys with which we are familiar would produce twelve recording stations. These transient data do show anomalous character away from the dry hole, but they give no depth control and concentrate on upper section variations alone. This upper section can have considerable geologic noise. An IP model would lock on the change modeled as a less resistive area, the area of the dry hole.

The comparison between the three approaches can be made more dramatically by comparing the results in map view. The controlled source EM resistivity map shows a north-south trending resistivity high through the area (Fig. 9). This trend is consistent with producing fields in the area.

The re-created magnetotelluric data echo the EM data but with considerably less detail (Fig.

10). There are fourteen MT soundings versus the 390 controlled source soundings on the same cost basis. Integrated time transient data also reveal an anomaly similar to the EM data (Fig. 11). The EM data exhibit more intense anomalies and more pronounced trends and are shown to be at the depth of the Strawn Formation.

FIELD TECHNIQUES

The most common source input is shaped in a "boxcar" fashion (Fig. 12). A generalized voltage response to this current input is illustrated in the lower half of the diagram. The steady state voltage between the current on and current off positions can be used to calculate DC resistivity. The DC resistivities are used in controlling the conversion from frequency to depth for EM data. Also, the DC data are used to define resistive layers that are somewhat transparent to EM waves. The main part of the decay curve represents the data shown in the case study just presented and is known as "early time," or as the "far zone," in the frequency domain. These far zones are theoretically plane waves and may be treated mathematically as such in modeling.

The transition between the early time decay and the steady state—DC levels is known as "late time," or "near zone" (in the frequency domain). These data are highly focused in that they represent data from a limited part of the geologic section. Near zone data reportedly have been used by Russian explorationists. These data hold promise for further controlling EM data and defining anomalously resistive portions of the section.

One of the most important aspects of EM prospecting is the recognition of noise in the data. Noise can most easily be identified by examining the data in the time domain. In Figure 13, the current input function is shown on the extreme left. Proceeding to the right, the next trace shows a normal time domain response. The next two traces illustrate the response of crossing a grounded conductor, such as a pipeline. Note the mirrored symmetry across the conductor. Time integration of these curves gives a much different result than integrating the normal response, and these are easily recognized as invalid readings. The curve on the right is degraded by the presence of 60 Hz and other noise. Examination of the basic data is critical for the recognition of recording problems and those data to be eliminated from the data set to be modeled.

A high signal-to-noise ratio is critical to electrical prospecting, as it is to all geophysical methods. Controlled source techniques can achieve a high signal-to-noise ratio by using a large current source. Indesco uses a 100 kw generator to provide the potential to generate up to 100 amps of current. The frequency output is limited by the capacity of the high speed switching circuitry to a 100 Hz maximum.

Current is to be conducted into the ground, limited by contact resistance between the earth and the electrodes. Generally, braided copper battery cable placed in a ditch that has been soaked with salt water will allow current on the order of 100 amps to flow into the earth. Metal objects that have been grounded for a long time, such as culverts or water-well casing that is in contact with the water table, will make excellent current electrodes.

Receiver electrodes must be nonpolarizing so that all the potential measurements result from electromagnetic fields and not from chemical interaction between the electrode and the soil. Indesco uses porous pots with saturated copper sulfate and copper electrodes. Receivers are connected to conventional multichannel seismic road cable. This cable allows multichannel recording and roll-along for the efficient collection of a high volume of data. The cables have to be specifically designed to eliminate cross feed. Because of cable capacitance problems which create changing frequency response and signal loss, roll-along techniques cannot be used to the distance they are used in seismic prospecting.

The different roll-along switches have been designed to accommodate various recording geometries. From the switch box, the data are passed to a set of eight specially designed amplifiers. The amplifiers contend with long period telluric currents and electrode SP through automated bucking circuits. Because the receivers are grounded, 60 Hz noise is always a problem, and the amplifiers contain efficient active 60 Hz and harmonic reject filters. The amplifiers operate in a fixed mode with gain range selections spanning 96 db. Each amplifier is also equipped with a saturation warning indicator.

Data are converted to digital (12 bit) format for summing on a microprocessor. Raw incoming data are monitored on an analog strip chart recorder while the summed data are displayed on the oscilloscope. Summing is continued until there is no further visual improvement of the signal-to-noise ratio seen on the oscilloscope display. The summed data along with the header information is then recorded onto floppy discs for further processing in a separate computing center.

SUMMARY

1. Controlled source electrical methods provide an increase in efficiency over passive source EM methods (magnetotellurics).

2. The controlled source methods promise increased resolution and reliability through the integration of DC and near-zone data as well as the plane-wave data.

3. Existing electrical techniques can solve a number of exploration problems, including penetrating volcanic and granitic cover, identifying sand development in many geologic sections, defining carbonate structures, serving as a reconnaissance basement mapping tool in remote, relatively unexplored areas, and providing an exploration tool in areas of poor seismic data quality.

FOR ELECTRICAL METHODS TO PROGRESS

1. Signal-to-noise enhancement techniques similar to the CDP stacking of seismic data must be implemented if the resolution of thinner layers is to be achieved.

2. Multimethod recording, processing, and interpretation using plane wave and near-zone EM soundings and DC resistivity sounding data will provide more complete information about the electrical sections, including information on rock anisotropy.

3. The development of "multichannel" telemetry type systems will allow more data to be recorded for statistical data processing. Such a system will also permit recording multiple components at each station that will further increase the reliability of the data.

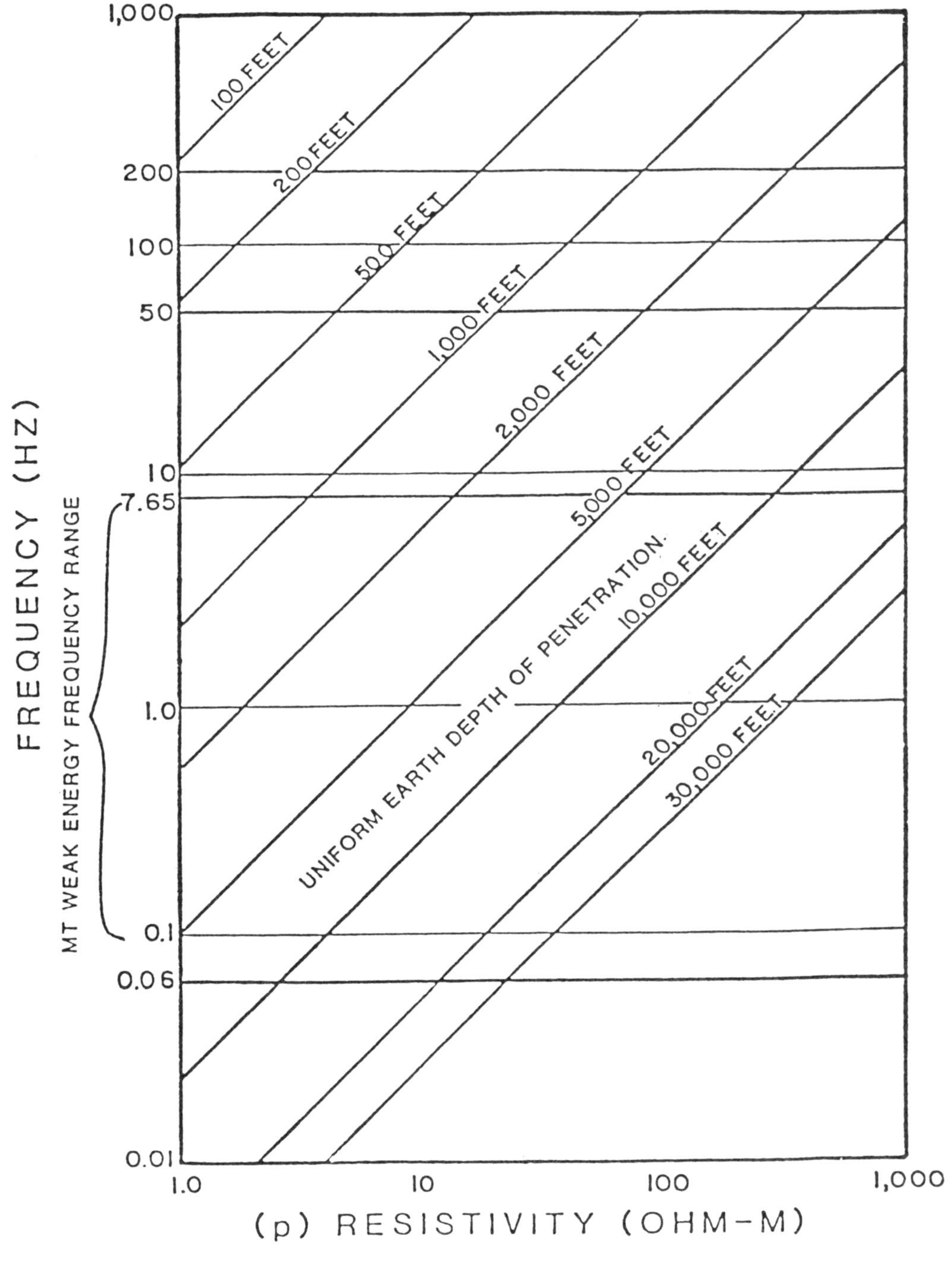

Fig. 1. Qualitative review of penetration depths for the magnetotelluric weak energy frequency range.

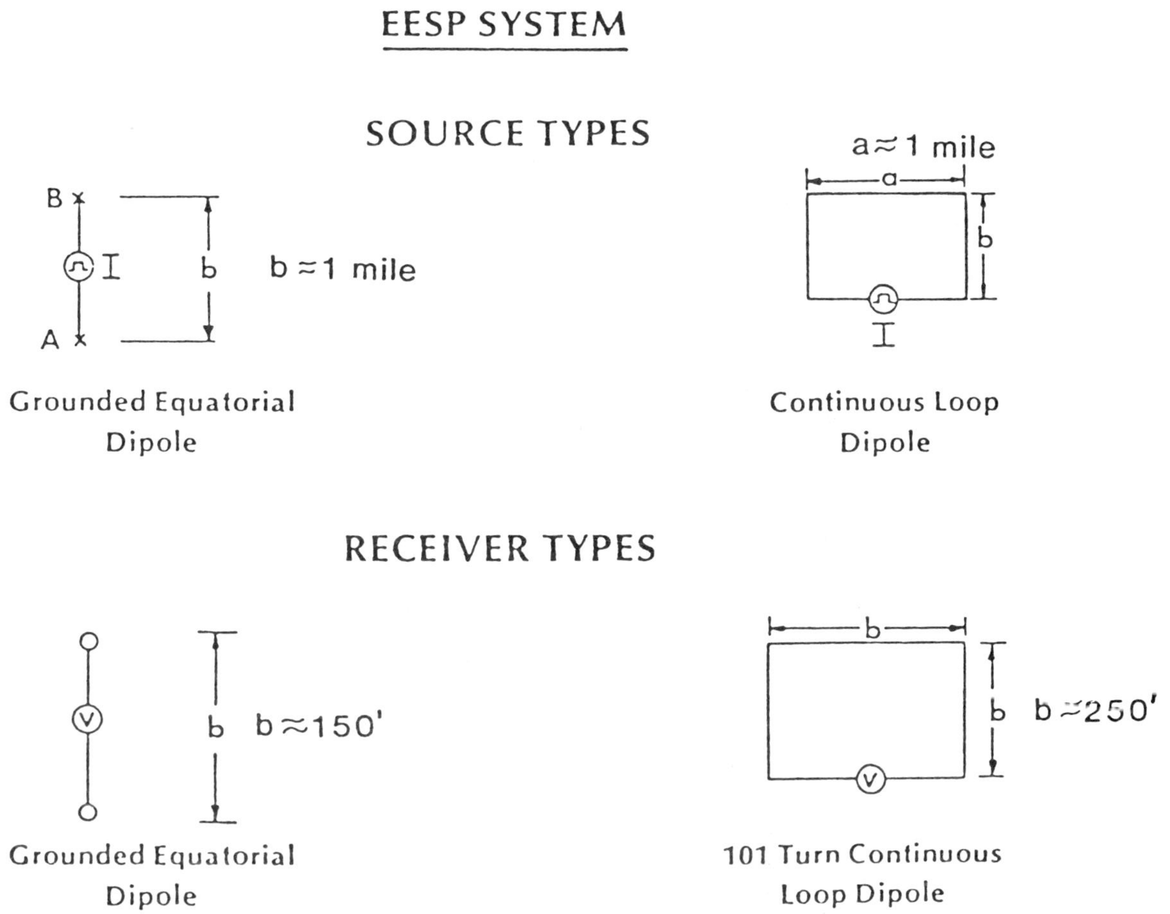

Fig. 2. Controlled EM sources and receivers.

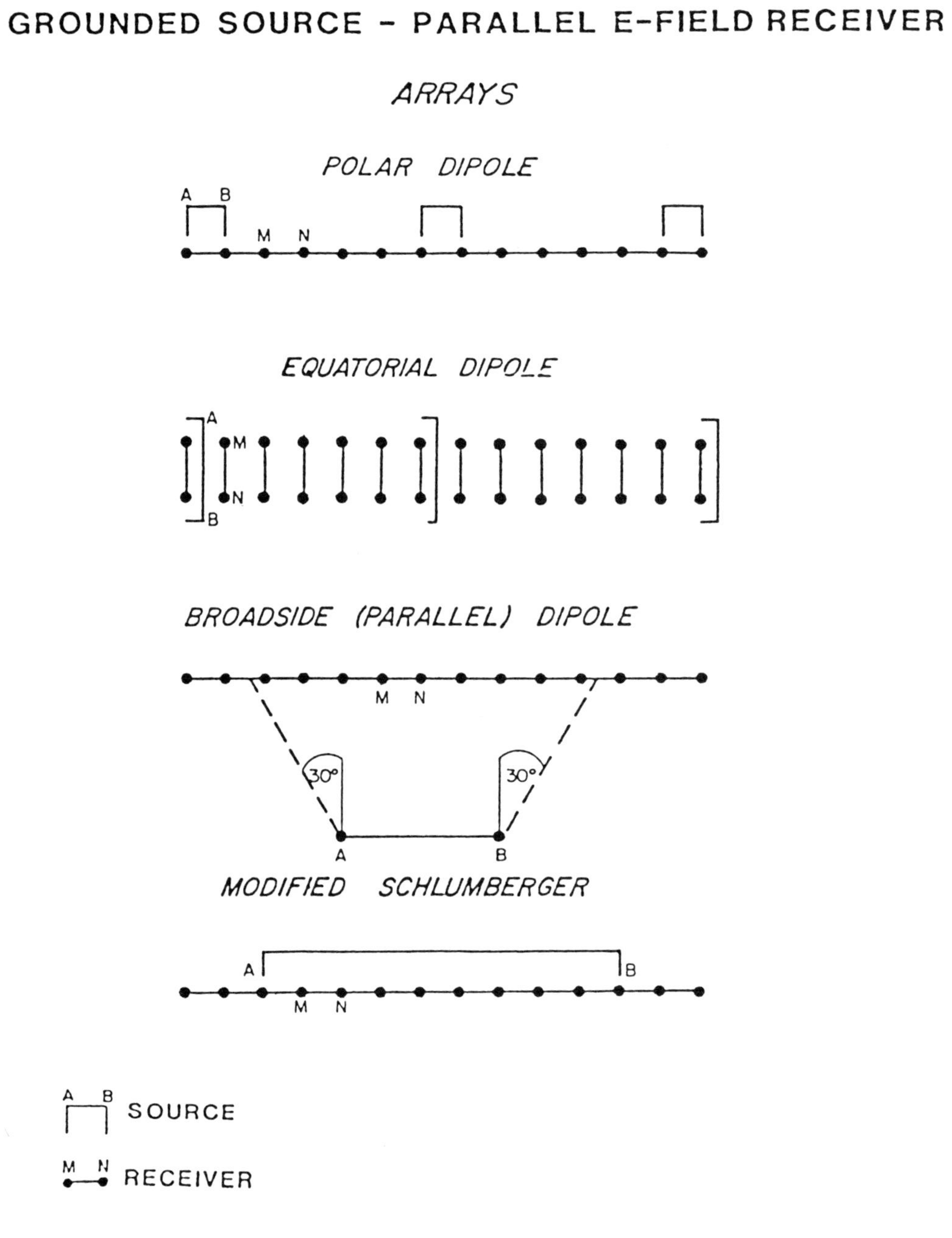
GROUNDED SOURCE - PARALLEL E-FIELD RECEIVER
ARRAYS
POLAR DIPOLE
A B
M N
EQUATORIAL DIPOLE
A
M
N
B
BROADSIDE (PARALLEL) DIPOLE
M N
30°
30°
A
B
MODIFIED SCHLUMBERGER
A
B
M N
A B
SOURCE
M N
RECEIVER

Fig. 3.

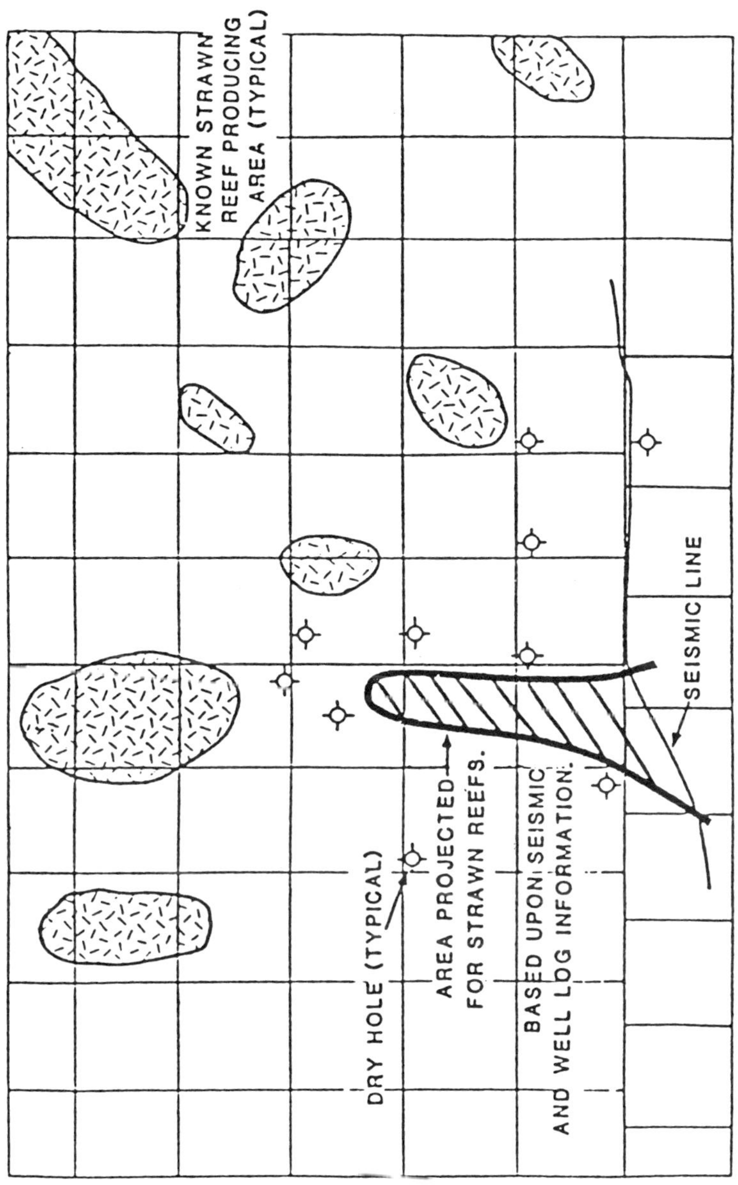

Fig. 4. East Shelf of the Permian Basin Project area.

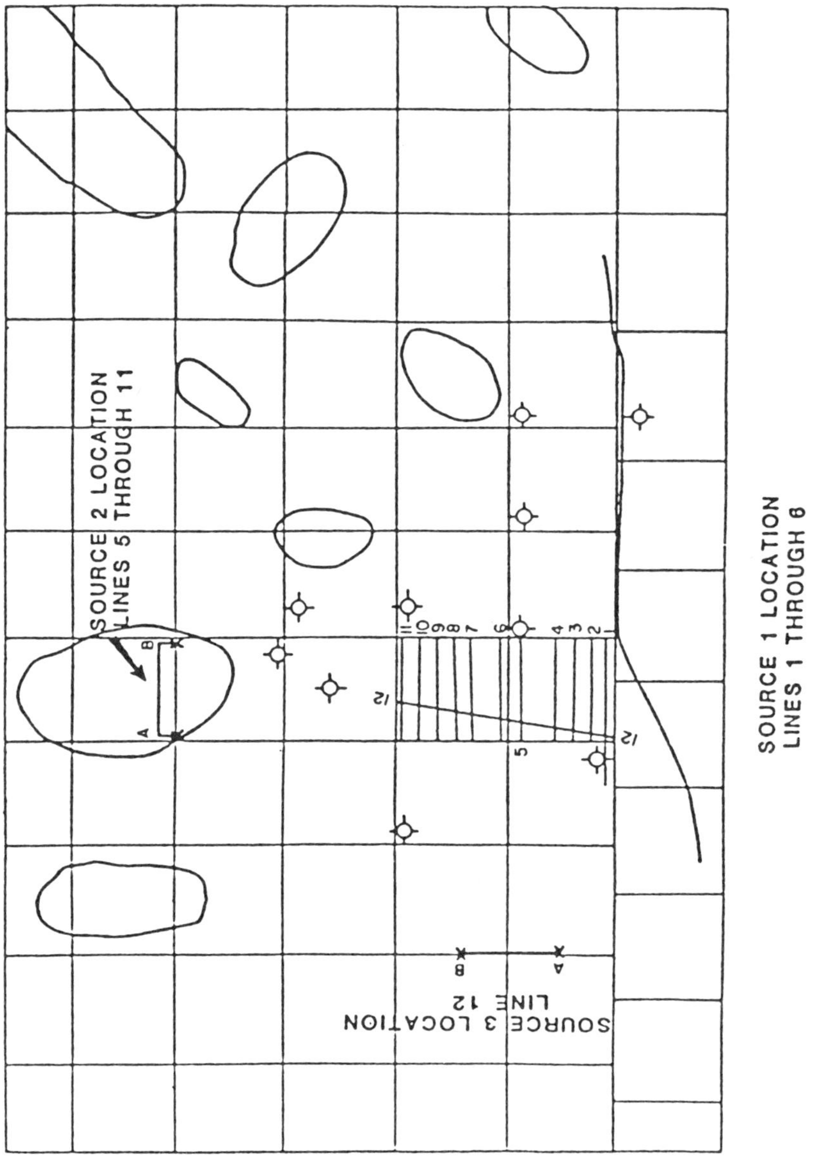

Fig. 5. Broadside array controlled source-parallel E-field receiver coverage.

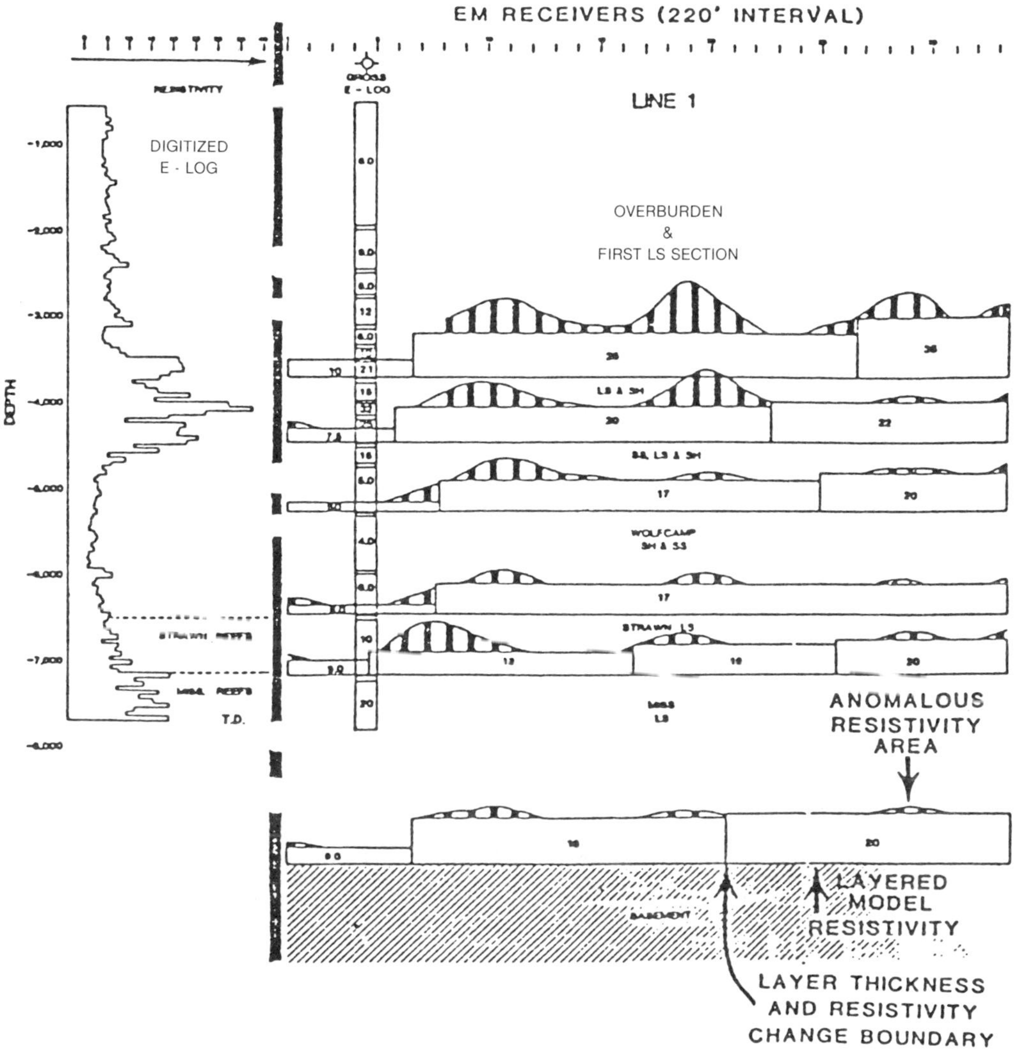

Fig. 6. Controlled source EM sounding and residual resistivity profile.

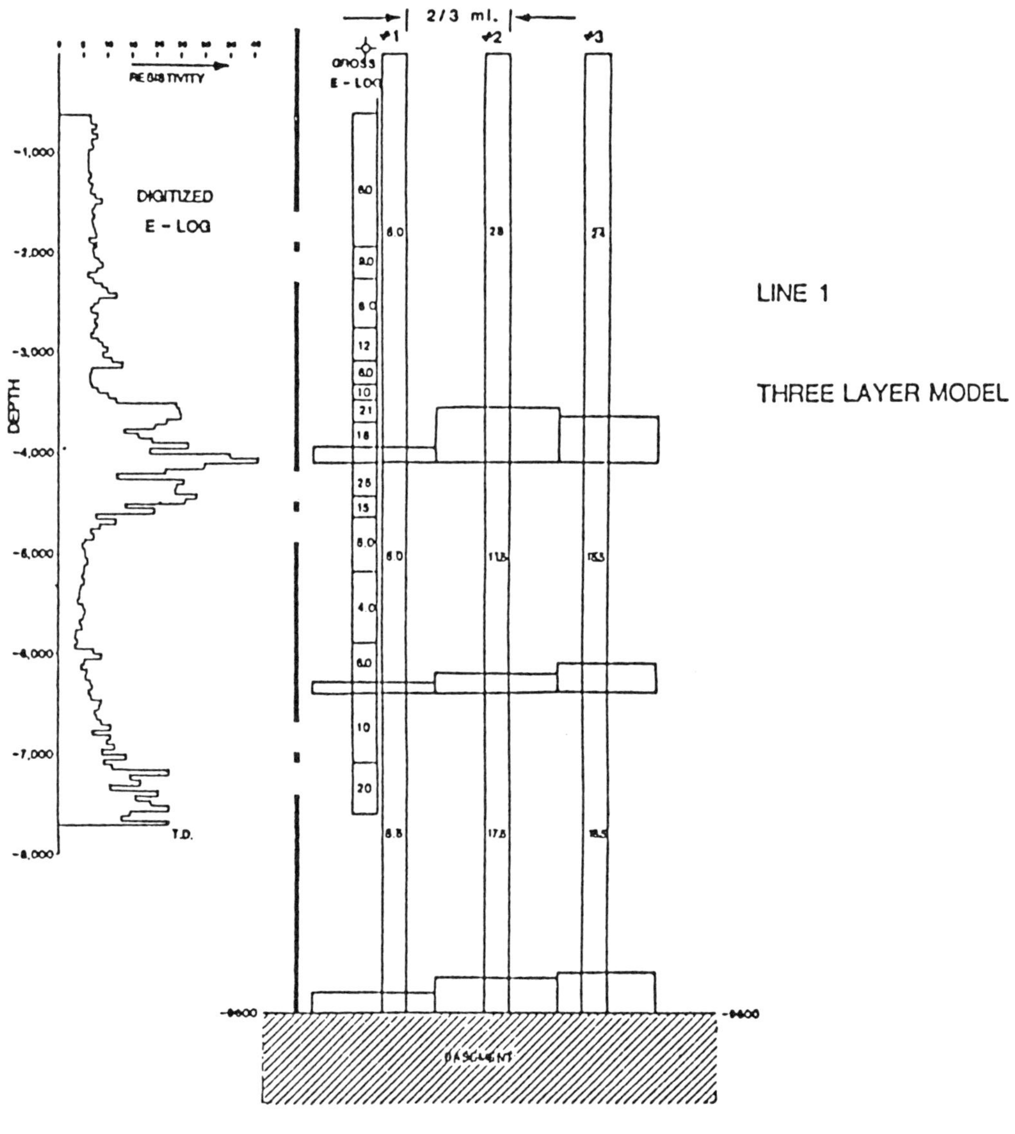

Fig. 7. Comparative MT sounding profile.

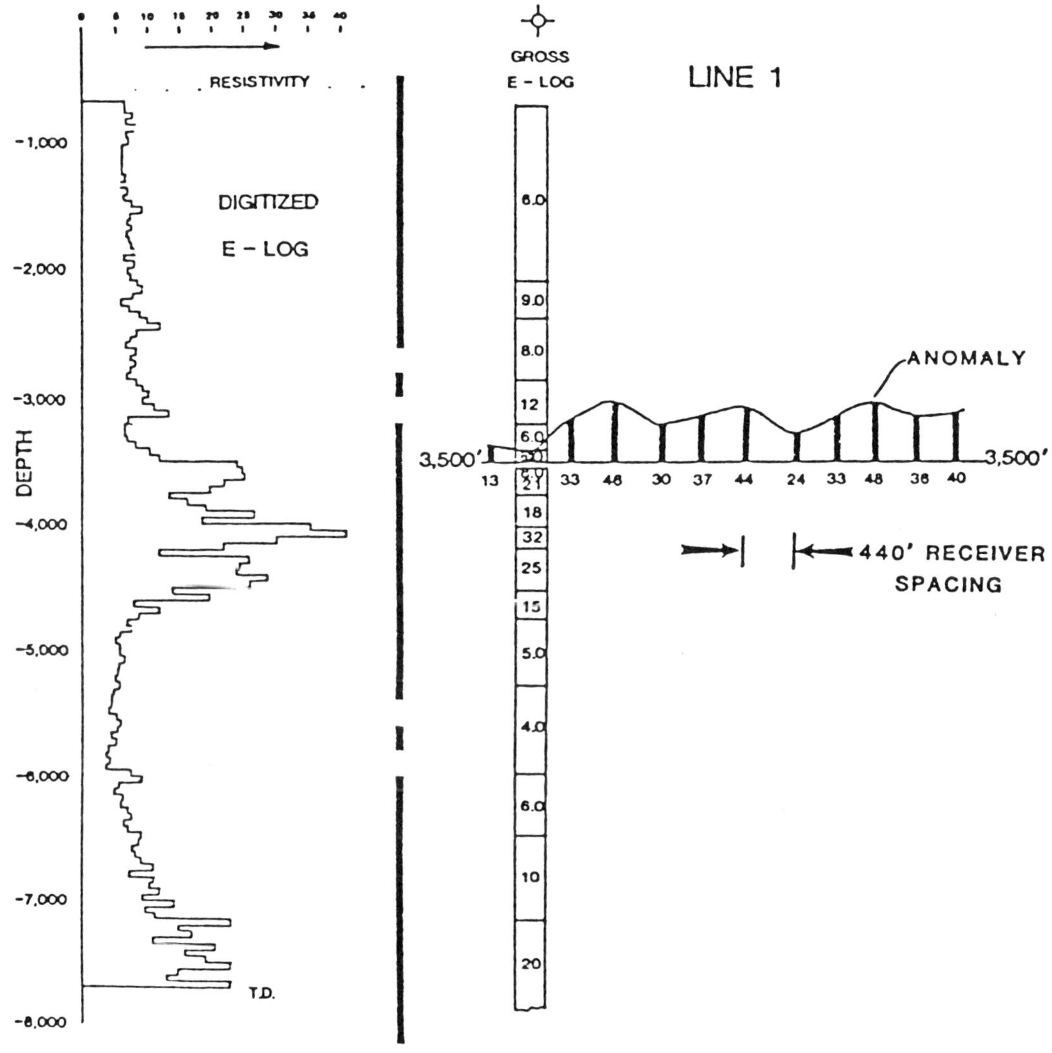

Fig. 8. Comparative integrated transient anomaly profile.

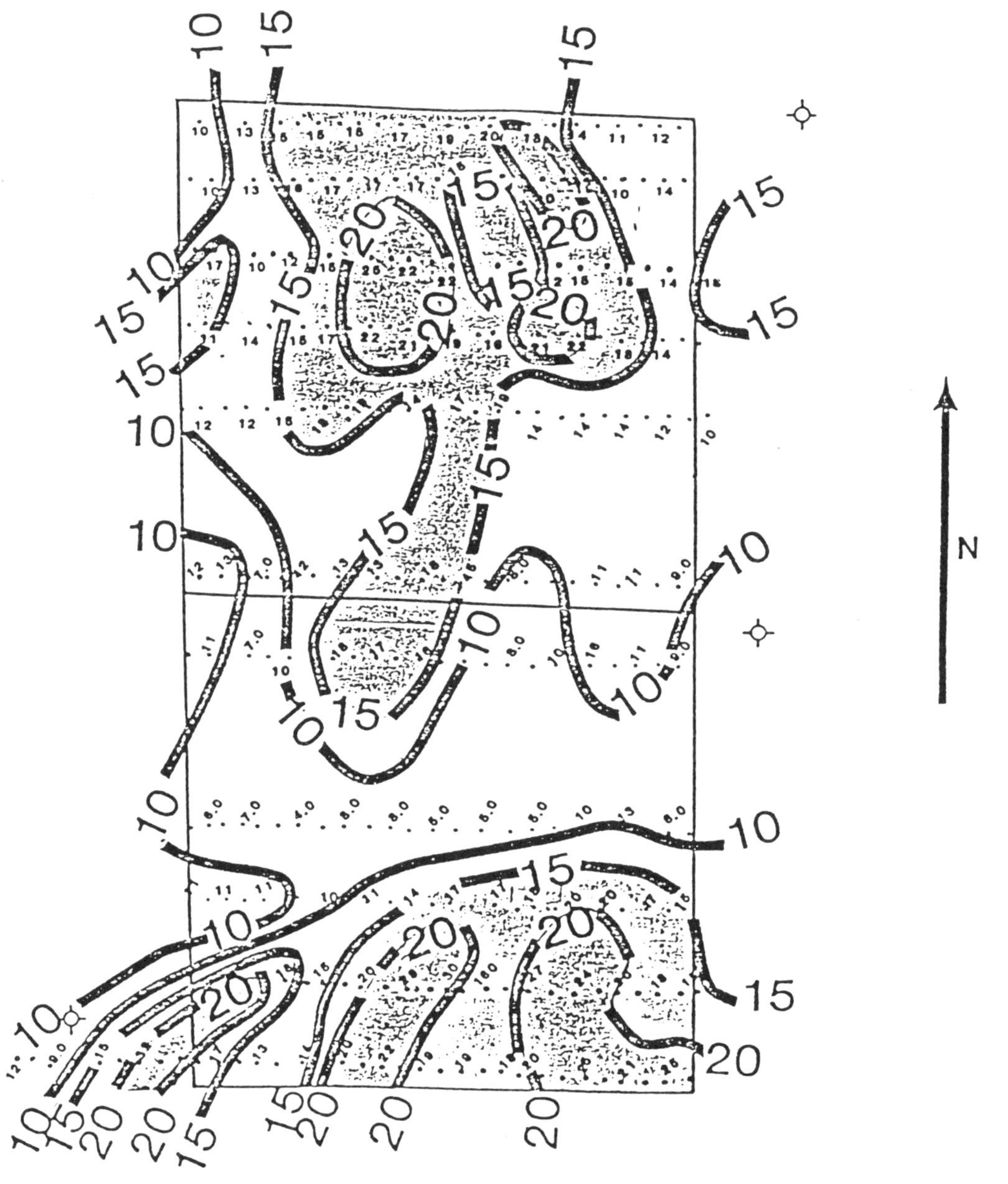

Fig. 9. Controlled source EM resistivity map of the Strawn Formation depth (layer 5).

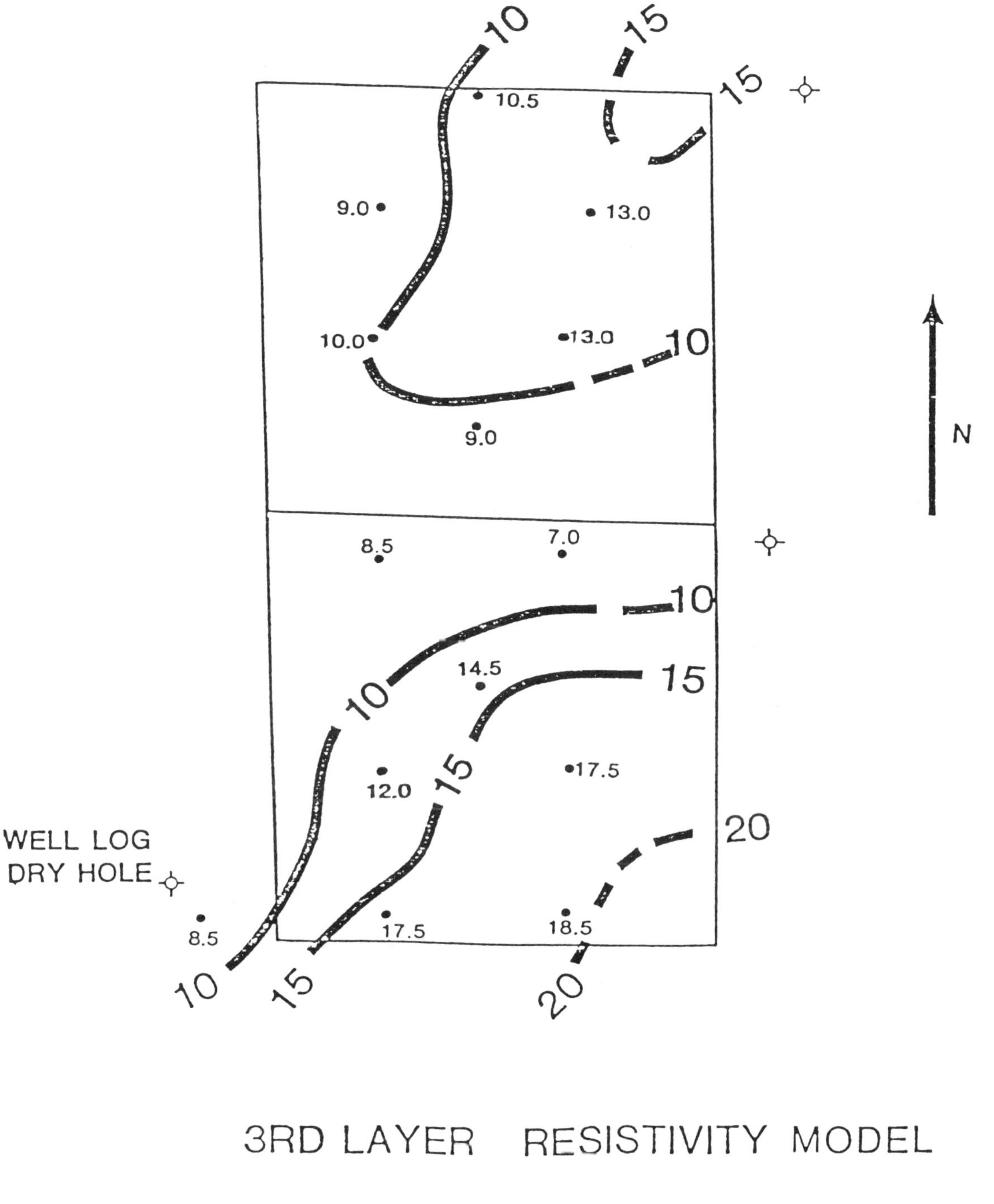

Fig. 10. Comparative MT resistivity map of the Pennsylvanian Formations depth (layer 2).

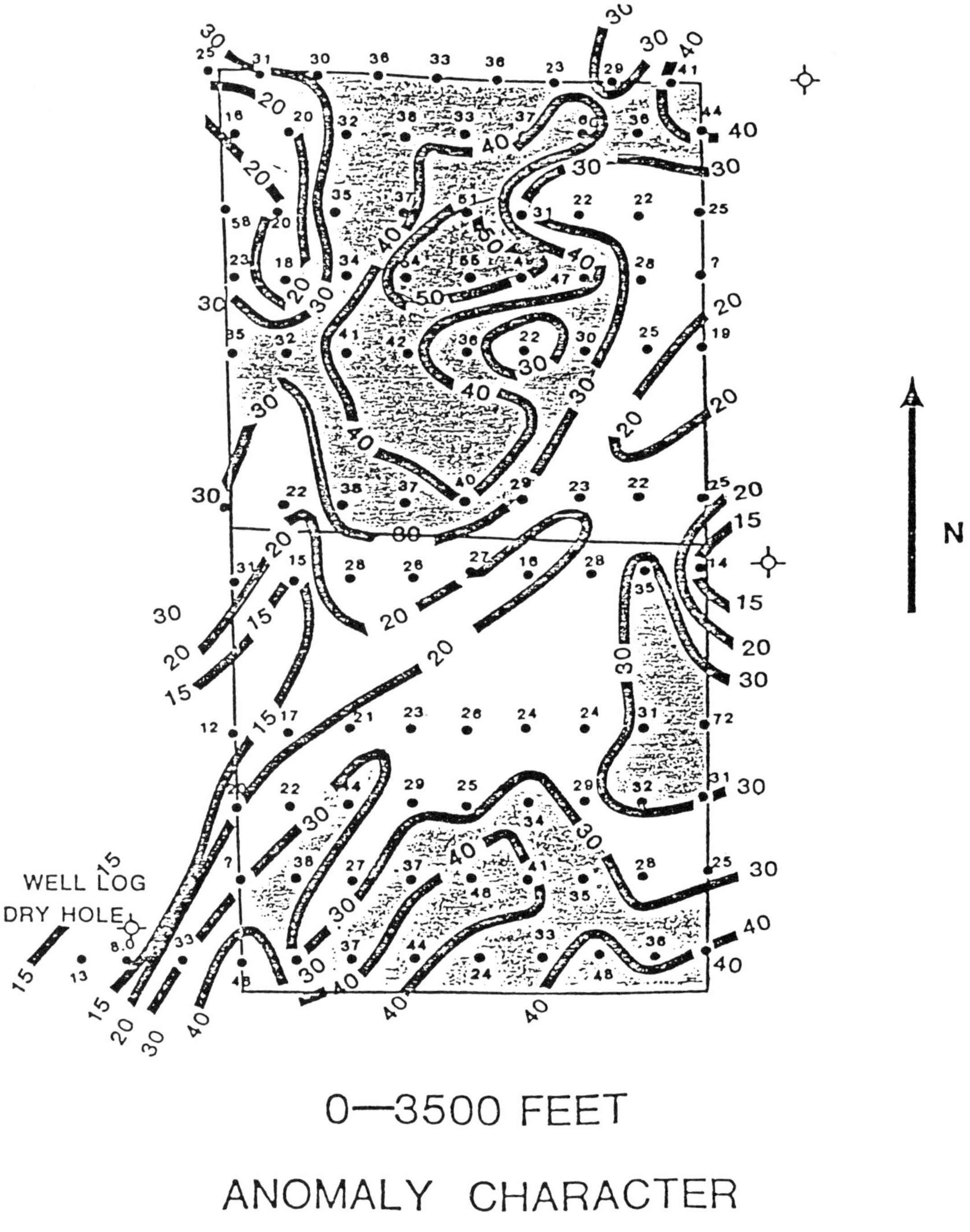

Fig. 11. Comparative integrated transient anomaly map (depth in the Upper Permian formation).

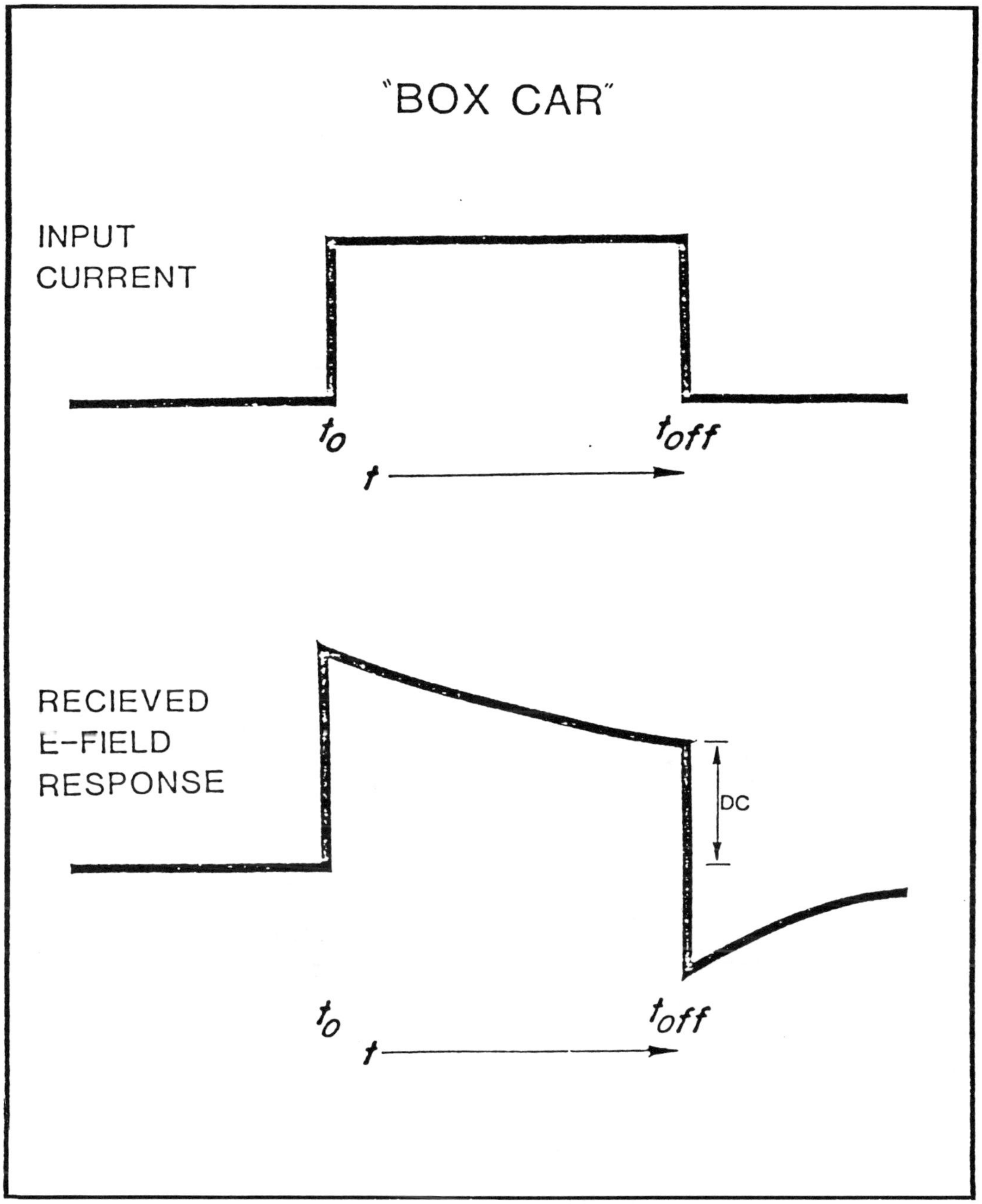

Fig. 12. Controlled source "boxcar" input-receiver transient feature.

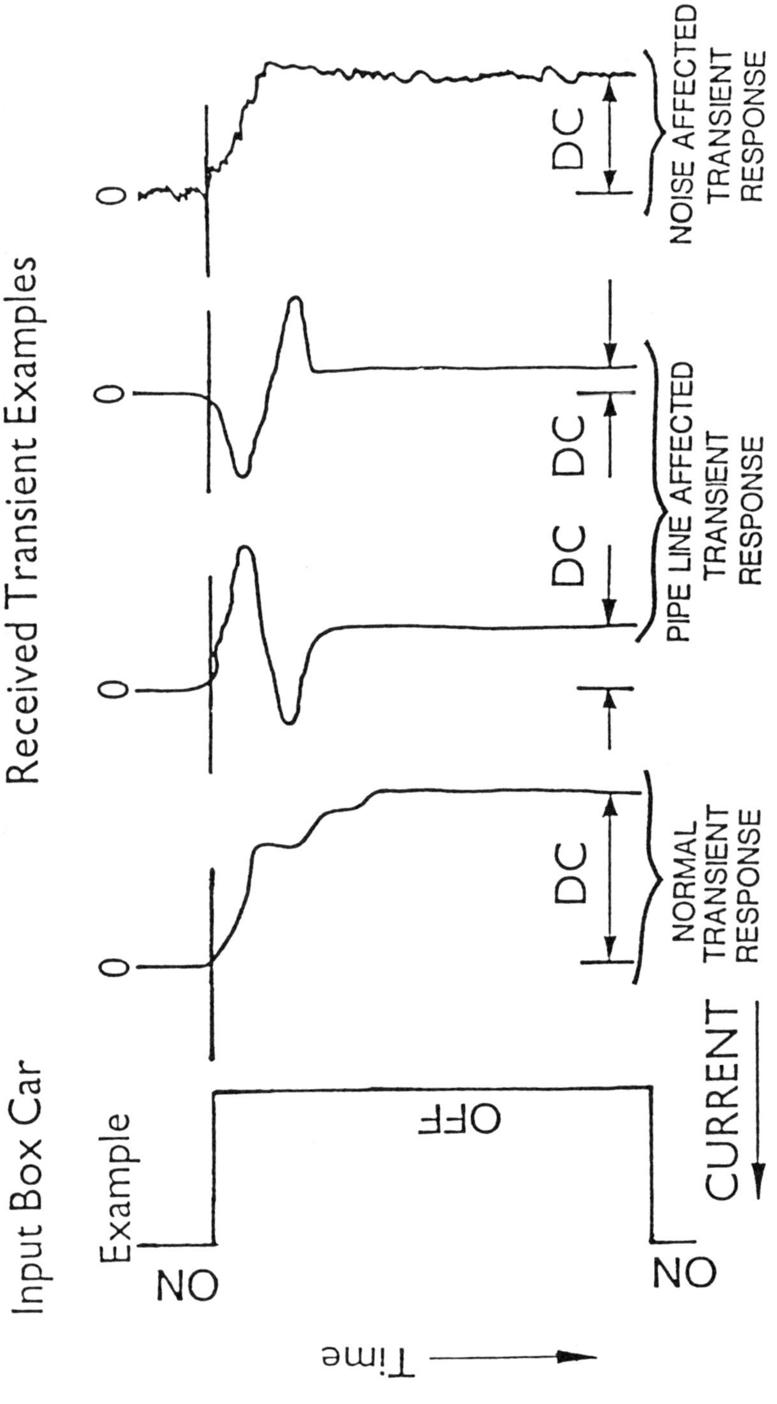

Fig. 13. E-field transient response types.

REMOTE DETECTION OF SOLAR-STIMULATED LUMINESCENCE: A POTENTIAL PETROLEUM PROSPECTING METHOD

Mitchell E. Henry

U.S. GEOLOGICAL SURVEY

ABSTRACT

The association of petroleum seeps with buried petroleum accumulations onshore and offshore is well known, and its usefulness in indicating the presence of petroleum in undeveloped basins is well established. Differences in chemical and physical properties between water and oil allow many crude oils and refined petroleum products introduced into the marine environment to float on the water surface and form films of varying thickness. Remote sensing methods for the detection and identification of these surface accumulations have met with varying degrees of success. One potentially useful method for remotely monitoring oil slicks is based on the observation that most oils luminesce when excited by a suitable source. The Fraunhofer Line Discriminator* (FLD), an instrument designed to measure low levels of solar-stimulated luminescence, was evaluated as a possible petroleum prospecting and pollution monitoring tool in the marine environment.

Laboratory luminescence measurements of crude oils show little correlation between chemical composition and luminescence properties; however, they indicate that multispectral luminescence data can be used effectively to separate crude oils from most refined petroleum products.

An airborne field test of the FLD, conducted in the Santa Barbara Channel, California, showed that the data compare favorably with visual observations of the distribution of the oil by airborne observers. Measurements on known quantities of crude and refined petroleum products in outdoor tank experiments were made with the instrument. Of forty-three determinations, seven occurrences were detected by a positive luminescence value relative to the water. The other occurrences, of which twenty-seven were detected, yielded negative luminescence values compared to the water background. These results are believed to reflect the intense attenuation of the solar energy as it passes through the oil film into the water, and the weak luminescence emission of those oils. It is concluded that a new FLD with simultaneous multispectral capability would yield information that would complement data gathered from an oil slick by other remote-sensing tools.

INTRODUCTION

Because it is evident that the earth's store of commercially valuable hydrocarbons is not unlimited, and that these deposits are increasingly difficult to find, research into new and improved prospecting methods is important. Recent increases in offshore exploration and production activities are expected to continue into the future, as these areas appear to be among the last frontiers for large discoveries.

Petroleum exploration and development in the marine environment is expensive to conduct. Vessels have been fitted with (1) seismic equipment capable of delineating subsurface structures favorable for hydrocarbon accumulation, (2) high-frequency sounding devices to detect gas bubbles in the water column rising from the sea floor (Duncan and Pantin, 1969; Sweet, 1972; Tinkle, 1973; Tinkle et al., 1973), and (3) hydrocarbon "sniffers" that detect these gases in sea water (Dunlap et al., 1960; Swinnerton and Linnebom, 1967; Frank et al., 1970; Schink et al., 1971).

Because of differences in chemical and physical properties between water and oil (Table 1), the remote detection and identification of petroleum on the sea surface is an attractive possibility. Much of the work done to date on remote sensing of marine oil films has exploited differences in the reflective properties between water and oil. These investigations have included conventional light photography (Welch, 1968; Munday et al., 1971; Vizy, 1974); television techniques (Horvath et al., 1971; Millard et al., 1975; Bulban, 1979); airborne multispectral scanner (MSS) devices (Lowe and Hasell, 1969; Munday et al., 1971); and spaceborne MSS devices (Stumpf and Strong, 1974; Goldman and Horvath, 1975; Deutsch et al., 1977; Deutsch and Estes, 1980).

An interesting phenomenon not thoroughly investigated previously is the photoluminescence of crude and refined petroleum products. The intensity and frequency of emitted radiation depends, in part, on the type and relative amounts of organic compounds present in the oils. Two approaches have been taken. In one, the remote measurement of luminescence has been accomplished by use of an airborne laser fluorosensor (Measures and Bristow, 1971; Fantasia and In-

*Use of trade names in this paper is for descriptive purposes only and does not constitute an endorsement by the U.S. Geological Survey.

grao, 1974). A second approach uses a system designed to measure solar-stimulated luminescence with a Fraunhofer Line Discriminator (FLD) (Stoertz et al., 1969; Hemphill et al., 1969; Watson and Hemphill, 1976).

The present study was initiated to evaluate the potential of the FLD as a marine petroleum prospecting tool. The project involved laboratory chemical analyses of twenty-eight crude oils; laboratory luminescence measurements of these crude oils, nineteen pure organic compounds, and fourteen commercially available refined petroleum products; outdoor tank experiments in which FLD measurements were made of thin films of selected crude oils and refined petroleum products; an airborne flight test with the FLD over the Santa Barbara Channel, California; and an evaluation of the most effective methods of digital image enhancement to aid in data interpretation.

THE FLD SYSTEM

Fraunhofer lines are dark lines in the solar spectrum observed from the earth, caused by selective absorption of particular wavelengths by the relatively cool photosphere of the sun. In order to measure the low levels of luminescence under daylight conditions, the FLD is designed to isolate these Fraunhofer lines by means of narrow band-pass filters. A complete description of the operation of the FLD is given in Watson and Hemphill (1976). Figure 1 shows the relationship between energy and wavelength across a Fraunhofer line for a luminescent and nonluminescent target. The relationships used by the FLD computer to compute the reflectance (ϱ) and luminescence (ξ) are $\varrho = \frac{d-c}{a-b}$ and $\xi = \frac{d}{a} - \varrho$. These values are displayed on the FLD front panel as 4 digit numbers ranging from 0000 to 9999, and are recorded on magnetic tape.

The FLD MKII was designed as a spot-reading radiometer capable of monitoring three Fraunhofer lines: 486.1, 589.0, and 656.3 nanometers (nm). The device (see Fig. 2) has since been modified by the addition of a scanning mirror, a dc standard lamp matched to the solar constant by means of fixed resistors in the FLD circuitry in order to eliminate the need for a sky-looking telescope, a mini-computer to carry out bookkeeping tasks in real time, a digital tape recorder for the data, and a television system bore-sighted with the FLD that aids in flight line recovery where landmarks exist in the survey area. Navigation and position location were accomplished with an RNAV (Area Navigation) system over the Santa Barbara Channel.

MATERIALS AND METHODS

The crude oils studied (Table 2) come from reservoirs that range in age from Precambrian to Pliocene and cover the range from heavy, tarlike oils to thin oils resembling diesel fuel. Several refined petroleum products (Table 3) were selected in order to compare their luminescence properties with those of the crude oils. Luminescence measurements were also made on nineteen pure organic compounds generally common to crude oils (Table 4). Measurements in the laboratory were made with a Perkin Elmer MPF-3 spectrophotometer.

All samples were placed in 1-cm quartz cuvetts as neat liquids for luminescence measurements. The quantitative measure of luminescence intensity is given in terms of a luminescent standard, rhodamine WT dye at a concentration of 1,200 parts per billion (ppb). The formula (modified from Watson et al., 1973) used for calculating rhodamine dye equivalence is

$$\text{rhodamine WT equivalence (ppb)} = \frac{\text{Spl } \Sigma_{320}^{RSP}}{\text{rhWT } \Sigma_{320}^{560}} \times 1{,}200$$

where rh Σ_{320}^{560} is the integrated excitation intensity from 320 to 560 nm of the rhodamine standard measured at the 589.0 nm. The Fraunhofer emission line, Spl Σ_{320}^{RSP}, is the integrated excitation intensity of the sample from 320 nm to, but not including, the 20 nm interval containing the Rayleigh Scattering Peak (RSP) at that particular Fraunhofer line. In this equation, 1,200 is the concentration in ppb of the dye standard.

Chemical analyses of the crude oils consisted of asphaltene removal, followed by open-column chromatographic separations to determine the weight percent of aliphatic, aromatic, and polar compounds present. (See Henry, 1982, for a description of procedures.)

The aromatic fraction from twenty-one of the crude oils was analyzed by High Performance Liquid Chromatography (HPLC) to determine the relative amounts of one to five-ring molecules present and their luminescence properties. The HPLC reference mixture contained 7.5μg/ml of benzene, napthalene, anthracene, pyrene, and perylene in hexane. A luminescence detector was added to the column down stream from the UV detector.

Experiments were conducted outdoors with the FLD suspended over a water-filled tank with selected oils floating on the water surface. These experiments were conducted to determine the effects of oil type and film thickness at the three Fraunhofer lines measured.

RESULTS AND DISCUSSION

The results of the chemical analyses are shown in Table 5. These data permit classification of the oils into two general groups: aliphatic oils, which contain over 50 percent aliphatic compounds, and aromatic oils, which contain less than 50 percent aliphatic compounds (Fig. 3).

Luminescence traces of the aromatic fraction of the crude oils are remarkably similar, in spite of the wide range in age and API gravity displayed by the samples (Henry, 1982). Little correlation was found between luminescence intensity of the bulk crude oils and the percentage of 1, 2, 3, or 4 ring aromatic compounds. The only filter combination that produced measurable luminescence was 340 nm excitation/486.1 nm emission.

Laboratory luminescence measurements of the crude oils are shown in Table 6. These data show that the aliphatic crudes generally display higher values of luminescence at the 486.1 nm line that at the 589.0 nm line. Aromatic crudes, those that show high luminescence at the 486.1 nm line, generally display high luminescence at the 589.0 nm line. Luminescence intensity of the crude oils is lowest at the 396.8 nm line for all oils. This appears to result from strong absorption of the excitation beam at shorter wavelengths and from the solar correction applied to the laboratory luminescence measurements.

Little correlation exists between luminescence intensity at the four Fraunhofer lines studied and the percentage of the four major groups of compounds present in the crude oils. Positive correlation was found to exist, however, between API gravity and the logarithm of the luminescence intensity (Figs. 4, 5, 6, and 7). These plots show an increase in luminescence with an increase in API gravity. This is a complex relationship resulting, in part, from the opacity of the oils; the low API gravity crudes are generally less transparent than the higher API gravity oils. The shift of luminescence intensity between aromatic and aliphatic crude oils is shown in Figures 4, 5, 6, and 7 by a decrease in the slope of the regression line with an increase in emission wavelength.

Results of laboratory luminescence measurements of the refined products are shown in Table 7. The major difference between crude oils and refined products is a large shift toward higher luminescence values at shorter wavelengths in the refined products. This results from removal during the refining process of the asphaltene fraction that, in addition to giving rise to long wave luminescence (Riecker, 1962), may effectively attenuate shorter wavelength excitation energy necessary to excite shorter wave (396.8 nm) luminescence.

The combined results from the laboratory luminescence measurements of bulk crude oils and refined products suggest a classification scheme using multispectral ratios. Such a scheme is shown in Figures 8 and 9. The ratio of luminescence intensity measured at the 396.8 and 656.3 nm lines (Fig. 8) shows a sharper separation of the two groups than does the ratio formed for measurements made at the 486.1 and 656.3 nm lines (Fig. 9).

Results from the outdoor tank experiments are summarized in Table 8. Although the oil-filled pan was detected in most of these experiments (86 percent), it registered as a higher signal in only seven instances. In all other cases, the oil generates a luminescence signal lower than or indistinguishable from the adjacent clear water. Figure 10 is an example of an enhanced image created from motor-oil luminescence data showing a higher luminescence signature. A crude oil sample showing a low luminescence signature is shown in Figure 11.

The low luminescence signatures for oil samples compared to water is believed to be the result of tap water luminescence not detected in the laboratory, but detected in the outdoor tank work. Additionally, the oil absorbs a fraction of the incoming solar energy available for excitation of the water column and some of the luminescence that is emitted by the water.

Many of the images display dark areas near the inside perimeter of the pan with values in the center of the pan similar to values from clear water (Fig. 11). This display is caused by a meniscus effect in which some of the oils tended to adhere to the pan walls giving rise to thick peripheral accumulations. A model is shown in Figure 12 that explains the observed results from the tank work.

Field testing of the modified FLD was done over the Santa Barbara Channel, California (Watson et al., 1978). The extent of the slick, its estimated film thickness, and the flight coverage are shown in Figure 13. FLD data were superimposed on the flight line map to show the oil distribution as determined from luminescence (Fig. 14). A good correlation exists between those FLD values above background and the visual observations of the slick position and size. These data points represent luminescence values above those for clear water. No significant difference exists between luminescence intensity and oil thickness (Watson et al., 1978). It has been reported, however, that oils with an API gravity less than about 30° show little or no change in luminescence as a function of film thickness over the range studied (Watson and Hemphill, 1976). The oils from selected seeps in this area display an API gravity of about 10° (Allen et al., 1970).

SUMMARY AND CONCLUSIONS

This study has demonstrated several limitations for FLD applications to marine petroleum prospecting. The present instrument cannot measure multiple Fraunhofer lines simultaneously, a necessary requirement for use of the instrument as an identification tool as well as a detection tool. However, the technique of remote sensing of luminescent materials by the Fraunhofer line-depth method, using an updated instrument capable of recording data simultaneously in multiple channels, has potential application to petroleum exploration in the marine environment, particularly in establishing the presence of naturally occurring seeping hydrocarbons in unexplored areas.

REFERENCES

Allen, A.A., Schlueter, R.S., and Mikolaj, P.G., 1970, Natural oil seeps at Coal Oil Point, Santa Barbara, California: Science, v. 170, p. 974-976

Bulban, E.J., 1979, Airborne sensor detects fish at night: Aviation Week and Space Technology, v. 110, no. 9, p. 60-63.

Deutsch, Morris, and Estes, J.E., 1980, Landsat detection of oil from natural seeps: Photogrammetric Engineering and Remote Sensing, v. 46, p. 1313-1322.

———— Strong, A.E., and Estes, J.E., 1977, Use of Landsat data from the detection of marine oil slicks: Ninth Annual Offshore Technology Conference, Preprints, v. 1, paper no. 2763, p. 311-318.

Duncan, A.R., and Pantin, H.M., 1969, Evidence for submarine geothermal activity in the Bay of Plenty: New Zealand Journal of Marine and Freshwater Research, v. 3, p. 602-606.

Dunlap, H.F., Bradley, J.S., and Moore, T.F., 1960, Marine seep detection—a new reconnaissance exploration method: Geophysics, v. 25, p. 275-282.

Fantasia, J.F., and Ingrao, H.D., 1974, Development of an experimental airborne laser remote sensing system for the detection and classification of oil spills: Ninth International Symposium on Remote Sensing of Environment, Proceedings, University of Michigan, Ann Arbor, p. 1711-1745.

Frank, D.J., Sackett, W.M., Hall, R., and Fredericks, A., 1970, Methane, ethane, and propane concentrations on the Gulf of Mexico: American Association of Petroleum Geologists Bulletin, v. 54, pt. 1, p. 1933-1938.

Goldman, G.C., and Horvath, Robert, 1975, Oil-pollution detection and monitoring from space using ERTS-1: Final Report, Environmental Research Institute of Michigan, Ann Arbor, 57 p.

Hemphill, W.R., Stoertz, G.E., and Markle, D.A., 1969, Remote sensing of luminescene materials: Sixth International Symposium on Environment Proceedings, University of Michigan, Ann Arbor, p. 565-585.

Henry, M.E., 1982, Marine petroleum prospecting and pollution monitoring with an airborne Fraunhofer Line Discriminator: Ph.D. thesis, Texas A&M University, 201 p.

Horne, R.A., 1969, Marine chemistry: New York, John Wiley and Sons, 568 p.

Horvath, Robert, Larrowe, V., and Stewart, S., 1971, Measurement and analysis program for oil-slick characteristics: Quarterly Progress Report 27660-15-P, Willow Run Laboratories, University of Michigan, Ann Arbor, 26 p.

Klemas, Vytautas, 1972, Detecting and measuring oil on water: Instrumentation Technology, v. 19, p. 54-59.

Lowe, D.S., and Hasell, P.G., 1969, Multispectral sensing of oil pollution: Proceedings, Sixth International Symposium on Remote Sensing of Environment, University of Michigan, Ann Arbor, p. 755-765.

Measures, R.M., and Bristow, Mike, 1971, The development of a laster fluorosensor for remote environmental probing: Joint Conference on Sensing of Environmental Pollutants, Palo Alto, California, American Institute of Aeronautics and Astronautics Paper 71-1121, 7 p.

Millard, J.P., Arvesen, J.D., and Lewis, P.L., 1975, Development and test of video systems for airborne surveillance of oil spills: U.S. National Aeronautics and Space Administration Technical Memorandum, NASA TM X-62, 429, 25 p.

Munday, J.C., MacIntyre, W.G., Penney, M.E., and Oberholtzer, J.D., 1971, Oil slick studies using photographic and multispectral scanner data: Seventh International Symposium on Remote Sensing of Environment, Proceedings, University of Michigan, Ann Arbor, p. 1027-1043.

Riecker, R.E., 1962, Hydrocarbon fluorescence and migration of petroleum: American Association of Petroleum Geologists Bulletin, v. 46, p. 60-75.

Schink, D.R., Guinasso, N.L., Sigalove, J.J., and Cima, N.E., 1971, Hydrocarbons under the sea — a new survey technique: Third Annual Offshore Technological Conference, Preprints, v. 1, paper no. 1339, p. I-131-I-142.

Stoertz, G.E., Hemphill, W.R., and Markle, D.A., 1969, Airborne fluorometer applicable to marine and esturine studies: Marine Technology Society Journal, v. 3, p. 11-26.

Stumpf, H.G., and Strong, A.E., 1974, ERTS-I views an oil slick: Remote Sensing Environment, v. 3, p. 87-90.

Sverdrup, H.U., Johnson, M.W., and Fleming, R.H., 1942, The oceans, their physics, chemistry, and general biology: Englewood Cliffs, New Jersey, Prentice-Hall, 1087 p.

Sweet, W.E., 1972, Data evaluation analysis, cruise 72-C-2; study of naturally occurring hydrocarbons in the Gulf of Mexico: Department of Oceanography, Texas A&M University, 15 p.

Swinnerton, J.W., and Linnebom, V.J., 1967, Determination of the C_1 to C_4 hydrocarbons in sea water by gas chromatography: Journal of Gas Chromatography, v. 5, p. 570-573.

Tinkle, A.R., 1973, Natural gas seeps in the northern Gulf of Mexico: a geological investigation: Master's thesis, Texas A&M University, 93 p.

Tinkle, A.R., Antoine, J.W., and Kuzela, R., 1973, Detecting natural gas seeps at sea: Ocean Industry, v. 8, p. 139-142.

Tissot, B.P., and Welte, D.H., 1978, Petroleum formation and occurrence: New York, Springer-Verlag, 538 p.

Vizy, K.N., 1974, Detecting and monitoring oil slicks with aerial photographs: Photogrammetric Engineering, v. 40, p. 697-708.

Watson, R.D., and Hemphill, W.R., 1976, Use of an airborne Fraunhofer line discriminator for the detection of solar stimulated luminescence: U.S. Geological Survey Open-File Report 76-202, 109 p.

Watson, R.D., Hemphill, W.R., and Hessin, T.D., 1973, Quantification of the luminescence intensity of natural materials: Symposium for the Management and Utilization of Remote Sensing Data, Proceedings, Sioux Falls, South Dakota, p. 364-376.

Watson, R.D., Henry, M.E., Theisen, A.F., Donovan, T.J., and Hemphill, W.R., 1978, Marine monitoring of natural oil slicks and man-made wastes utilizing an airborne Fraunhofer line discriminator: Fourth Joint Conference on Sensing of Environmental Pollutants, Proceedings, p. 667-671.

Weast, R.C., and Astle, M.J., eds., 1980, CRC handbook of chemistry and physics (60th ed.): Boca Raton, Florida, CRC Press, 2447 p.

Welch, R.I., 1968, Remote sensing in California resource management: American Society of Photogrammetry, Proceedings, p. 330-344.

TABLE 1
SELECTED PROPERTIES OF NATURAL WATERS AND OILS

Property	Material			
	Clear Seawater	Clear Freshwater	Crude Oil	Refined Products
Density gm/cc, 25°C	1.024[a]	1.003[a]	0.7-1.007	0.7-> 1.0[b]
Refractive index ratio to air	1.34[c]	1.33[a]	1.44[d]	1.44-1.58[d]
Dielectric constant ratio to air	20-78[d]	78-80[a]	range 1.85-2.94[d] avg 2.0	1.7-> 2.2[d]
Luminescence	Not determined	None detected	As much as 60% compared to a 1,200 ppb rhodamine WT standard	Some over 100% compared to a 1,200 ppb rhodamine WT standard

[a] From Horne (1969)
[b] Calculated from density values of individual hydrocarbons given in Weast and Astle (1980)
[c] Calculated from data given in Sverdrup et al (1942)
[d] From Klemas (1972)

TABLE 2
CRUDE OILS USED

Sample No.	Chevron ID No.	Reservoir Age	Producing Horizon or Unit	Location	Specific Gravity	°API Gravity
1	19327	Devonian	Dundee Fm.	Michigan	0.8327	38.4
2	17394	Late Cretaceous	Parkman S. Mbr.	Wyoming	0.7866	48.4
3	31473	Cretaceous	Muddy S. Mbr.	Wyoming	0.8098	43.2
4	24294	Miocene	Ala[1]	Nigeria	0.8163	41.8
5	24563	Late Cretaceous	Lakota[1]	Colorado	0.8172	41.7
6	17636	Middle Miocene	8700' sand	Louisiana	0.8834	28.7
7	29350-5	Eocene	Zone IV	Denmark	0.7790	50.1
8	24461	Late Cretaceous	Stevens[1]	Mississippi	0.7932	46.9
9	31316	Miocene	1st[1] Carneros	California	0.8532	34.3
10	28212	Upper Cretaceous	Padgett[1] sand	Mississippi	0.8455	35.9
11	31317	Miocene	3rd[1] Carneros	California	0.8377	37.4
12	26241	Middle Miocene	Cruse	Trinidad	0.8318	38.5
13	30534-1	Jurassic	N.A.	N.A.	0.7317	61.9
14	32219	Oligocene	Vaqueros Fm.	California	0.7713	52.0
15	24569	Pennsylvanian	Weber Ss.	Colorado	0.8574	33.5
16	27358	Paleocene	Harash	Libya	0.8350	38.0
18	27458-2	Mississippian	Pekisko	Canada	0.7735	51.4
19	31851	Cretaceous	N.A.	Bahrain	0.8695	31.2
20	30229-1	Jurassic	N.A.	Alaska	0.8654	32.0
21	13588	Permian	Phosphoria Fm.	Wyoming	0.9110	23.8
22	25968-2	Permian-Pennsylvanian	Minnelusa Fm.	Wyoming	0.8430	36.4
23	25179	Miocene	Asmari	Iran	0.8854	28.3
25	29619	Pliocene	Repetto[1]	California	0.8639	32.3
26	27360	Precambrian	N.A.	Libya	0.8481	35.3
27	14525	Eocene	N.A.	Venezuela	0.8752	30.2
28	17112	Ordovician	Sparkyzone	Canada	0.9677	14.7

N.A. — Not available
[1]Of local or subsurface usage

TABLE 3
REFINED PETROLEUM PRODUCTS

Sample No.	Product	Supplier
31	Air conditioner pump oil	Everco Industries
32	Power steering fluid	Sta-Lube Inc.
33	Hydraulic jack oil	Sta-Lube Inc.
34	Brake fluid	Raybestos/Manhattan
35	Pennzoil motor oil—30 weight	Pennzoil Co.
36	Aero-Shell aviation oil—50 weight	Union Oil Co.
37	Regular gasoline	Union Oil Co.
38	Premium gasoline	Union Oil Co.
39	Unleaded gasoline	Union Oil Co.
40	Chevron solvent	Union Oil Co.
41	Kerosene	Union Oil Co.
42	Diesel fuel	Union Oil Co.
43	Heating oil	Union Oil Co.
44	Menhaden oil	Zapata Haynie Corp.

TABLE 4
PURE ORGANIC COMPOUNDS*

Sample No.	Product	Grade
45	Pentane	PHOTREX
46	Hexanes	BAKER ANALYZED
47	n-Heptane	PHOTREX
48	2,4-Dimethylpentane	PHOTREX
49	2, 2, 4-Trimethylpentane	PHOTREX
50	n-Nonane	BAKER
51	n-Decane	PRACTICAL
52	Cyclopentane	BAKER
53	Cyclohexane	PHOTREX
54	Methylcyclohexane	PHOTREX
55	1-Propanol	BAKER ANALYZED
56	2-Propanol	PHOTREX
57	Isobutyl Alcohol	PHOTREX
58	Ethylene Glycol	BAKER ANALYZED
59	Oleic Acid	FOOD CHEMICALS CODEX
60	Benzene	PHOTREX
61	Toluene	PHOTREX
62	O-Xylene	PHOTREX
63	1, 2, 4-Trimethylbenzene	BAKER

*J.T. Baker Chemicals Company

TABLE 5
RESULTS OF ASPHALTENE REMOVAL AND CHROMATOGRAPHIC SEPARATION OF CRUDE OILS

Sample No.	Weight (mg)	Aliphatics (mg)	Aromatics (mg)	Polar (mg)	Asphaltene (mg)*
1	25.7	13.7	5.7	2.9	3.4
2	29.2	15.5	3.5	1.5	8.7
3	28.5	10.9	3.8	3.3	10.5
4	29.9	13.5	3.8	2.2	10.4
5	29.3	14.2	4.6	5.2	5.3
6	37.0	22.8	8.0	5.1	1.1
7	22.1	13.2	2.4	0.5	6.0
8	29.5	20.9	4.2	4.0	0.4
9	27.3	9.9	3.9	0.9	12.6
10	35.0	21.7	7.2	3.3	2.8
11	27.8	14.3	7.0	2.8	3.7
12	28.5	15.7	6.7	3.6	2.5
13	25.1	15.4	2.1	0.2	7.4
14	22.4	11.6	6.4	3.5	0.9
15	31.3	18.6	10.1	1.5	1.1
16	25.3	17.5	5.6	1.6	0.6
18	13.9	2.8	3.7	0.7	6.7
19	35.0	12.8	15.3	4.5	2.4
20	33.0	9.6	11.3	10.2	1.9
21	35.0	14.7	17.0	2.4	0.9
22	28.9	12.8	9.3	2.7	4.1
23	29.0	12.4	10.4	4.3	1.9
25	30.3	11.4	10.2	7.3	1.4
26	30.5	13.9	6.5	2.8	7.3
27	28.2	9.0	6.8	1.7	10.7
28	26.3	6.9	10.0	2.6	6.8

*The weight of asphaltene was determined by difference between the original sample weight and the sum of the weights of aliphatic + aromatic + polar material as the asphaltene material was discarded before weighing.

TABLE 6
INTEGRATED EXCITATION INTENSITY OF CRUDE OILS*

		Emission Wavelength (nm)									
Sample	Oil Type	396.8	486.1	589.0	656.3	486.1 / 656.3	589.0 / 656.3	486.1 / 396.8	589.0 / 396.8	656.3 / 396.8	(396.8 + 486.1) / 656.3
1	Aliphatic	59.1	749.3	709.3	428.2	1.75	1.66	12.68	12.00	7.25	1.89
2	Aliphatic	37.6	316.9	287.5	139.7	2.27	2.06	8.43	7.65	3.72	2.54
3	Aromatic	59.4	498.4	443.9	206.0	2.42	2.15	8.39	7.47	3.47	2.71
4	Aromatic	12.6	311.7	406.9	208.3	1.50	1.95	24.74	32.29	16.53	1.56
5	Aromatic	2.5	78.9	128.9	78.8	1.00	1.64	31.56	51.56	31.52	1.03
6	Aliphatic	20.8	325.5	352.6	193.3	1.68	1.82	15.65	16.95	9.29	1.79
7	Aliphatic	78.9	425.4	329.7	164.2	2.59	2.01	5.39	4.18	2.08	3.07
8	Aliphatic	40.9	385.6	279.3	140.1	2.75	1.99	9.43	6.83	3.43	3.04
9	Aromatic	6.9	92.0	101.3	52.1	1.77	1.94	13.33	14.68	7.55	1.90
10	Aliphatic	14.2	137.8	104.8	52.8	2.61	1.98	9.70	7.38	3.72	2.88
11	Aliphatic	14.2	164.1	136.3	64.0	2.56	2.13	11.56	9.60	4.51	2.79
12	Aliphatic	13.6	157.2	149.1	81.9	1.92	1.82	11.56	10.96	6.02	2.09
13	Aliphatic	58.0	445.5	263.1	85.6	5.20	3.07	7.68	4.54	1.48	5.88
14	Aliphatic	11.0	110.9	94.4	44.1	2.51	2.14	10.08	8.58	4.01	2.76
15	Aliphatic	1.7	51.8	69.4	46.2	1.12	1.50	30.47	40.82	27.18	1.16
16	Aliphatic	1.9	57.1	81.0	45.8	1.25	1.77	30.05	42.63	24.11	1.29
17	N.D.†	3.9	81.6	125.8	88.0	0.93	1.43	20.92	32.26	22.56	0.97
18	Aromatic	29.3	123.8	85.4	41.0	3.02	2.08	4.23	2.91	1.40	3.73
19	Aromatic	8.1	84.1	88.2	56.6	1.49	1.56	10.38	10.89	6.99	1.63
20	Aromatic	5.5	56.7	66.0	39.0	1.45	1.69	10.31	12.00	7.09	1.59
21	Aromatic	8.1	81.9	88.4	52.8	1.55	1.67	10.11	10.91	6.52	1.70
22	Aromatic	1.0	20.2	36.6	27.4	0.74	1.34	20.20	36.60	27.40	0.77
23	Aromatic	5.4	54.4	58.3	28.4	1.92	2.05	10.07	10.80	5.26	2.11
24	N.D.†	5.6	46.2	48.5	30.4	1.52	1.60	8.25	8.66	5.43	1.70
25	Aromatic	3.3	40.4	40.9	21.2	1.91	1.93	12.24	12.39	6.42	2.06
26	Aromatic	6.0	36.7	32.3	17.6	2.09	1.84	6.12	5.38	2.93	2.43
27	Aromatic	6.2	32.4	28.3	11.7	2.77	2.42	5.23	4.56	1.89	3.30
28	Aromatic	2.0	19.4	26.2	19.1	1.02	1.37	9.7	13.10	9.55	1.12

*Values shown are corrected for nonlinearity of the excitation lamp, detector response and solar intensity as a function of wavelength, and given in parts per billion (ppb) rhodamine WT equivalence.
†N.D. = Not determined.

TABLE 7
INTEGRATED EXCITATION INTENSITY OF REFINED PRODUCTS*

Sample	Emission Wavelength (nm) 396.8	486.1	589.0	656.3	486.1/589.0	486.1/656.3	589.0/656.3	486.1/396.8	589.0/396.8	656.3/396.8	(396.8 + 486.1)/656.3
Air conditioner pump oil	1,074.0	2,211.9	91.5	5.0	24.17	442.38	18.30	2.06	0.09	0.004	657.18
Power steering fluid	1,367.3	1,216.7	81.3	25.5	14.97	47.71	3.19	0.89	0.06	0.02	101.33
Jack oil	828.3	3,638.6	503.5	76.5	7.23	47.56	6.58	4.39	0.61	0.09	58.39
Brake fluid	2,074.7	293.3	18.2	2.9	16.12	101.14	6.28	0.14	0.01	0.001	816.55
Pennzoil 30 weight	313.5	4,154.0	2,880.4	687.6	1.44	6.04	4.19	13.25	9.19	2.19	6.50
Aviation oil Aeroshell 50 weight	630.8	5,039.7	1,033.3	163.5	4.88	30.82	6.32	7.99	1.64	0.26	34.68
Regular gasoline	469.8	412.6	29.3	10.5	14.08	39.29	2.79	0.88	0.06	0.02	84.04
Premium gasoline	229.8	212.5	126.1	2.2	1.69	96.59	57.32	0.92	0.55	0.01	201.05
Unleaded gasoline	549.1	479.6	17.2	44.4	27.88	10.80	0.39	0.87	0.03	0.08	23.17
Chevron solvent	18.0	1.5	0	2.2	∞	0.68	0	0.08	0	0.12	8.86
Kerosene	12.9	2.3	0	2.3	∞	1.00	0	0.18	0	0.18	6.61
Diesel fuel	972.7	709.6	29.8	0	23.81	∞	∞	0.73	0.03	0	∞
Heating oil	173.3	45.2	3.0	3.0	15.07	15.07	1.00	0.26	0.02	0.02	72.83
Menhaden oil	29.2	10.0	47.0	2,245.0	0.20	0.004	0.02	0.34	1.60	76.80	0.017

*Values shown are corrected for nonlinearity of the excitation lamp, detector response and solar intensity as a function of wavelength, and are given in parts per billion (ppb) rhodamine WT equivalence.

TABLE 8
LUMINESCENCE CHARACTER OF OIL FILM COMPARED TO TAP WATER MEASURED WITH THE FLD DURING OUTDOOR TANK EXPERIMENTS

Sample	Ideal Film Thickness (μm)	Qualitative Characteristics
		Emission Wavelength 486.1 nm
1	1	Positive luminescence near pan edge
1	10	Positive luminescence near pan edge
1	100	Positive luminescence over most of pan surface
7	1	Generally less than water
7	10	Pan not distinguishable
7	100	Pan not distinguishable
25	1	Pan not distinguishable
25	10	Pan not distinguishable
25	100	Pan not distinguishable
Pennzoil	1	Positive near pan edge
Pennzoil	10	Positive near pan edge
Pennzoil	100	Positive over most of pan surface
Menhaden oil	1	All three film thicknesses display positive
Menhaden oil	10	and negative values randomly distributed
Menhaden oil	100	within pan boundaries
		Emission Wavelength 589.0 nm
1	1	Slightly negative
1	10	Some positive values near pan edge, mostly negative values inside pan
1	100	Slightly positive over much of pan surface
7	1	All three film thicknesses display negative
7	10	values near pan edge. Values in center of pan
7	100	comparable to clear water values
25	1	All three film thicknesses display negative
25	10	values near pan edge. Values in center of pan
25	100	comparable to clear water values
Pennzoil	1	Negative values near pan edge. Values in center of pan comparable to clear water values
Pennzoil	10	Poor image quality, pan not distinguishable
Pennzoil	100	Negative values over most of pan surface
Menhaden oil	1	All three film thicknesses display negative
Menhaden oil	10	values over most of pan surface
Menhaden oil	100	

TABLE 8 *(continued)*

Sample	Ideal Film Thickness (μm)	Qualitative Characteristics
		Emission Wavelength 656.3 nm
1	1	Both film thicknesses display negative values
1	100	over most of pan surface
7	1	All three film thicknesses display negative values
7	10	over most of pan area with values in center of
7	100	pan comparable to those of clear water
25-C	1	Both film thicknesses display negative values over
25-C	100	most of pan area with values in center of pan
		comparable to those of clear water
Pennzoil	1	All three film thicknesses display negative values
Pennzoil	10	over most of pan area with values in center of
Pennzoil	100	pan comparable to those of clear water
Menhaden oil	1	All three film thicknesses display negative values
Menhaden oil	10	over most of pan area with values in center of
Menhaden oil	100	pan comparable to clear water

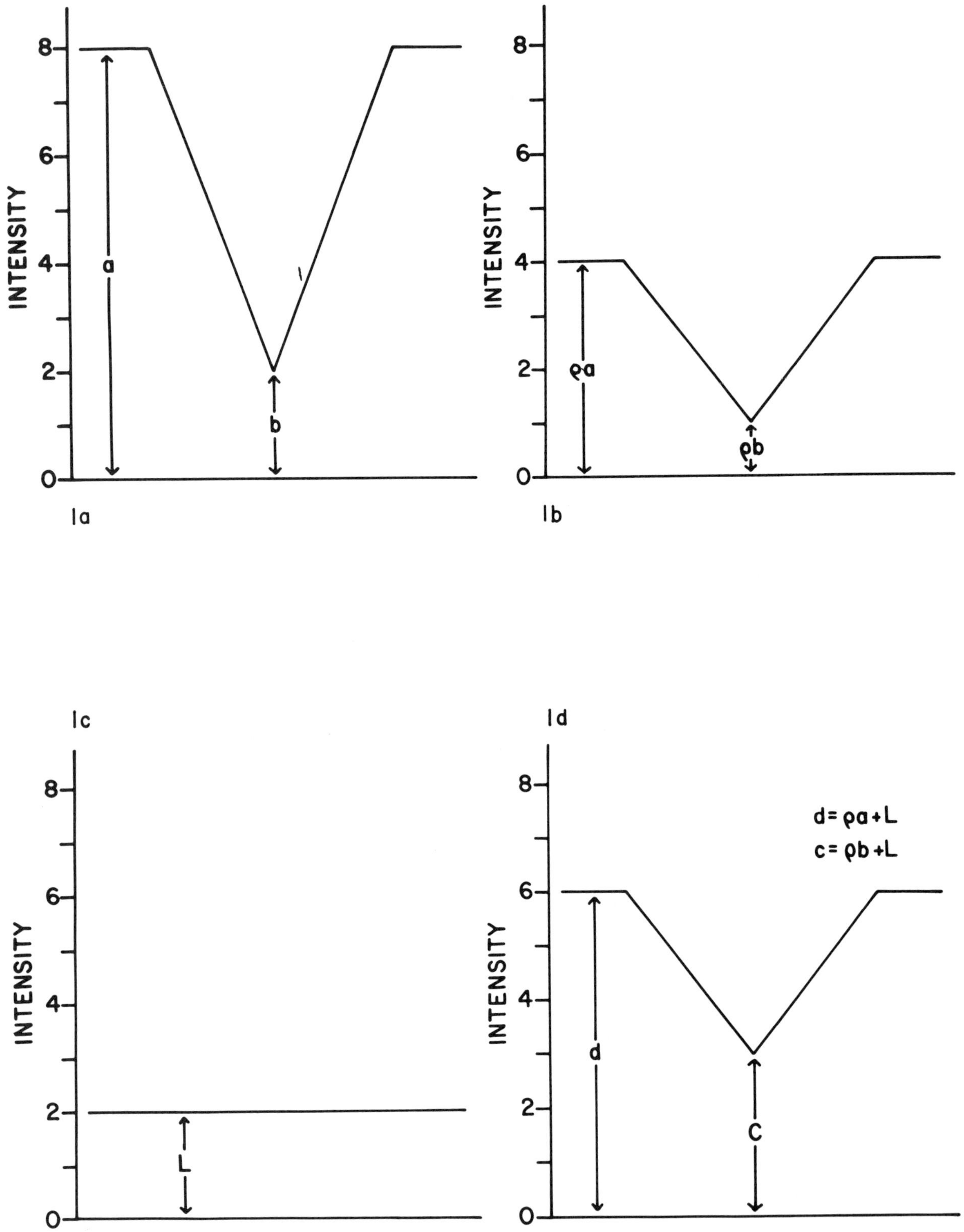

Fig. 1. Diagrammatic representation of the intensity, in arbitrary units, across a Fraunhofer line from: (1a) the solar spectrum; (1b) a non-luminescent, reflective target; (1c) the emitted luminescent component above; and (1d) a luminescent, reflective target (modified from Watson and Hemphill, 1976)—a represents the energy level adjacent to the Fraunhofer line, b represents the energy at the Fraunhofer line center, p represents the reflectivity coefficient, L represents the luminescent component, and c and d are defined in Fig. 1d.

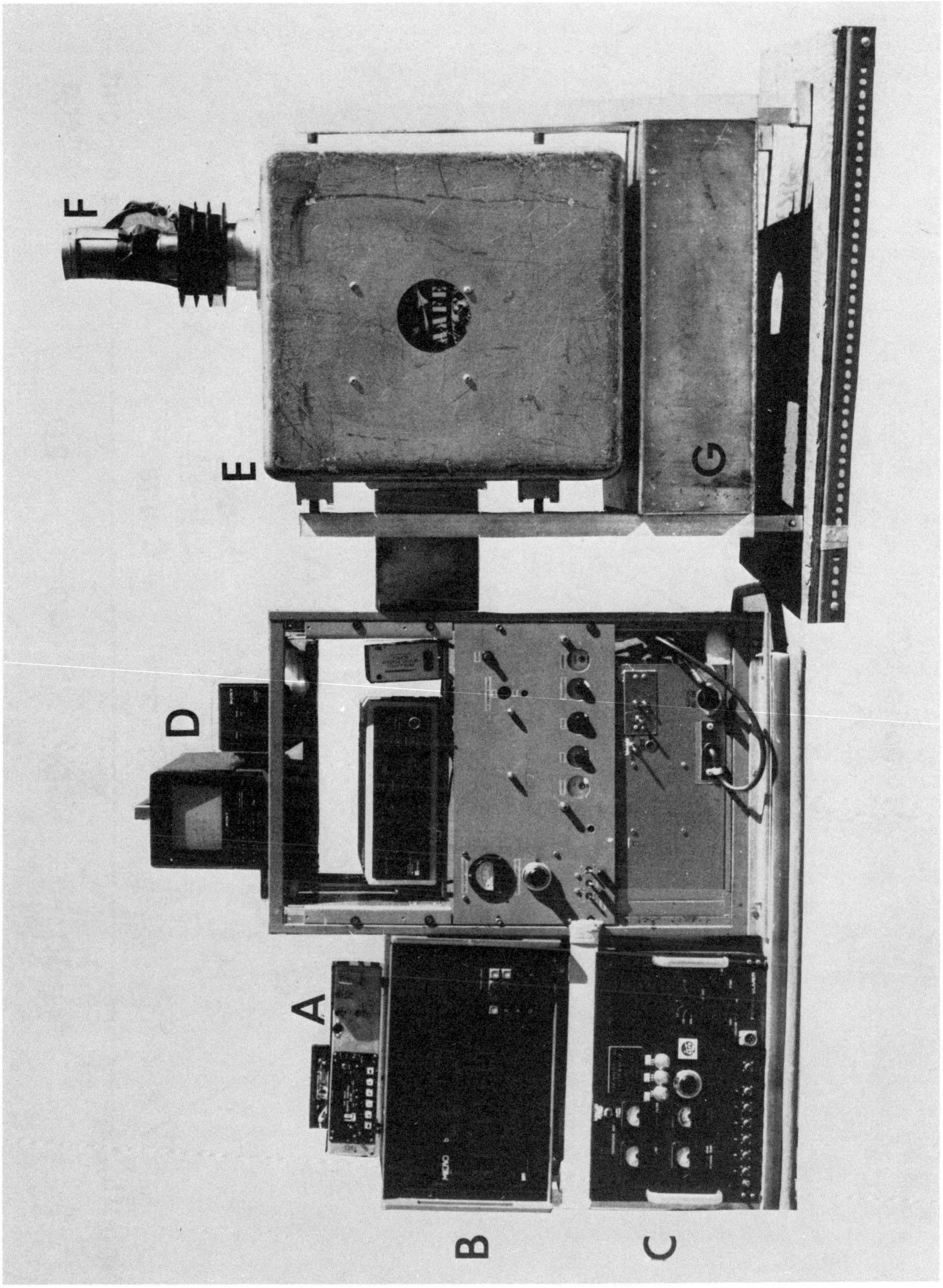

Fig. 2. Fraunhofer line discriminator system. A is the magnetic cassette recorder. B shows the microprocessor. C is the FLD computer. D is the television package. E is the optical head. F is the housing that contains the standard lamp. G is the housing for the scanning mirror and the television camera.

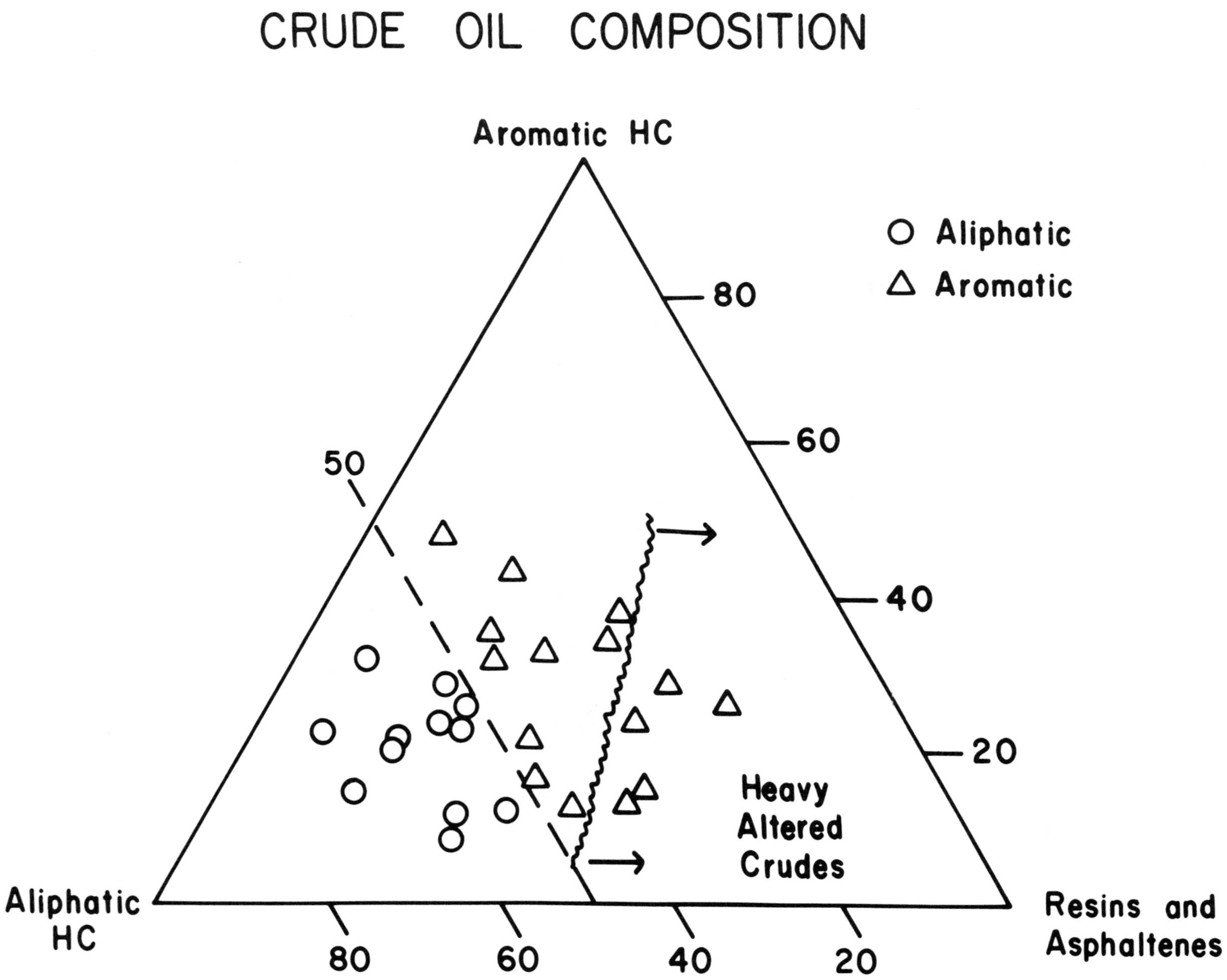

Fig. 3. Triangular diagram showing the composition of the crude oils used in this study (modified from Tissot and Welte, 1976). Dashed line indicates biodegraded crudes (to the lower right of line).

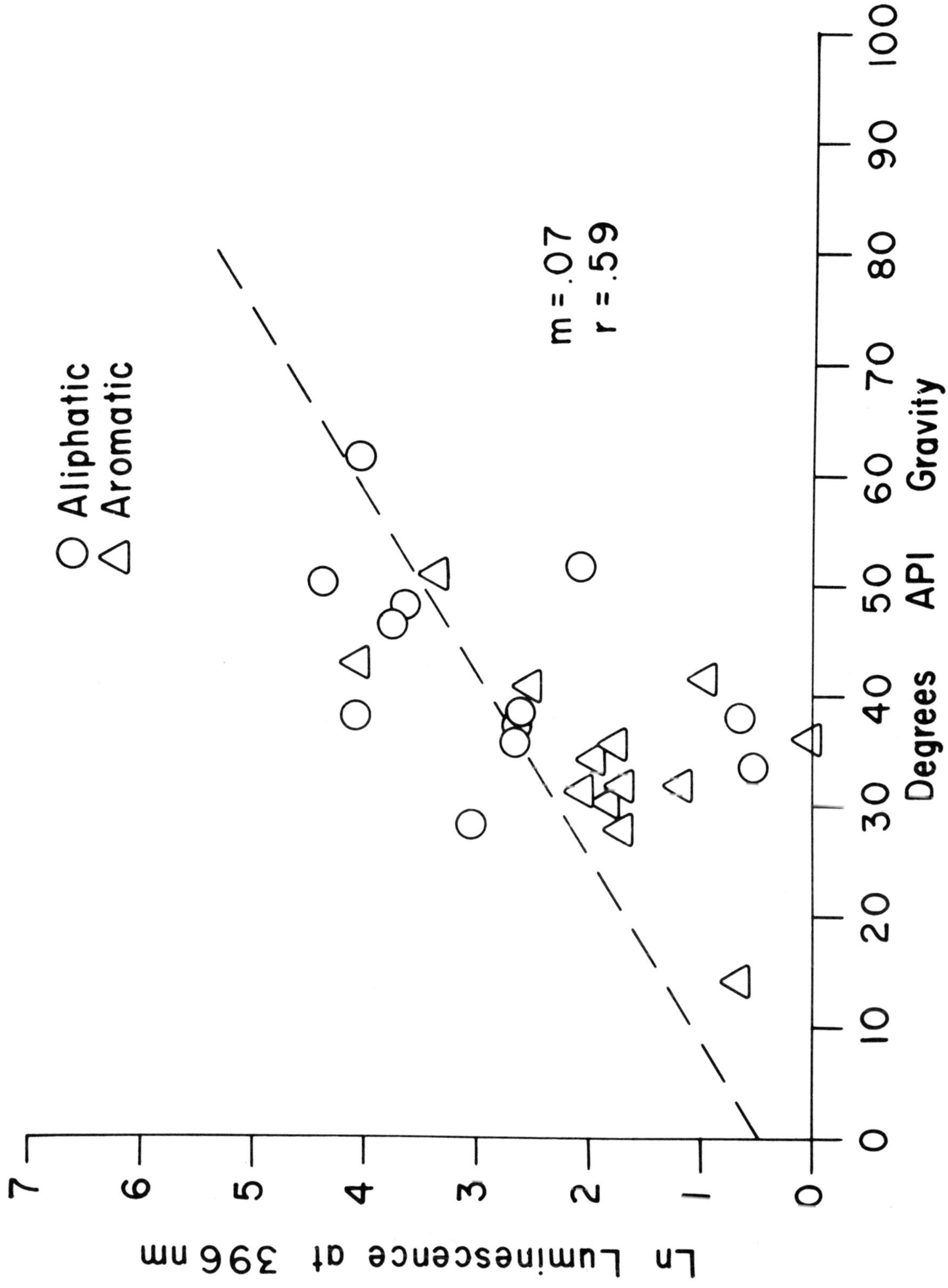

Fig. 4. Ln of luminescence intensity measured at the 396 nm Fraunhofer line plotted atainst API gravity of the crude oils.

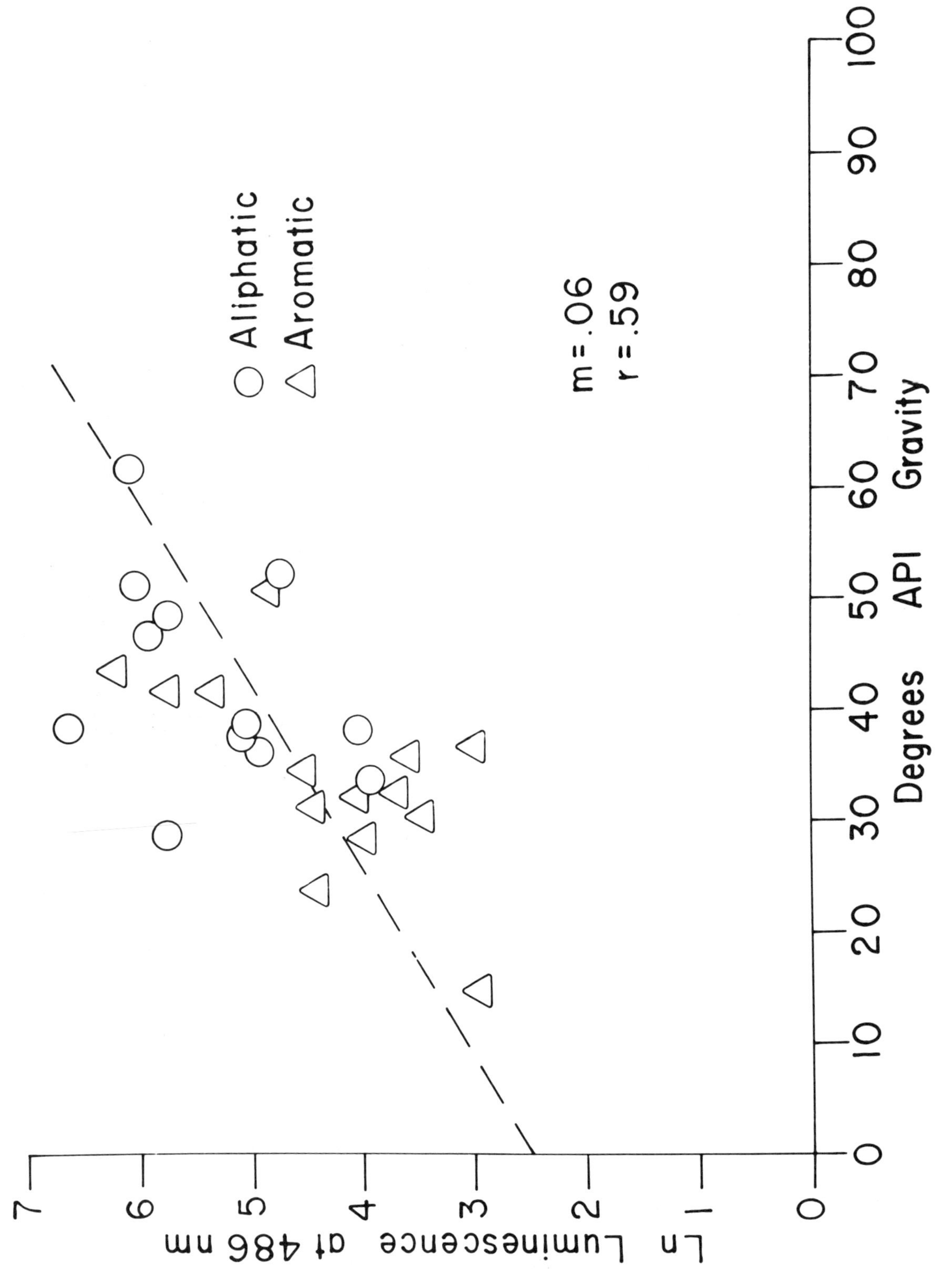

Fig. 5. Ln of luminescence intensity measured at the 486 nm Fraunhofer line plotted against API gravity of the crude oils.

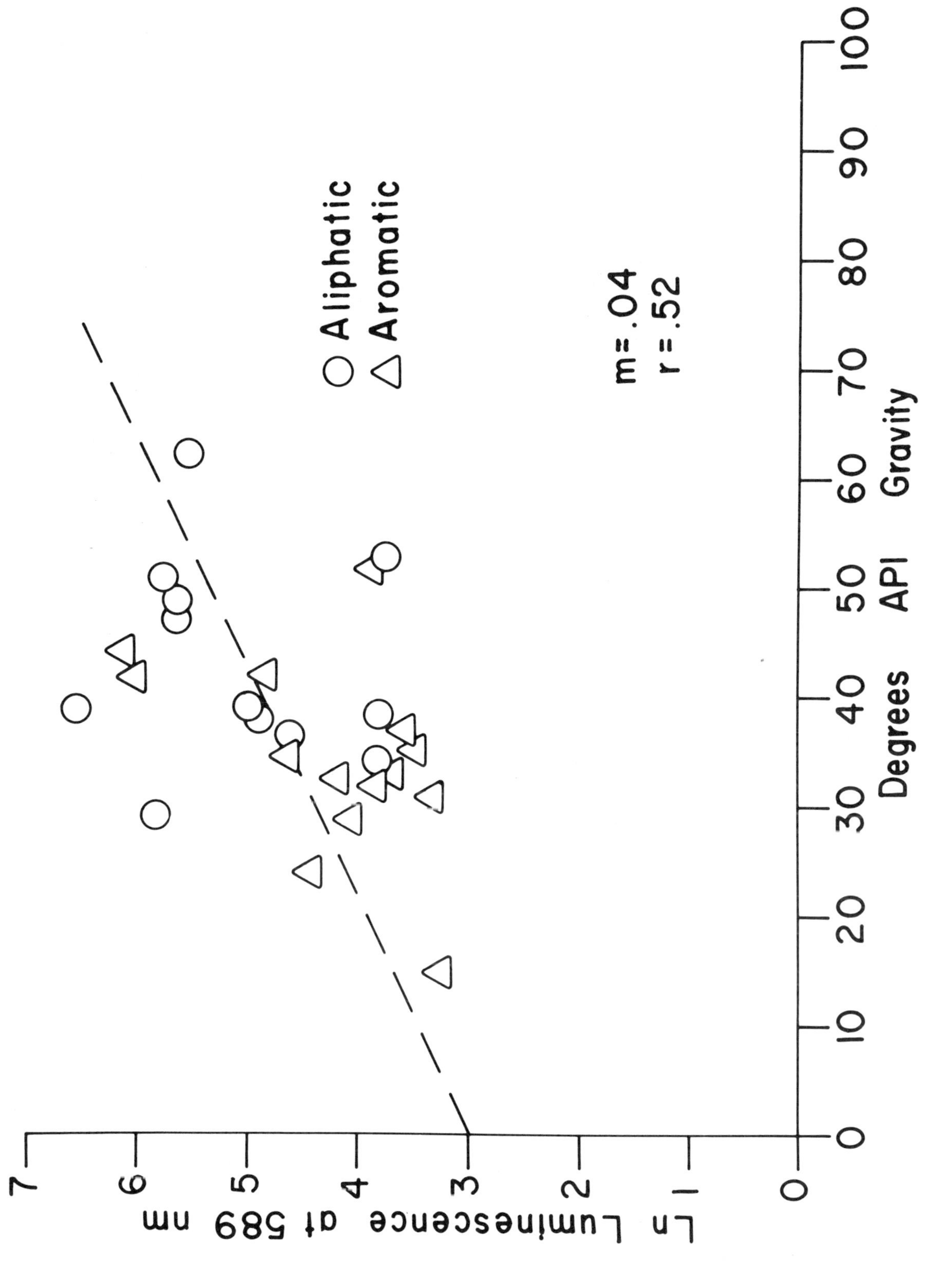

Fig. 6. Ln of luminescence intensity measured at the 589 nm Fraunhofer line plotted against API gravity of the crude oils.

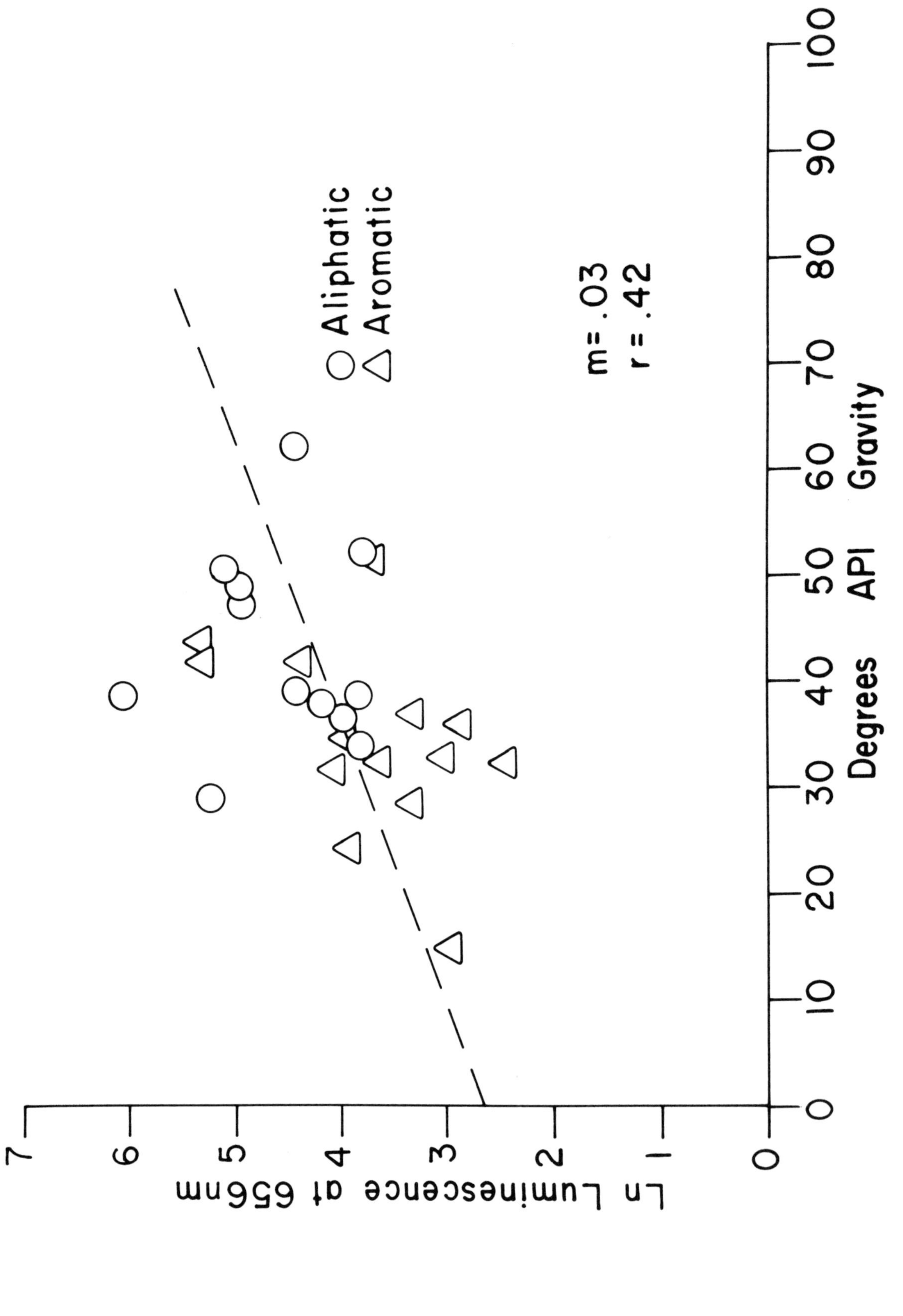

Fig. 7. Ln of luminescence intensity measured at the 656 nm Fraunhofer line plotted against API gravity of the crude oils.

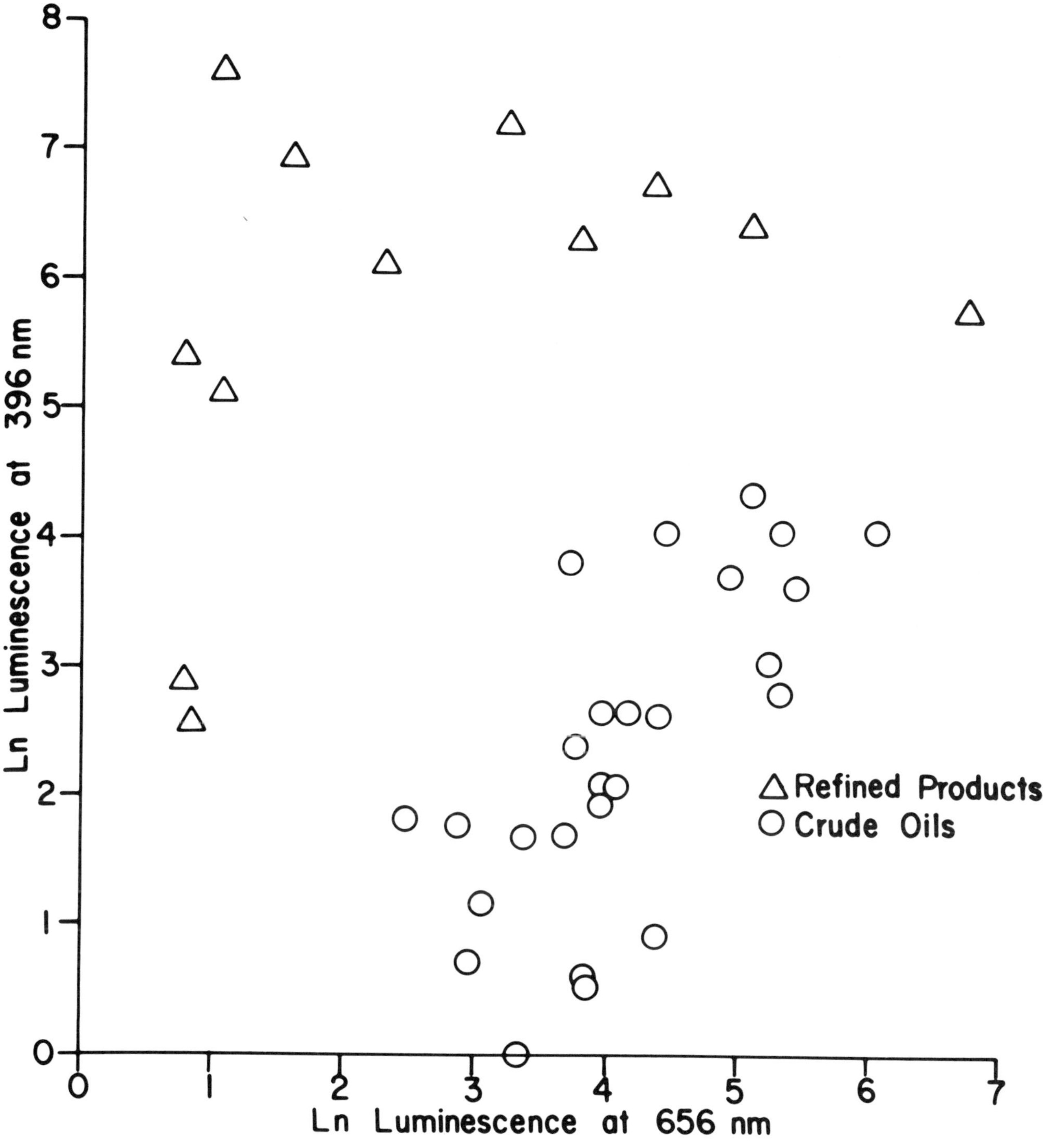

Fig. 8. Ln of luminescence intensity measured at the 396 nm Fraunhofer line plotted against the ln of luminescence intensity measured at the 656 nm Fraunhofer line.

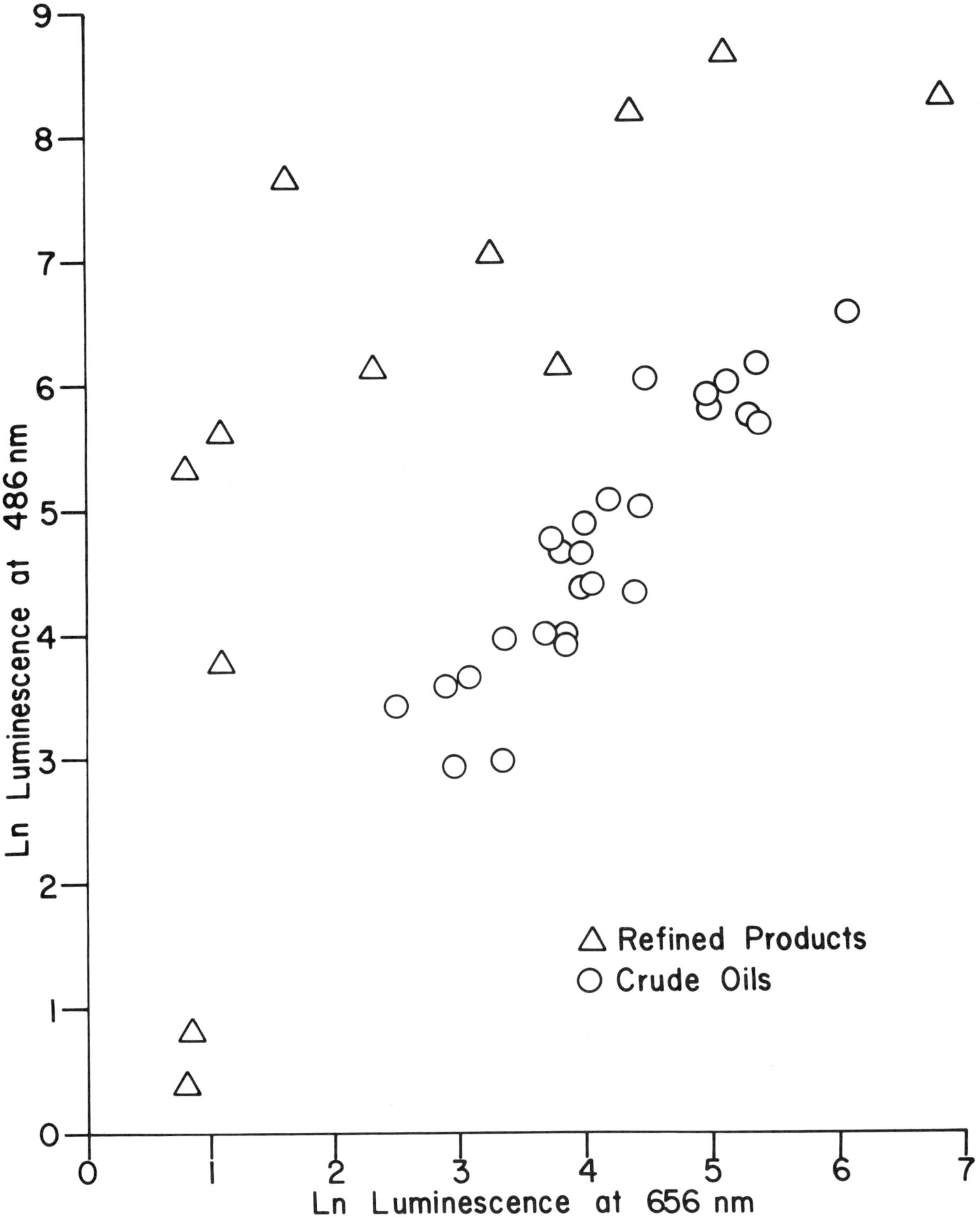

Fig. 9. Ln of luminescence intensity measured at the 486 nm Fraunhofer line plotted against ln of luminescence intensity measured at the 656 nm Fraunhofer line.

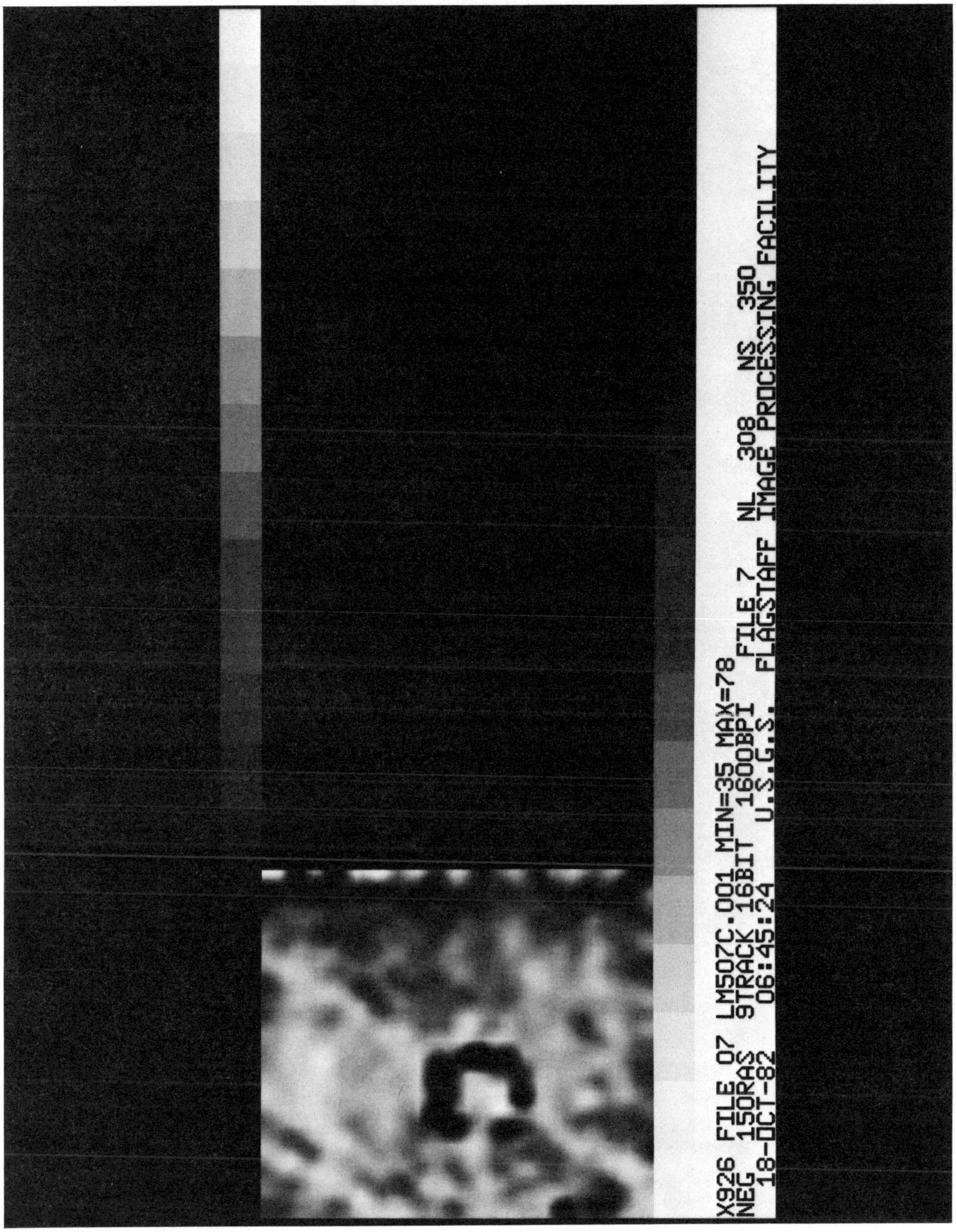

Fig. 10. Computer generated image created from FLD data. Pennzoil motor-oil film shown as light rectangular area in center of image. Effective film thickness is 100μm, data gathered at the 486.1 nm Fraunhofer line.

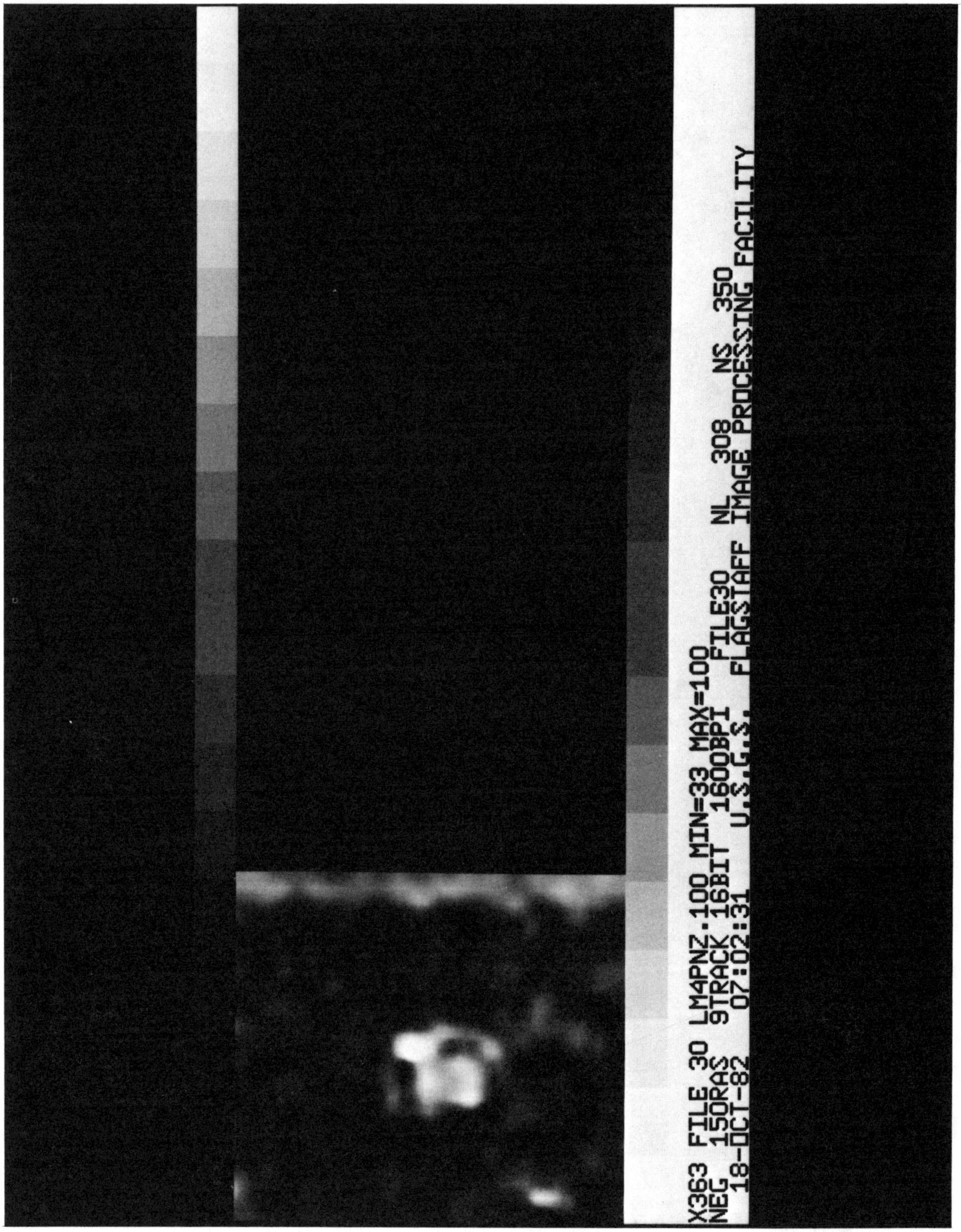

Fig. 11. Computer-generated image created from FLD data. Crude oil sample no. 7 shown as dark rectangular outline in center of image. Effective film thickness is 1 μm, data gathered at the 589.0 nm Fraunhofer line.

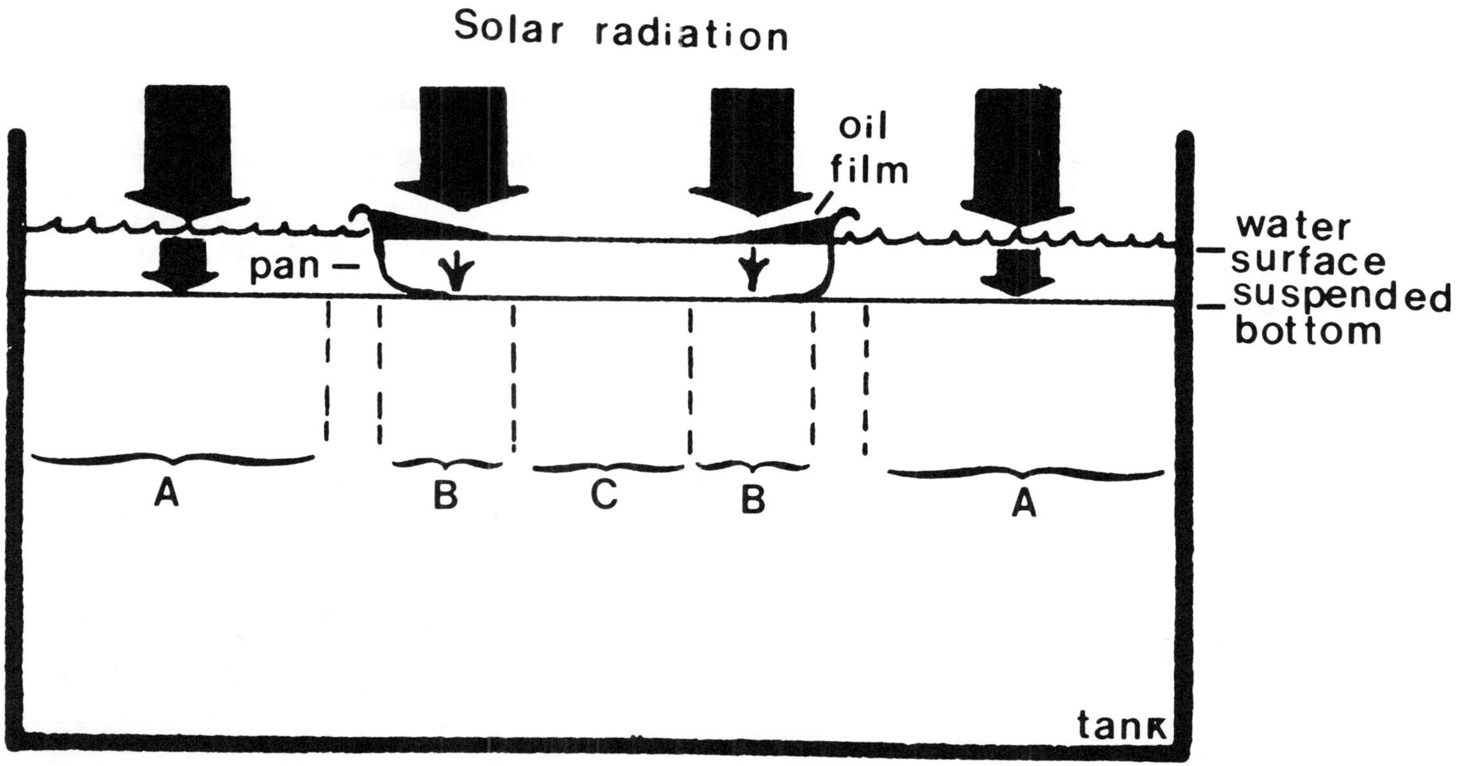

Fig. 12. Proposed model explaining the results of the outdoor tank experiments. Area A indicates clear water; incident solar radiation undergoes little attenuation, and penetrates entire water column exciting luminescence from that column.

Area B indicates a thick accumulation of oil due to a "meniscus effect." Incoming solar radiation is strongly attenuated by thick oil film near pan edge. Upwelling luminescence excited by the sunlight that penetrates the film is further absorbed by the oil film.

Area C indicates an area where the oil film is thin or absent. Luminescence values recorded from C are similar to those recorded for areas A.

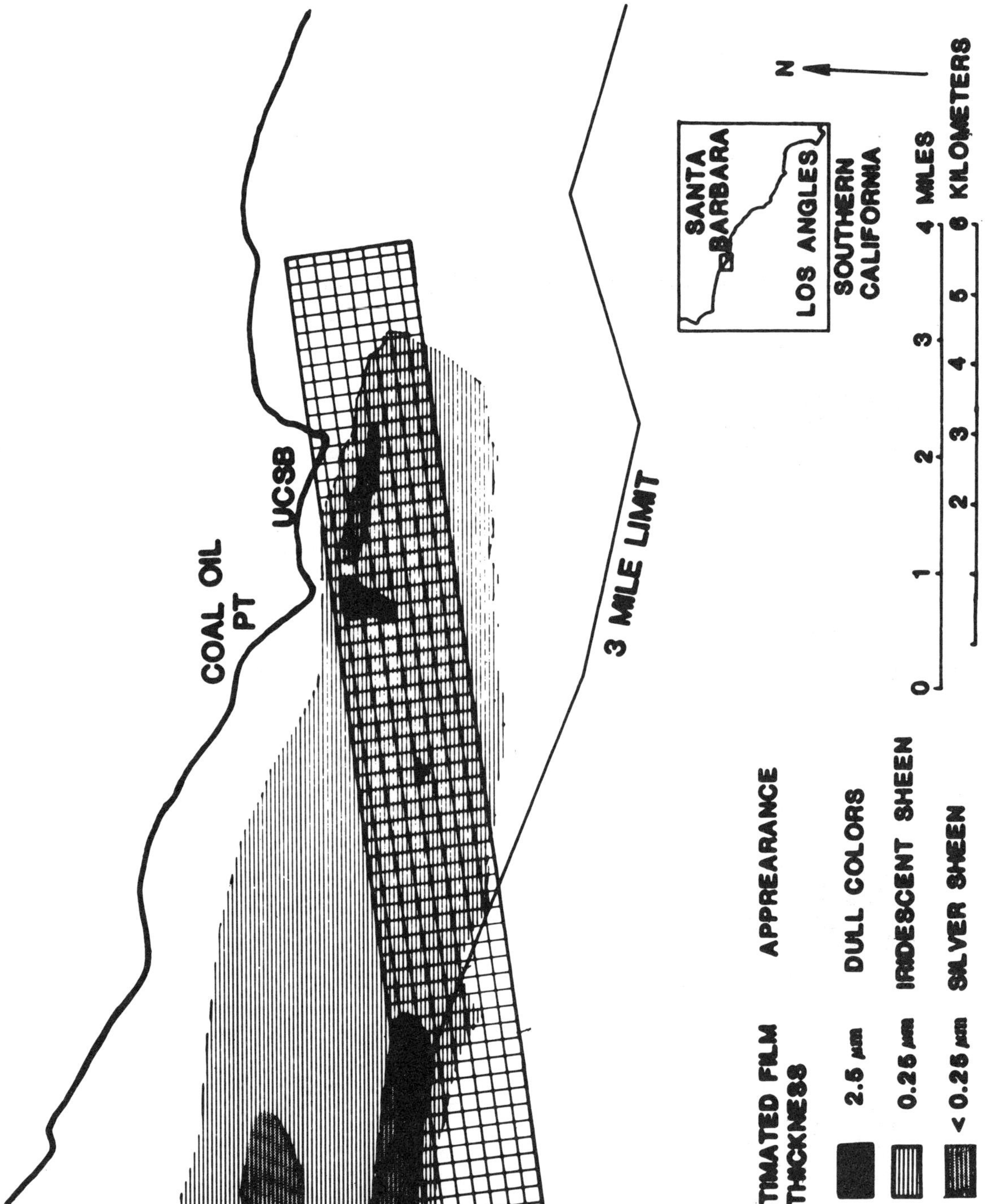

Fig. 13. Field test site. Gridded rectangle represents flight coverage. Oil coverage as determined by observers during FLD data collection shown as different tones of shading (modified from Watson et al., 1978). Small rectangle in inset map denotes survey area.

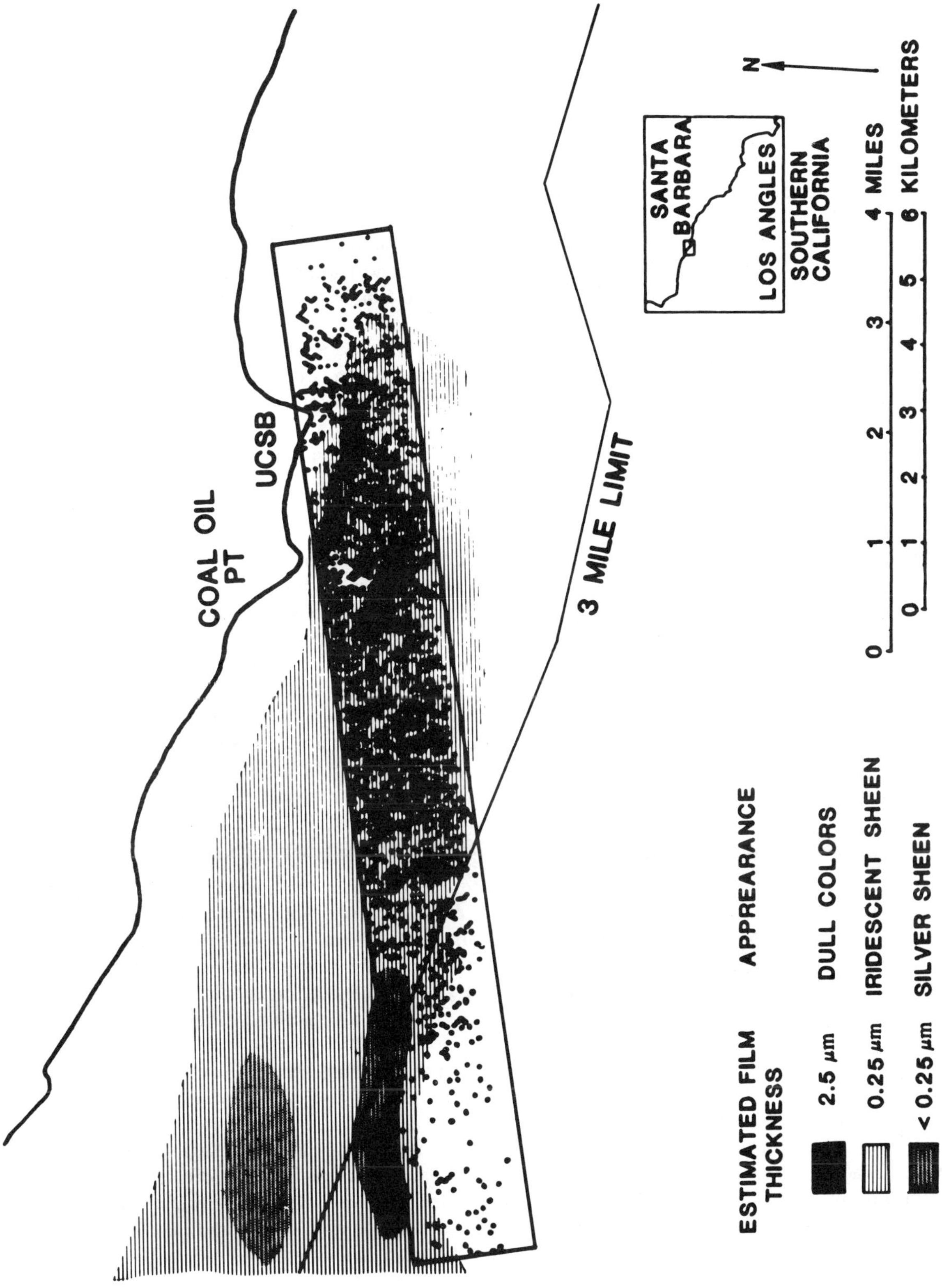

Fig. 14. FLD data superimposed on visual observational data of slick extent and thickness (modified from Watson et al., 1978). Small rectangle in inset map denotes survey area.

IODINE: A PATHFINDER FOR PETROLEUM DEPOSITS

Alton V. Gallagher

GALLAGHER RESEARCH & DEVELOPMENT CO.

ABSTRACT

The relationship between high concentrations of iodine with oil and gas field waters is well documented (Collins, 1975). Iodine is a good, indirect indicator of areas favorable for oil and gas accumulations. Initially humic substances in subsurface waters are the major sources of iodine (Kudel'skiy, 1977). At the high temperatures achieved during burial, structural degradation of large molecules of unsaponified organic matter from insoluble residues and bitumens also provides a source of iodine. Although the iodine background content in formation waters of different geologic ages varies widely, it does not preclude iodine as a pathfinder for potential oil and gas deposits.

Soil depth, mineralogical composition, humic content, and moisture content influence the extent of iodide fixation and iodine retention. Therefore, identification of sand-to-clay ratios in the 170-mesh sieved fraction of the soil sample is imperative. Samples taken from the top two to four in of the soil are analyzed for their total iodine content. Total iodine content—iodine which is firmly retained by humic soils, and iodine which is contained in soluble iodide form—is represented as a function of weight percent (i.e., parts per million, ppm).

High iodine concentrations occur about the perimeter of an oil or gas accumulation. This surface expression of the reduction chimney is the classic "halo effect." Anomalous concentrations of iodine also occur in an apical distribution directly over restricted petroleum deposits. Background levels for iodine generally range from less than 1 ppm to more than 15 ppm. A typical anomaly exhibits iodine values greater than two standard deviations above the statistical mean. Values of greater than 40 ppm iodine have been observed in association with oil and gas anomalies.

Iodine is an effective pathfinder in surface prospecting because (1) the surface samples require noncritical handling and are easily collected; (2) a detail survey may be integrated into earlier reconnaissance surveys; (3) relatively low-cost analysis permits greater sampling density; and (4) iodine is almost universally present in geochemical anomalies associated with oil and gas. As with any oil and gas exploration tool, iodine should be used in conjunction with other geochemical techniques (e.g., radiometrics) and the results cross-correlated for optimum confidence level.

INTRODUCTION

Conducting soil analyses for trace elements associated with petroleum deposits is a common geochemical practice. The relationship between high concentrations of iodine with oil field waters and oil accumulations is well documented (Collins, 1975). Field studies indicate high iodine concentrations about the perimeter of an oil or gas field, the "halo effect," and over an oil or gas field, the apical anomaly. Iodine is a good, indirect indicator of areas favorable for oil and gas accumulations.

Why iodine and not bromine, nickel, vanadium, cobalt, chromium, manganese, etc.? Following a series of geochemical surveys in several basins of the Rocky Mountains and in the San Joaquin Valley of California, in which over twenty-five elements were analyzed, the empirical results were clear-cut. Iodine was consistently present over oil and gas fields in anomalous concentrations. Bromine, also a good indicator, does not appear to be as universal as iodine. However, it is a good back-up trace element. The other trace elements may be excellent pathfinders in one area, but undetectable in another. In a thorough geochemical program, there is a great deal of merit in running a suite of elements in the detail phase; but it is not always cost or time effective in the reconnaissance phase.

ORIGIN AND OCCURRENCE

Kudel'skiy (1977) states that during the first stages of hydrocarbon generation from organic matter, the influx of iodine into subsurface waters is primarily associated with the breakdown of humic acids and humins. Structural degradation of large molecules of unsaponified organic matter from insoluble residues and bituminoids also produces iodine. This reaction is specific to the later high temperature stages (i.e., above 65° -100°C).

Although soil depth and moisture content influence iodine concentration in soils, the most important factors are the mineralogical composition and the humic content of the soils (Rozen, 1970). For example, Rao (1970) records that the clay mineral illite fixes more iodide than montmorillonite or kaolinite. Ermolenko (1966) notes that in any given soil, the colloidal (clay-size) fraction absorbs more iodine than the coarse (sand-size) fraction. Moreover, the adsorptive capacity of silica is two times lower than that of clays for iodine (Rozen, 1970). The strength with which iodine is retained by the soil depends on the nature of its bond with the soil elements.

Clayey soils are often richer in both humic and iodine content than are sandy soils. The distribution of iodine in soil profiles is directly proportional to that of humus (Aubert, 1977). In typical soil profiles, more than one horizon exhibits high accumulations of organic debris or humus.

SAMPLE COLLECTION

Field studies indicate that proper examination of the top 1/8 in of soil can identify oil and gas-related anomalies. Studies on the merits of sampling at depth (greater than 3 ft) indicate deeper samples generally achieve better resolution and delineation, but not necessarily better identification of anomalies. In particular, soil samples analyzed for hydrocarbons benefit from depth sampling.

More important than depth of sampling is the lithology, or mineralogical composition, of the soil samples. A well-developed clayey soil has sufficient cation exchange capacity for adsorption of trace elements and hydrocarbons so that deep samples are not required. Conversely, in a very sandy soil (e.g., a near-eolian environment), samples at 30 to 50 ft may be required for optimum resolution. The moisture content of the soil also influences fixation of trace elements. Aubert (1977) states that in partially or totally water-saturated soil clays, iodide fixation increases with increasing depth. In arid or humid-temperate soil clays, however, the extent of iodide fixation increases with decreasing depth.

Although the iodine background content varies widely in formation waters of different geologic ages and in soil types, it does not preclude the use of iodine as a pathfinder for potential oil and gas prospects. These changes in iodine background content occur on a regional basis. Most geochemical prospecting is in highly localized areas where *relative* changes are sought. Caution should be exercised when sampling densely drilled areas with many one- or two-well fields. A tightly spaced data grid and attention to possible sources of surface contamination may prevent misidentification of a halo. Continuous-profile radiometrics also works well with iodine in this situation. Samples on a quarter-mile reconnaissance survey along roads can be collected at the rate of two every five minutes. Analyses can be run and the data be ready for interpretation within two weeks. Because the elution rate for iodine varies widely, it is imperative that soil samples not be collected immediately following groundsoaking rain and snow falls.

SAMPLE PREPARATION AND ANALYSIS

In sample preparation for iodine analyses, use of the 170-mesh sieved fraction, rather than the conventional 80-mesh sieved fraction, minimizes effects of changing clay : sand ratios. It is not so much the composition of the field sample as it is the composition of the sieved fraction used for analysis that is critical. In some areas, particularly on larger surveys, it is necessary to establish more than one statistical background to compensate for soil type changes. This is also true for radiometrics and hydrocarbon soil-sampling.

Samples from the top 2 to 4 in of the soil are analyzed for their total iodine content. Analysis for the total iodine content includes iodine which is firmly retained by humic soils, and iodine which is contained in soluble iodide form. The proportion of iodine in the soil which can be eluted has been reported as ranging from 5 to 90 percent. Results are represented as a function of weight percent (i.e., parts per million, ppm) and are held to within 0.3 ppm.

Background levels generally range from less than 1 ppm to more than 15 ppm iodine. Observed values in excess of 40 ppm iodine have been noted in Texas in association with oil and gas fields. A typical anomaly exhibits iodine values greater than two standard deviations above the statistical mean in the high values of the halo, and as much as 0.75 to 1.0 standard deviations below the mean directly over the reduction zone.

ANOMALY TYPES

There are two basic types of surface soil geochemical signatures: the halo effect and the apical anomaly. The classic halo effect, in which various elements are concentrated about the perimeter of the reduction zone above an oil or gas deposit, is shown in Figure 1. The light shading denotes the oil or gas deposit and the dark shading, the halo area around the reduction zone. A profile of closely spaced geochemical data is also shown to indicate a typical response. Note the decrease in the geochemical response over the oil and gas reduction chimney. This reduction in the geochemical signature is not universally present over oil and gas fields. When present, the geochemical signature is usually abrupt and strongly supporting. This is the only instance where you can place significance on low iodine values. In reality, halos are rarely symmetrical. Surface expression of the halo is often restricted from two-thirds to three-quarters around the prospect. The halo anomaly is generally associated with a good, economic oil or gas reservoir. In practice, a good halo anomaly is present less than half of the time.

The second type of geochemical anomaly is the apical anomaly (Fig. 2). The apical anomaly is generally associated with linear reservoirs (e.g., channel sands and narrow porosity trends) or with one- to two-well fields. These fields often exhibit no halo because the halo geochemical parameters tend to coalesce into a point or linear anomaly. Even with the closely spaced sampling-program (one-tenth of a mile) in Figure 2, it would be easy to miss a 300-ft-wide channel. The size of the potential target in a geochemical survey program determines the density of data needed. It is best to run a few profiles of closely spaced data-points to identify an anomaly and to follow it up with a detailed grid to delineate anomalous areas. The apical anomaly is encountered more than half of the time.

INTERPRETATION

Geochemical data should be plotted on a topographic map base or on an aerial photo. This provides the interpreter with at least some measure of quality control over his data. Second, a good statistical analysis determines background and threshold values for the area of interest. In many instances, common geologic sense should prevail over a rigid statistical analysis. Third, consider the following questions:

1. Are adequate data available to establish the presence of an anomaly?
2. Is a reasonably stable background in the geochemical data apparent?
3. Is more than one background level present as a result of changes in soil types?
4. Is the anomaly restricted to a topographic high or low and, if so, why?
5. Is the anomaly apical or halo?
6. Are the data quantity and quality adequate to permit delineation (contouring)?
7. Is adequate information present to identify the age and type of potential production?

These questions presume that care, good judgment, and quality control have been exercised in the acquisition and analysis of the samples. The way in which these questions are answered is related to the risk factor that would be assigned to a prospect. For example, sampling only in the immediate vicinity of a suspected prospect anomaly, the interpreter will delude himself into thinking he has adequate data.

A stable background permits confident identification of anomalies. As we have seen, changes in background level may be attributed to soil type, moisture content, or formation variations. Quite definitive background level changes are often encountered when dropping onto a flood plain from the surrounding countryside. If an anomaly is restricted to one type of topographic or geomorphic feature, its validity should be questioned unless a rational geologic explanation is available.

The fifth question is the most critical of all: "Is the anomaly apical or halo?" Misinterpretation insures a dry hole.

If the sixth and seventh questions can be answered in a positive manner, data coverage is sufficient for a valid interpretation and recommendation of a drill site.

CASE STUDIES

The first case study of an iodine geochemical soil survey is of the Moulton Gas Field in LaVaca County, Texas (Fig. 3). The halo, represented by the 15 ppm contour on the northwest flank of the field, is long and narrow. This linear expression could be interpreted as being associated with a fault. The iodine low, over the productive portion of the field, is well developed in this survey and represents the central portion of an oil or gas field. One might question why the iodine low area northwest of the halo is not interpreted as being anomalous. There are two reasons: (1) no sharp break denoting good halo development is present farther to the northwest; and (2) no radiometric anomaly is associated with the anomalously low area northwest of the Moulton Field. The anomalously high iodine values of 20 and 26 ppm at the northern edge of the mapped area may be related to another anomaly. It is interesting to note that this survey represents data collected on a reconnaissance survey which were later filled in with data from a more detailed survey.

The next study is the Ryckman Creek Field area in the Overthrust Belt of southwestern Wyoming. Figure 4 shows the extent of the reconnaissance survey. Samples were taken along a county road at one-quarter-mile intervals. The sampling program begins on the eastern edge of the map east of the outcrop of a major thrust fault. It extends westward beyond the Ryckman Creek Field area to the eastern edge of the Whitney Canyon Field area. A line of samples was also taken down a road to the northern edge of the Painter Reservoir Field. This reconnaissance survey required approximately six hours of field time.

Data density is insufficient to contour on this type of survey. The object is to identify areas which merit further investigation. In this instance, different sizes of squares indicate increasingly anomalous values of iodine in ppm. Note the total absence of anomalous values on the eastern flank of Ryckman Creek. The west flank has several anomalous values. In particular, note the strong grouping of anomalous values around the northern edge of the Painter Reservoir Field. It is a good, relatively new field that has not experienced significant reservoir depletion and is an excellent analog for geochemical surveys.

The eastern flank of Ryckman Creek Field exhibits a poor anomaly. The Ryckman Creek Field has produced some 16 million barrels of oil and was on a repressurization program after some six years of production at the time of the survey. An identical survey, run one year after this survey, indicates that the Ryckman Creek Field no longer exhibits a significant iodine anomaly. When using an older field as an analog for unknown anomalies along similar stratigraphic or structural strike, consider the age and status of the known field. An area with a complex tectonic history, such as the Ryckman Creek Field of the Overthrust Belt, is elementary for geochemical exploration techniques.

Figure 5 shows the Sorrento Field area in southeastern Colorado. The Sorrento Field is productive from the Pennsylvanian Morrow Formation. It provides a valid analog for unknown anomalies within a thirty-mile radius. This particular reconnaissance survey combines radiometric data and iodine analyses. Radiometric lows or anomalous areas are indicated on the map by solid black lines. Iodine values are plotted in ppm. A general outline of the high and low values of iodine is indicated by light shading for anomalously low values and dark shading for anomalously high values. Iodine reasonably identifies the field limits, even though data coverage is insufficient to contour. A survey of this extent using a combination of radiometrics and iodine surface soil sampling can be accomplished by one day in the field.

Figure 6 is a map of the Edgemont Oil Field, South Dakota, in the extreme southeastern edge of the Powder River Basin. The dark shading represents iodine values above 5 ppm; the medium shading, values from 3 to 5 ppm; and the light shading, values of 1 ppm or less. The well within the light shading is the discovery well. The three subsequent dry holes and the one other producer could have been predicted by the geochemistry. The geochemistry indicates two other potentially productive locations in this field. This survey was a combination radiometric and iodine survey and required less than one day in the field. Note that this is a reconnaissance survey.

The final case study differentiates between the ages of production in a survey run in southern Kansas (Fig. 7). The area encompassed by the mapped square is 1 sq mi. The three wells near the center of the survey area produce from the Viola Formation. A fourth well producing from the Reagan is in the southwestern quarter of the map. On the basis of gas chromatograms, a comparison of butane, benzene, and toluene values was made. High values of butane and benzene represent Viola production in the area, while high values of toluene and low values of butane and benzene indicate Reagan production. Interpretation of this data indicated that the location marked ''After Survey''would be productive from both the Viola and the Reagan: it was. The hydrocarbon detail survey was run after an iodine and radiometric survey. Note that the hydrocarbon anomalies are coincident with the oil and gas deposit and are not in a halo position.

CONCLUSION

Geochemistry applied to finding oil and gas has matured greatly in the past decade. Analysis for iodine is a technique that has a place in the explorationist's arsenal of routine geochemical tools. It is an effective pathfinder in surface prospecting because (1) the soil samples require noncritical handling; (2) data are repeatable so that a detail survey may be integrated with earlier reconnaissance surveys; (3) relatively low-cost analysis permits a greater sampling density which provides better identification and definition of anomalies; and (4) iodine is near-universal in geochemical anomalies associated with oil and gas. The true strength of geochemistry as applied to petroleum exploration is in the use of several geochemical parameters and the cross-plotting of their results.

ACKNOWLEDGMENTS

The writer wishes to thank Petro-Labs, Inc., Golden, Colorado, for permission to use some of its data and also Traverse Oil Co., Wulf Oil Corp. et al., and Robert Shaw for permission to use data from surveys conducted for them. The writer is indebted to Brenda C. Gallagher and Lola Tapia for their assistance in the preparation of this paper. He also thanks Cheryl J. Dunn for typing the manuscript.

REFERENCES

Aubert, H., and Pinta, M., 1977, Trace elements in soils: Laboratoire de Spectrographie, Office de la Recherche Scientifique et Technique Outre-Mer, 93140 Bondy, France, p. 35-38.

Collins, Gene A., 1975, Geochemistry of oilfield waters: New York Elsevier Scientific Publishing Co., p. 164-166.

Ermolenko, N.F., 1966, Trace elements and colloids in soils: Academy of Sciences of the Belorussian SSR.

Kudel'skiy, A.V., 1977, Forecasting petroleum prospects on the basis of the amount of iodine in subsurface waters: Geoligiya Nefti i Gaza, no. 4, p. 45-49.

Rao, S.S., De, S.K., Tripathi, C.M., and Rai, C., 1971, Retention of iodide by soil clays: Indian Journal of Agricultural Chemistry, v. 4, p. 44-49.

Rozen, B. Ya, 1970, Geochemistry of bromine and iodine: Publishing House ''Nedra,'' Moscow, p. 1-144.

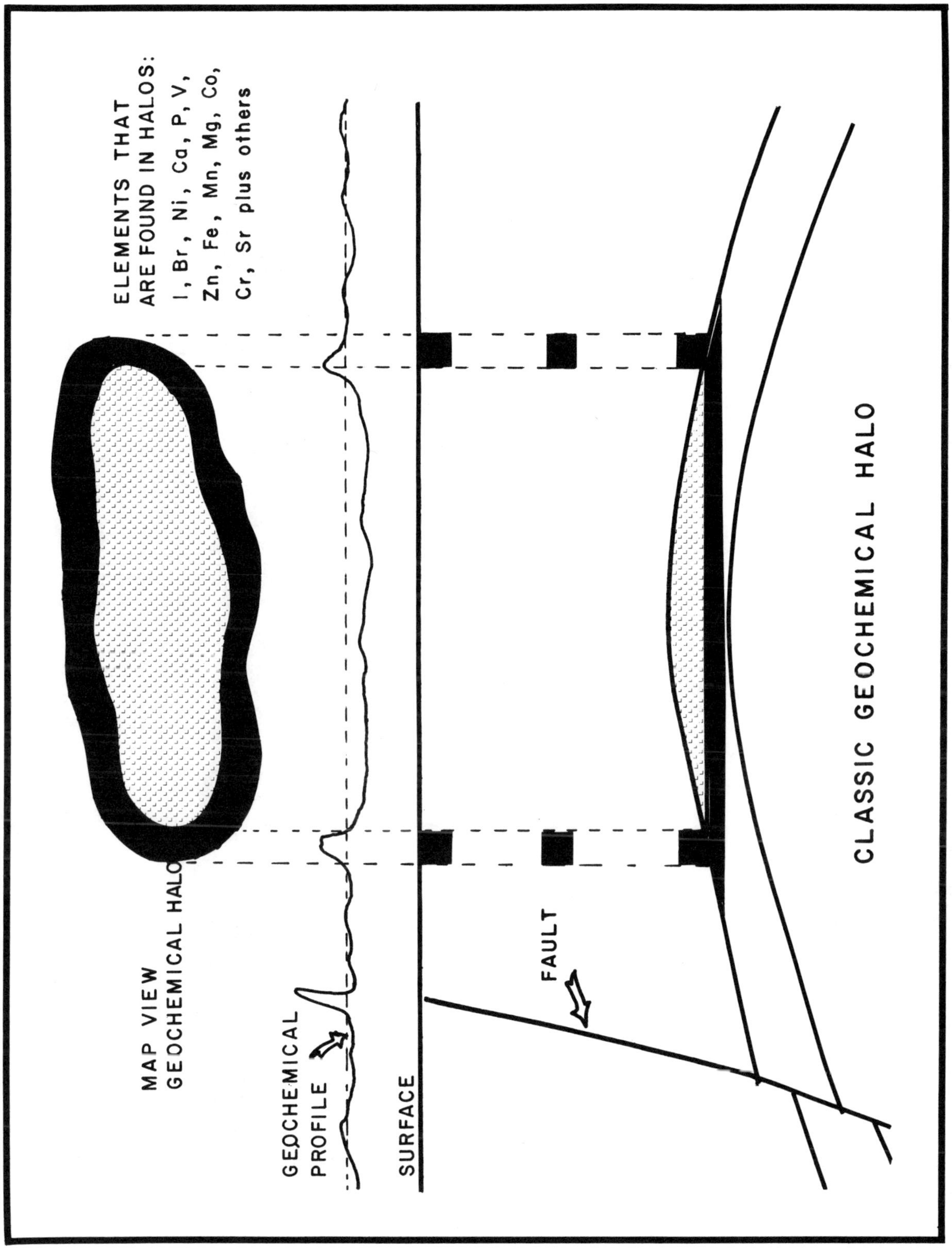

Fig. 1. The geochemical halo is a surface expression of the reduction chimney associated with oil and gas deposits.

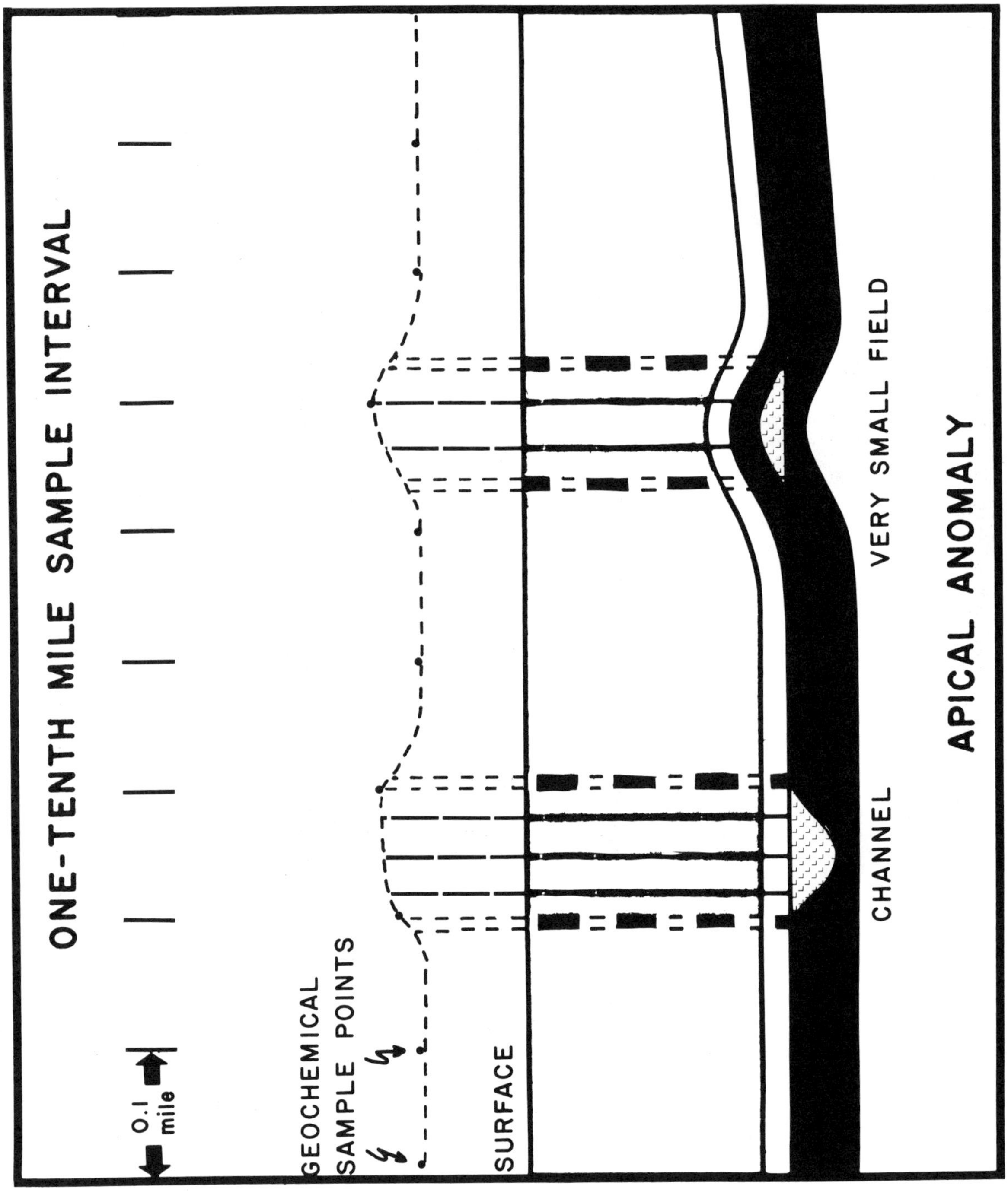

Fig. 2. Halo geochemical parameters may coalesce over narrow or small petroleum accumulations resulting in apical anomalies.

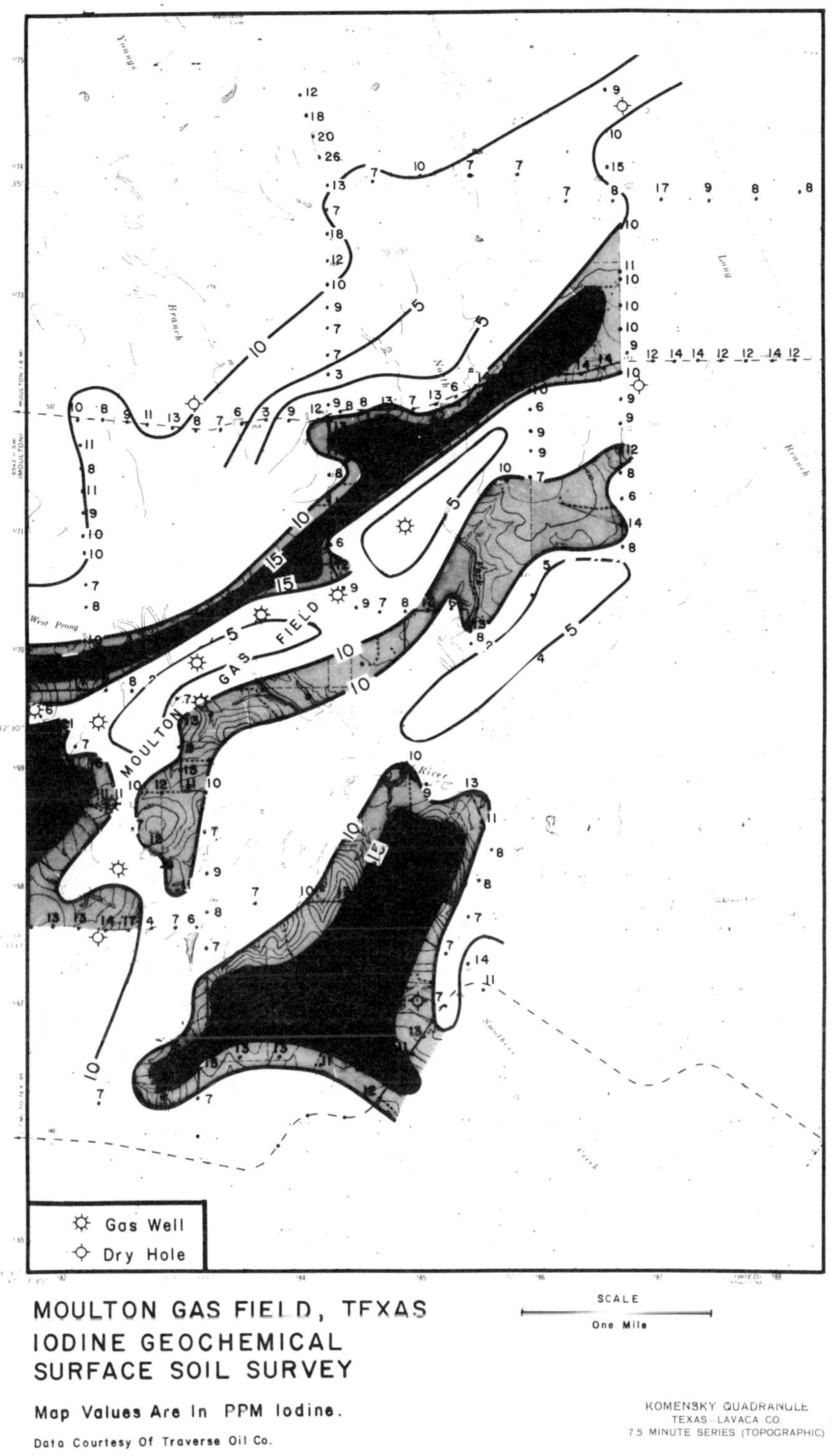

Fig. 3. Iodine soil survey in the Moulton Gas Field, La Vaca County, Texas.

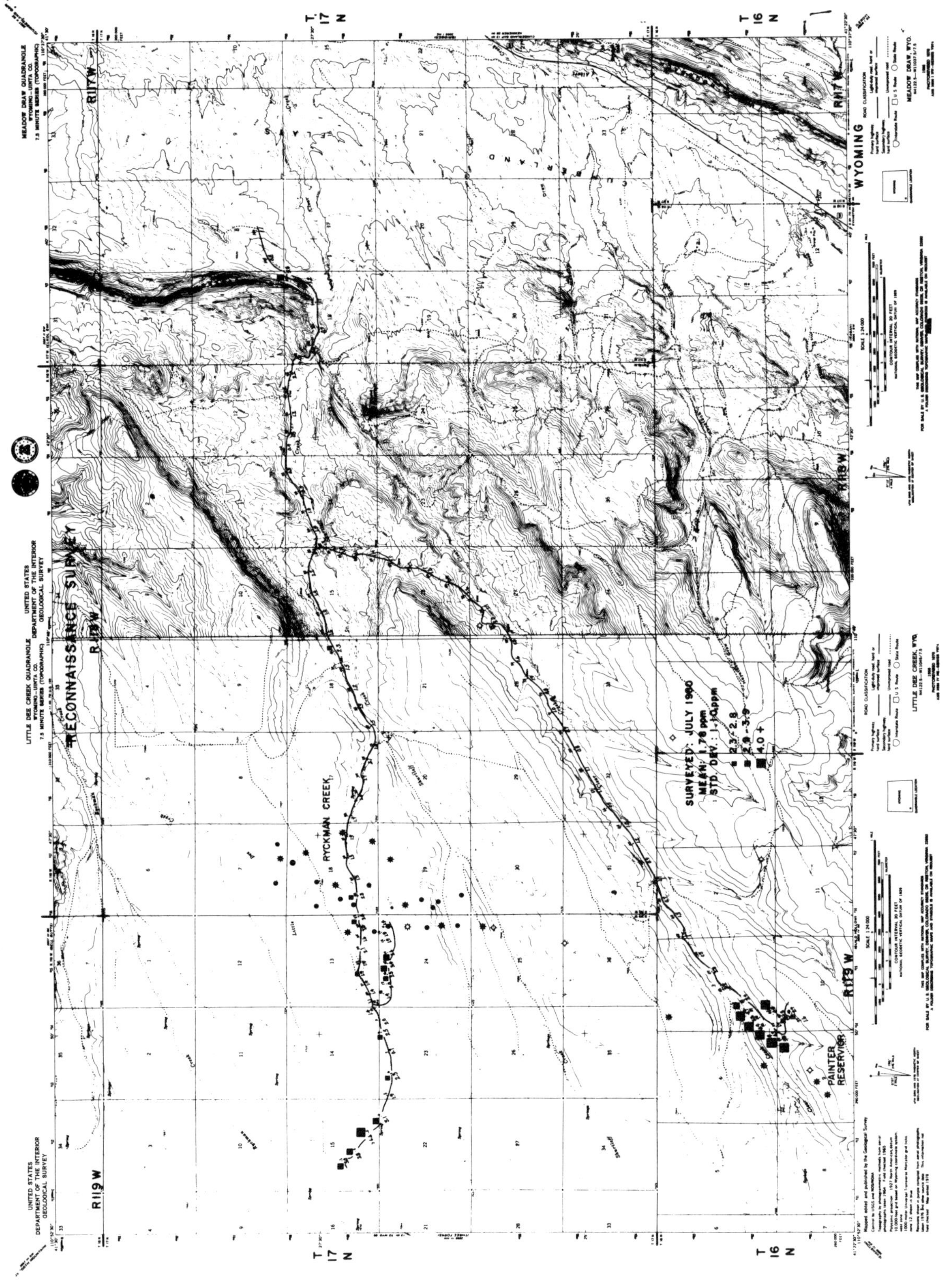

Fig. 4. Reconnaissance iodine soil survey in the Ryckman Creek Field area, Uinta County, Wyoming.

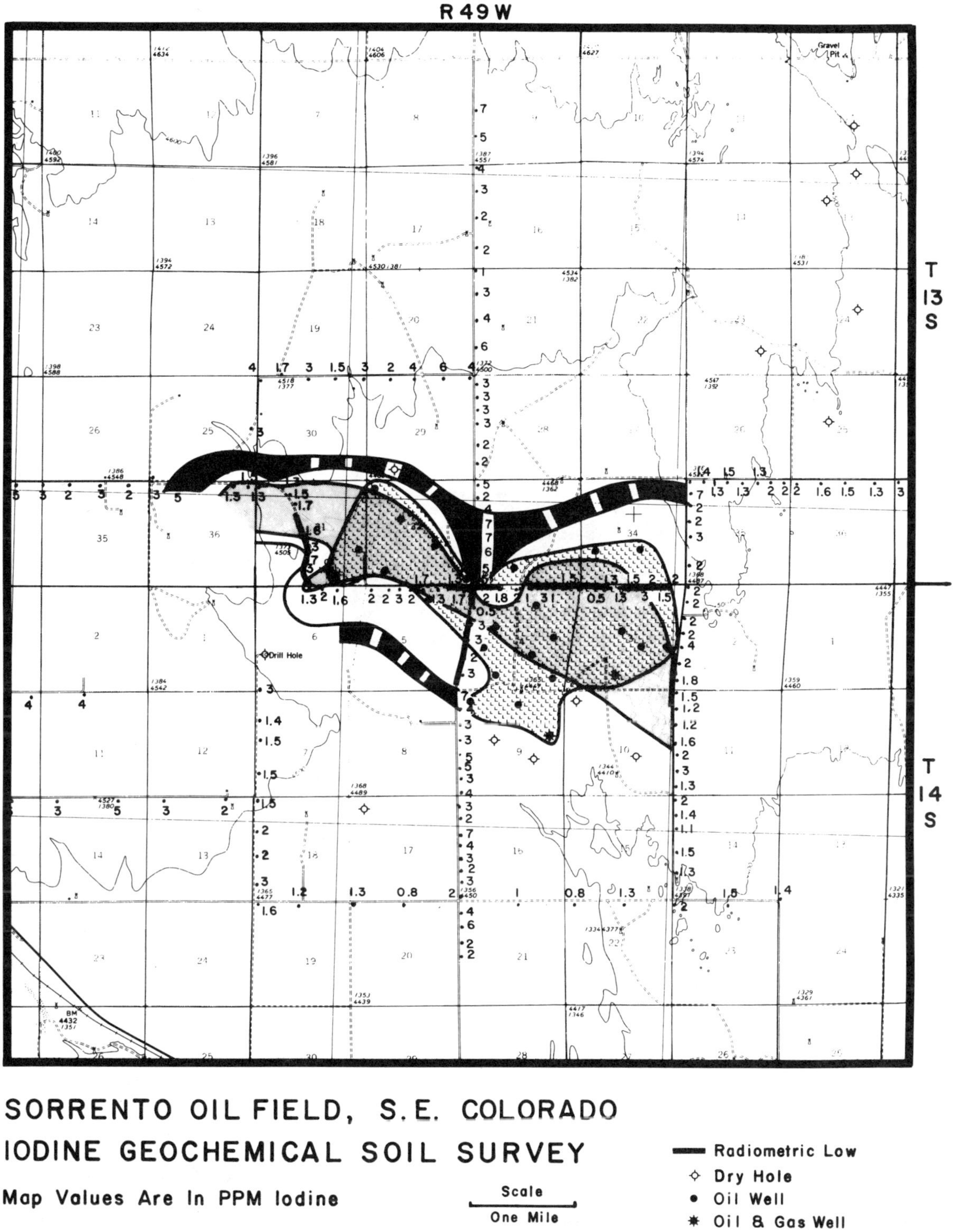

Fig. 5. Reconnaissance iodine soil survey and radiometric survey of the Sorrento Oil Field area, Cheyenne County, Colorado.

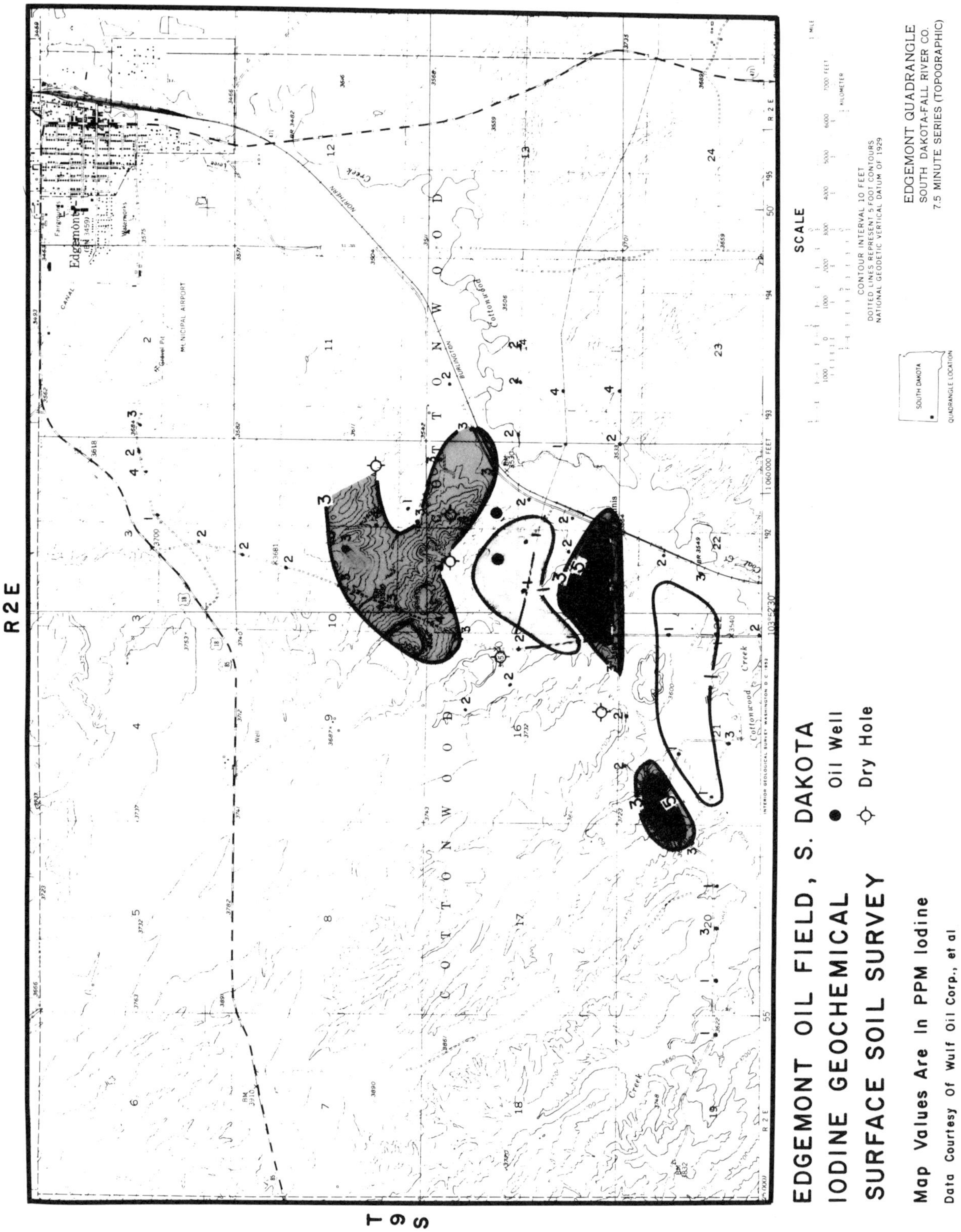

Fig. 6. Iodine, geochemical, surface soil survey in Edgemont Oil Field, Fall River County, South Dakota.

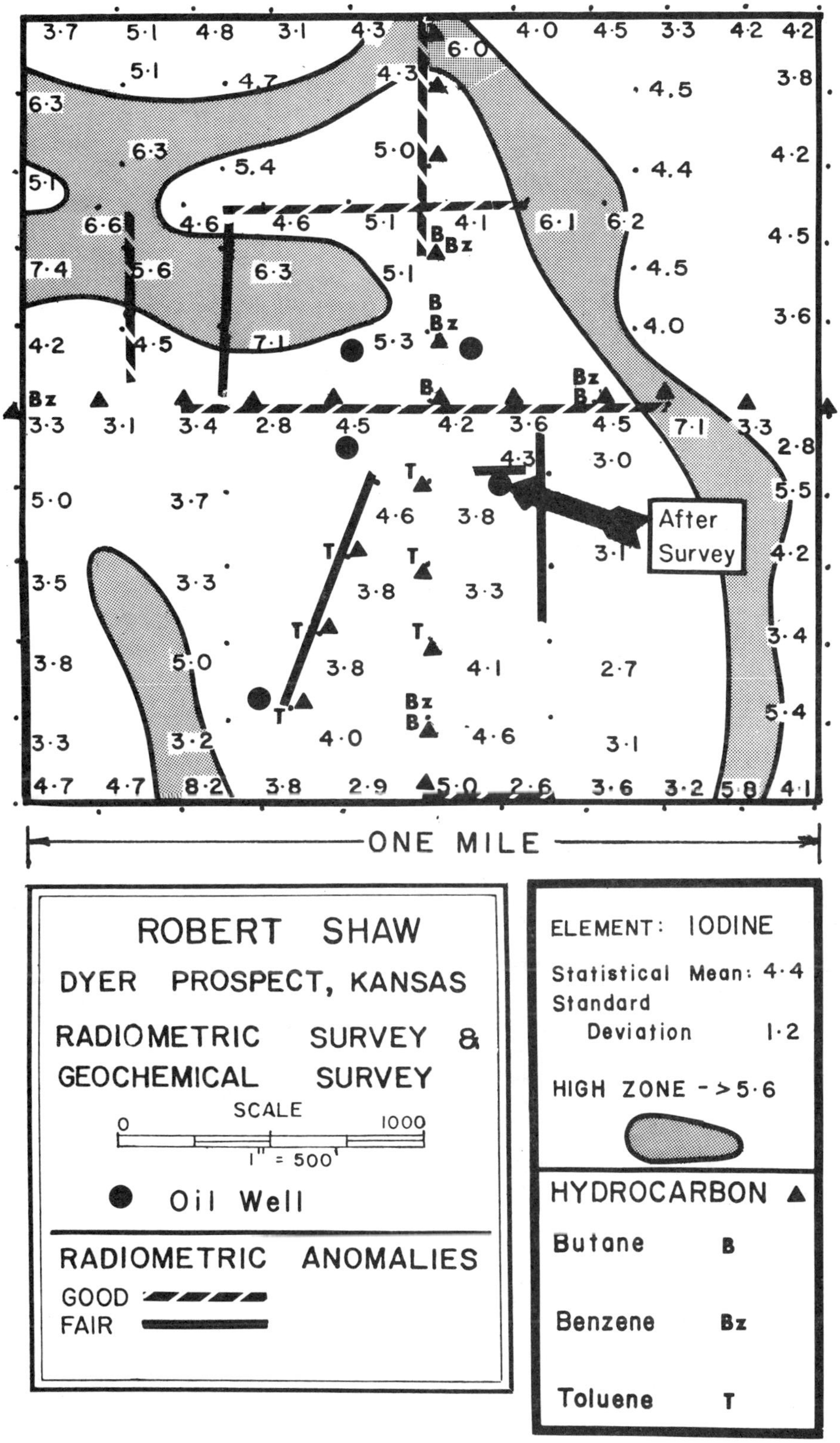

Fig. 7. Radiometric and gechemical surveys over the Dyer Prospect in southern Kansas.

THE ORIGIN AND EXPLORATION SIGNIFICANCE OF DILATION STRUCTURES

John Napier Monroe

MONROE COMPANY

One of the principal objects of theoretical research in any department of knowledge is to find the point of view from which the subject appears in its greatest simplicity.
—J. WILLARD GIBBS.

Letter to the American Academy of Arts and Sciences (1881).

ABSTRACT

Organic sediment buried under anaerobic conditions results in the generation of methane. This methane, both in solution and as gas bubbles within sediment, has two major effects: (1) it weakens the sediment to the point of fluidization by dilation, and (2) it alters the sediment density to the point of effecting isostatic readjustment, also by dilation. The weakness by fluidization results in the development of structures commonly attributed to submarine slumping, while the attendant reduction in density results in the development of larger structural features which are sometimes attributed to diastrophism.

Gas fluidization structures, as well as the larger structural features, are characteristically developed in the environment of petroleum and are there recognized in well data and seismic records as well as in surface exposures.

Fluidization of sands by dilation from intergranular gas makes sand masses behave much like gas itself; thus, "currents of sand" produce spiral rolls and other exotic forms, and masses of sand expand to invade newly formed cracks as clastic dikes. Thickening and thinning of beds result from pressure differentials; intraformational folding is common, and sediment failure by shear is singularly absent. Since the formation of these and other structures results from the distension of sediment by the expansion of interstitial gas, gas pressure exceeding hydrostatic pressure, and in many cases equal to the weight of overburden, may be inferred. As a result of this gas pressure, free communication to the surface by hydrocarbon fluids is denied. Thus the existence of a seal or trap, either breached or intact, is implied. In the absence of evidence of the breach, the intact trap should be sought.

Buoyancy effects of the dilation of gas-laden sediments are formidable. "Boils" of methane on the surface of the waters of continental shelves directly overlie oil structures. Gas-laden muds have risen in bulk over large acreages in many parts of the world. Frothy clay has come onto the surface from depths as great as 5,000 ft. Islands of gas-laden sediment have risen out of the sea in such far-flung locations as offshore Burma, Borneo, and Venezuela. High gas content of sediment appears to be genetically related to diapirism as is evidenced by attendant mud volcanoes and gas-fluidization structures. The direct cause of all these vertical movements is not gas pressure in the dynamic sense, but is the buoyancy or isostatic readjustment caused by the innocuous gas content which results in density reduction and sediment weakening.

These same mechanisms, isostatic readjustment due to gas content and sediment weakening through fluidization, both resulting from dilation, help explain structures of many of the world's oil areas.

INTRODUCTION

Dilation structures form in sediments which have been distended or volumetrically expanded by the growth of gas bubbles (usually of methane) within intergranular spaces. Only through this dilation process can sands, for example, behave like elastic fluids: that is, flow, crumple, roll into balls, and squirt into cracks of contiguous rocks to form clastic dikes and sills (Monroe, 1957, 1969). True, these fluidity structures have already been "explained," for like most natural phenomena they do not escape explanation unless they escape notice. Space does not allow the examination of such prior explanations as iceberg drag, glacial override, submarine slumping, air-heave, and gravity sliding. As this discussion develops, however, it will be seen that these explanations do not address the problem.

In addition to the fluidization effects of interstitial gas, there is also the effect of density reduction which leads to isostatic readjustment for the formation of oil structures. Accordingly, these gas-generated buoyancy structures are developed in areas of maximum petroleum generation; that is, they develop where the hydrocarbon gases are being maximally produced.

THE FLUIDITY STRUCTURES

Dikes of clean sand are sometimes injected hundreds of feet into overlying indurated shales (see Fig. 1). Associated sills must *lift* the overlying rocks in the process. Dikes of chalk derived from coccolithophorid or foraminiferal oozes are also injected into contiguous shales and marls as shown in Figure 3. Marl dikes are hard to recognize, but the one shown in Figure 4 was confirmed as such by the identification of microfossils derived from the parent formation, stratigraphically distant from the invaded rocks.

Angular chips of shale which are carried along with dike sands during injection are shown in Figure 5. Also, angular projections of delicate points and edges from the containing walls of dikes are often undisturbed. Fragile pelecypod valves are also transported without being broken. Therefore, we know that during injection dike sands are as weak as a low viscosity fluid. In the case of intrastratal folding, the sands and silts are again more mobile than the the enclosing clays. Contiguous beds are not affected. Often all stratal lines within the folded bed remain unbroken (Fig. 6). Instead of breaking they become thinner and thereby remain continuous. Usually no slip planes at the top and bottom are in evidence except by implication. Fluidity of the clastic material is again required. Since the crumpled sands and silts are overlain by the beds at the time of deformation, John Lyon Rich (1950) has likened their environment to that within a "closed container" in which the containing beds are not visibly affected by the deformation. Also, the overlying strata, or the heavy "lid" to this "container," provides confining pressure. The behavior of clastic materials in "closed containers" and under confining pressure is examined in pages to follow.

Sometimes associated nonfluidized beds separate to form Boudinage (Fig. 7). Also, balled-up rolls of sand or silt are often found intercalated with other fluidity structures where they either appear to spiral off from a layer of sand or silt, or appear as isolated, concentrically layered balls of clastic material. One such isolated roll is shown within a sandstone dike of the Pepper Shale equivalent of the Woodbine Sand in Central Texas (Fig. 8). The photograph in Figure 9 shows the sheer wall of this dike as it forces a turn in the creek. The long axis of the roll is horizontal and parallel to the dike walls. This roll exposed in the end of the dike testifies to the identity of the environment of formation for both structures. The roll appears to have been formed at the time of dike injection, presumably as an eddy of fluidized sand within the gaseous dike slurry. Stratigraphic continuity of beds on opposite sides of the dike proves that the formation of sand rolls does not require differential movement of enclosing rock surfaces or beds. Apparently, simple sand-slurry injection can result in roll formation.

Other fluidity structures include sealing-wax flow, injection lenses, intrastratal brecciation, isolated sand bodies, pinches and swells, and even load casts (see Figs. 1 and 2).

Rhodes W. Fairbridge (1946) called attention to the association of fluidity structures, which he called "slumping structures," with oil areas of the world in his 1946 paper entitled, "Submarine Slumping and the Location of Oil Bodies." His list of examples includes the oil fields of the Russian Caucasus, the Carpathians, Alberta, Iraq, Iran, Burma, Venezuela and Trinidad, Ecuador, Peru, Sumatra, New Guinea, and California. Today this list can be formidably extended to include the Gulf Coast, the Niger Delta, the MacKenzie Delta, the Prudhoe Bay Off-shore, and the Permian Basin of West Texas and New Mexico. Yet except in oil areas and some other very special euxinic environments, fluidity structures are singularly absent.

THE MECHANISM

The mode of failure of clastic sediment is a function of the degree of dilation of that sediment. When deformed, granular materials must expand to permit grains to pass over one another. If dilation or expansion of the mass is prevented, then the mass cannot be deformed except by the crushing of the constituent grains (Reynolds, 1881). In the case of extreme dilation such that individual grains are totally separated (as by methane bubbles), complete loss of rigidity results, and the mass assumes the mechanical characteristics of the interstitial fluid.

If dilation is merely permitted as a result of deforming stress, then the sediment will fail along shear planes, since this mode of failure requires the least expansion of the mass (see Fig. 10). In order to flow, the mass must first be dilated throughout by some means other than the stress itself (Mead, 1925).

Since the environment of intrastratal folding has been correctly likened to a "closed container" under confining pressure, it is appropriate to examine experimentally the behavior of clastic materials within this environment. Our "closed container" is a balloon (see Fig. 11). Sand is poured into the balloon and the air is exhausted.

Thus, normal atmospheric pressure serves as the confining pressure. Since the balloon has been previously fitted inside a cardboard tube, the loose sand is formed into the shape of a cylinder. When the cardboard tube is peeled off, the remaining cylinder of sand is firm: it feels like a core of quartzite coated with a thin rubber film.

A solid cast iron weight is balanced on the cylinder. The shear strength of the sand mass (in terms of differential pressure) is approximately equal to the confining pressure. If the balloon of sand were in deep sea water or buried under a thick column of sediment, the strength in shear could approach the crushing strength of many clastic materials. Obviously, sand under confining pressure and in a closed container will not flow.

If the balloon is punctured with a pin, air enters slowly through the pinhole. When enough air enters the balloon, such that the mass may be conveniently deformed, then a firm push of the finger results in shear of the sand mass. As more air enters the balloon, the confining pressure furnished by the atmosphere disappears because of pressure equalization. The balloon of sand collapses and then looks and feels like any other toy balloon with sand inside. This experiment was performed with both wet and dry sand. The results were the same.

The manner in which sediment fails depends on the allowable manner of packing of the constituent grains in accordance with the amount of dilation resulting from interstitial gas. Thus, we shall briefly discuss packing modes. A more exhaustive treatment of packing modes is furnished by Graton and Fraser (1935).

Four modes of packing are represented in Figure 12. These packing modes include rhombohedral, monoclinic, orthorhombic, and cubic. In this figure, each mode is illustrated by use of four solid circles, each representing a bottom layer of spherical grains. An additional dotted circle represents an appropriate position of a grain of the next higher layer. The position of this single grain of the higher layer illustrates the arrangement of grains for a particular packing mode in the third dimension.

For rhombohedral packing, the arrangement for minimal void space requires a porosity of only 25.95% for uniform spherical grains. Cubic packing results in a porosity of 47.64%; formation from rhombohedral packing requires a sediment expansion of 21.69%. This sediment expansion also results in a density reduction of 17.82%, or to 82.18% of that for rhombohedral packing. Orthorhombic packing shows a porosity of 39.54% and a sediment expansion over rhombohedral packing of 13.59%, with a density reduction from rhombohedral of 11.96%. Monoclinic packing shows a porosity of 30.18%, a sediment expansion over rhombohedral packing of 4.23%, and a density reduction of 4.06%. For this packing mode, note that the grain (represented by a dotted line in the figure) of the upper layer lies halfway between two grains of the lower layer.

Figure 13 shows two cross-sectional views illustrating the inherent instability of monoclinic packing. The top grain (note cross section on the right in the figure) is free to move in at least one dimension. Assuredly it does not remain in a "balanced" condition, and if it moves at all it can result in rhombohedral packing. If dilation remains, however, it is free to move farther on. Similar movement of other grains can result in a mobile mass of granular material.

The shaded areas in Figure 14 represent example routes of free movement for grains (in monoclinic packing) without the requirement for additional dilation. This freedom of movement is available, however, only in one plane. For three-dimension flowage, a small amount of additional dilation is necessary to allow directional rearrangement, even if intergranular friction is allowed. Thus flowage appears to be a probable mode of failure for sands of well-rounded and spherical grains if dilated in excess of about 5%, the volume required for monoclinic packing.

Relations between sediment expansion, packing mode, and mode of deformation are confirmed experimentally. A steel box with a 2-in-thick plexiglass window is used. It is so constructed to provide a view of the development of structures and also to contain gas under pressure. Also, a screw, a hand crank, and piston are provided for sediment deformation (see Fig. 15).

Because of the difficulties anticipated in dispersing small bubbles of methane within intergranular spaces, high pressure carbon dioxide and carbonated water are provided. Hoses are arranged to deliver gas, carbonated water, or tap water. So that experimental results can be measured against mathematically determined predictions, natural sand is not used. Rather, glass beads of uniform size (between 90 and 100 microns diameter) are employed. Dilation of sediment is achieved by first pressurizing the chamber containing sediment with CO_2 and then filling the sediment with carbonated water from below while maintaining constant pressure. To activate expansion of the sediment, pressure is released by opening the top-side stopcock. Gas bubbles then form between sediment grains.

For the first experiment, the beads were covered with tap water; no CO_2 was used. The piston was backed off to allow slumping. A view through the window confirms failure by shear at depth by the offset layers of iron filings. (With iron filings marking strata, the glass-bead sand could be easily cleaned with a magnet for reuse (see Fig. 16). The experiment is repeated for the production of both normal and thrust faults after inducing sediment expansion in various amounts. No folding occurs until the sand has been expanded or dilated by 5% or 6%.

With a 5% expansion, aligned folds representing an incipient shear plane develop (see Fig. 17). Then the incipient shear plane becomes real. Based on the observation of this final shear plane photograph alone, one might identify drag folds as opposed to the incipient shear folds which developed before the shear itself. In any case, this mode of failure represents the transition stage between failure by shear and failure by folding. This 5% expansion also represents the dilation requirement for monoclinic packing for the case of uniform spheres.

With 8% expansion prior to deformation, failure is totally by folding; no shearing develops. Note in Figure 18 that the folds are parallel, there being no thickening or thinning of beds. Stresses are well transmitted within the mass during failure. Beds yield by bending, not by totally free flowage, and not at all by shear. Even for this mode of failure, however, the sediment must be expanded by more than the 5% to accommodate monoclinic packing. Thus, if a buried bed of clean sand folds at all, it must *lift* all of the overlying strata in order to accommodate the required dilation.

The structures thus far discussed are *allogenous* dilation structures, so termed because the forces responsible for the formation of these structures are of external origin—in this case, the piston. Other dilation structures are self-generated, formed through forces derived from within the deformed material itself. These are called *autogenous* dilation structures. The production of these autogenous structures generally requires dilation approaching that of cubic packing, or an expansion beyond rhombohedral packing of more than 20% in the case of well-sorted and equidimensional grains.

The series of photographs shown in Figure 19 selected from a movie strip shows the development of autogenous dilation structures in slightly tilted beds of clean Woodbine Sand interstratified with magnetite sand. The photographs are negatives; therefore, the Woodbine sand is black and the magnetite is white. No piston is used.

The sand was dilated by about 25%, again by the use of carbonated water. Three aspects of this photograph sequence are of note: (1) the broad warp of the upper beds in the right of the picture in the first of the sequence develops into an injection lens, then into a dome without roots, and finally into a diapir; (2) the genesis of a spiral roll is shown in the upper beds at the left, showing good development in the last picture; and (3) the underlying beds remain undisturbed. The last photograph of this series was published in prior literature (Monroe, 1969). Since the dome and diapir developed at the top rather than at the bottom of the slope, they are shown to be buoyancy structures and not of compressional origin. Note the similarity of this anticlinal development of Figures 19c or 19d to the structural cross section of the Fandango Field, a high-pressure field of Zapata County, Texas (see Fig. 20).

Since it is known that clastic dikes are closely associated with these other dilation structures, it is appropriate to demonstrate the injection of a sandstone by use of the same equipment.

In this experiment a lower orifice of the box is fitted with a vertical glass tube, the top of which can be either pinched off or connected to a vacuum pump. Then the box of sand (sand obtained from the Woodbine Dike) is filled with carbonated water. The lower stopcock between the box and the evacuated tube is then opened. This is functionally the same as the opening of a crack in overlying strata. Sand immediatedly fills the tube. This experiment was repeated using noncarbonated tap water. No sand entered the tube in spite of the differential pressure. Clear water passed into the tube instead.

THE BUOYANCY STRUCTURES

Structures resulting from the isostatic readjustment of gas dilated sediments are subject to straightforward analysis. First we must remember that undilated sands or silts under high confining pressure are formidable structural members, in shear strength more like granite than like clay or shale. We must also remember that this same clastic material when adequately dilated by methane gas bubbles becomes as fluid as water and as elastic as the gas which forms the bubbles. Thus, when sediments are dilated they are both weakened and reduced in density. For example, a sediment expansion of the conservative 6% results in a density reduction of 5.66%. This reduction in density, along with the sediment weakening, results in isostatic readjustment in accordance with the diagram shown in Figure 21.

The second column represents the result of a 6% sediment expansion. The weight of the sediment remains the same, but the water depth is reduced by the amount of sediment expansion: that is, from 500 to 440 units. Therefore, the column is lighter than the first, having a weight of only 1,940 units.

The third column represents isostatic equilibrium for the weakened and less dense dilated sediment. Accordingly, the sediment moves up 145 units, reducing the water depth to 295 units. (We shall ignore the effects of the additional growth of gas bubbles caused by the lower pressure attendant upon this lessening water depth.) The last column represents the sediment of column three after enough fluid has escaped so that gas is no longer expanding the sediment through dilation. As gas is lost, the grains interlock and the sediment stands firm and out of isostatic equilibrium. The remaining closure relative to surrounding sediment is 137 units. Since all gas is not lost from pore spaces (some water is lost instead of all of the gas), the weight of the final column is something less than 2,069 units.

In the fifteen-year period between 1897 and 1911, the gas dilation process produced new islands in the petroleum areas of the northwest coast of the island of Borneo, off the Arakan coast of Burma, and off the southwest coast of Trinidad (Stigand, 1925). As one would expect, domal and anticlinal buoyancy structures are much more common in the underwater environment.

The Law of Uniformitarianism requires myriads of these buoyancy structures for the Tertiary and myriads more for older strata. These structures are seen in many of the oil and gas fields of today. In areas including and surrounding many of these oil and gas fields are the clastic dikes. Within this clastic dike zone is the area of allogenous fluidity structures, and well within this second zone is the area of autogenous fluidity structures along with the buoyancy features which so often function as traps.

APPLICATION TO EXPLORATION

As an illustration of the occurrence and significance of these dilation structures, the South Texas area of Oligocene sedimentary volcanism was chosen for investigation (Freeman, 1968), partly because of the surface indications of sediment-dilation conditions and partly because it is presently an active exploration area. This area lies between Webb and Live Oak counties and includes part of the counties of Webb, Zapata, Duval, McMullen, and Live Oak. At the surface we find sandstone dikes and mud-flow layers.

Some subsurface data are as follows:

1. From electric logs and seismic data we see abnormal thickening and thinning of beds; from seismic we see pinches and swells. Also, seismic gives us intrastratal crumpling and injection lenses.

2. Both dip meter and seismic data show abnormal dips up to 40°; the electric logs and seismic show subsurface diapirs. Drilling proves high-pressure gas. in the case of the Laredo Gas Field, autogenous crumpling is found right in its middle.

Several points, then, are clear: (1) Clastic dikes telegraph gas pressure information many thousands of feet into overlying strata; (2) all fluidity structures require high-pressure gas, with the autogenous ones requiring more dilation than the allogenous; (3) in turn, high-pressure gas requires both a seal and a trap; (4) all of these things combine to make an oil or gas field. Fairbridge was quite right. His so-called "slumping structures" are indeed closely related to oil bodies.

Accordingly, proof of high-pressure gas is an important piece of knowledge for the explorationist even in the absence of additional interpretation. With the additional capability to make judgment with regard to comparative gas pressures and probable fluid motion, again by use of dilation structures, these features become doubly valuable.

The first generation of application of the knowledge of dilation structures requires neither the development of new instrumentation nor the acquisition of new data. Instead, the reevaluation of existing data in the light of recognition and understanding of dilation structures should bring marked scientific progress and important petroleum discoveries.

ACKNOWLEDGMENTS

The writer expresses appreciation to Dibler Seismic Service for the seismic data of the Fandango Field (Fig. 20). He is especially grateful for the encouragement furnished toward this work by Dr. Claude C. Albritton, Jr., and Dr. James E. Brooks, both of the Institute for the Study of Earth and Man. The writer also appreciates the help of his son, John N. Monroe, Jr., who was of great help in the performance of the experimental work involved.

REFERENCES

Fairbridge, R.W., 1946, Submarine slumping and location of oil bodies: American Association of Petroleum Geologists Bulletin, v. 30, p. 84-92.

Freeman, P.S., 1968, Exposed middle Tertiary mud diapirs and related features in South Texas: Memoir 8, Diapirism and Diapirs, Tulsa, Oklahoma, American Association of Petroleum Geologists, p. 162-182.

Graton, L.C., and Fraser, H.J., 1935, Systematic packing of spheres—with particular relation to porosity and permeability: Journal of Geology, v. 43, no. 8, pt. 1, p. 785-909.

Mead, W.J., 1925, The geological role of dilatancy: Journal of Geology, v. 33, p. 685-689.

Monroe, J.N., 1957, Organically derived gases as a controlling factor in subaqueous slumping: Geological Society of America Bulletin, v. 68, no. 12, pt. 2, p. 1770.

__________, 1969, Slumping structures caused by organically derived gases and sediments: Science, v. 164, p. 1394-1395.

Reynolds, O., 1886, Dilatancy: Nature, v. 33, p. 429-430.

Rich, J.L., 1950, Flow markings, groovings, and intrastratal crumplings, etc.: American Association of Petroleum Geologists Bulletin, v. 34, p. 717-741.

Stigand, I.A., 1925, Outlines of the occurrence and geology of petroleum, London: Charles Griffin & Co.

TABLE 1

Dilation Parameters for Uniform Spheres (Relative to Rhombohedral Packing)

Packing Mode	Porosity (in %)	Expansion From Rhombo-Hedral (in %)	Fluid Expansion (in %)	Ratio of Gas to Total Fluid in Poor Space	Ratio of Gas to Total Sediment	Density Reduction From That of Rhombohedral Packing (in %)
RHOMBOHEDRAL	25.95	—0—	—0—	—0—	—0—	—0—
MONOCLINIC	30.18	4.23	16.3	0.14	0.0406	4.06
ORTHORHOMBIC	39.54	13.59	52.37	0.34	0.1196	11.96
CUBIC	47.64	21.69	83.58	0.46	0.1782	17.82

TABLE 2

Webb to Live Oak County Area

DATA SOURCE	OBSERVATIONS
SURFACE OBSERVATIONS	OLIGOCENE MUDFLOWS SANDSTONE DIKES
ELECTRIC LOGS AND SEISMIC	ABNORMAL THICKENING AND THINNING OF BEDS
SEISMIC	PINCHES AND SWELLS
SEISMIC	INTRASTRATAL FOLDING
SEISMIC	INJECTION LENSES
DIP METER AND SEISMIC	ANOMALOUS DIPS (UP TO 40 DEGREES)
ELECTRIC LOGS AND SEISMIC	SUBSURFACE DIAPIRS
DRILLING	HIGH-PRESSURE GAS

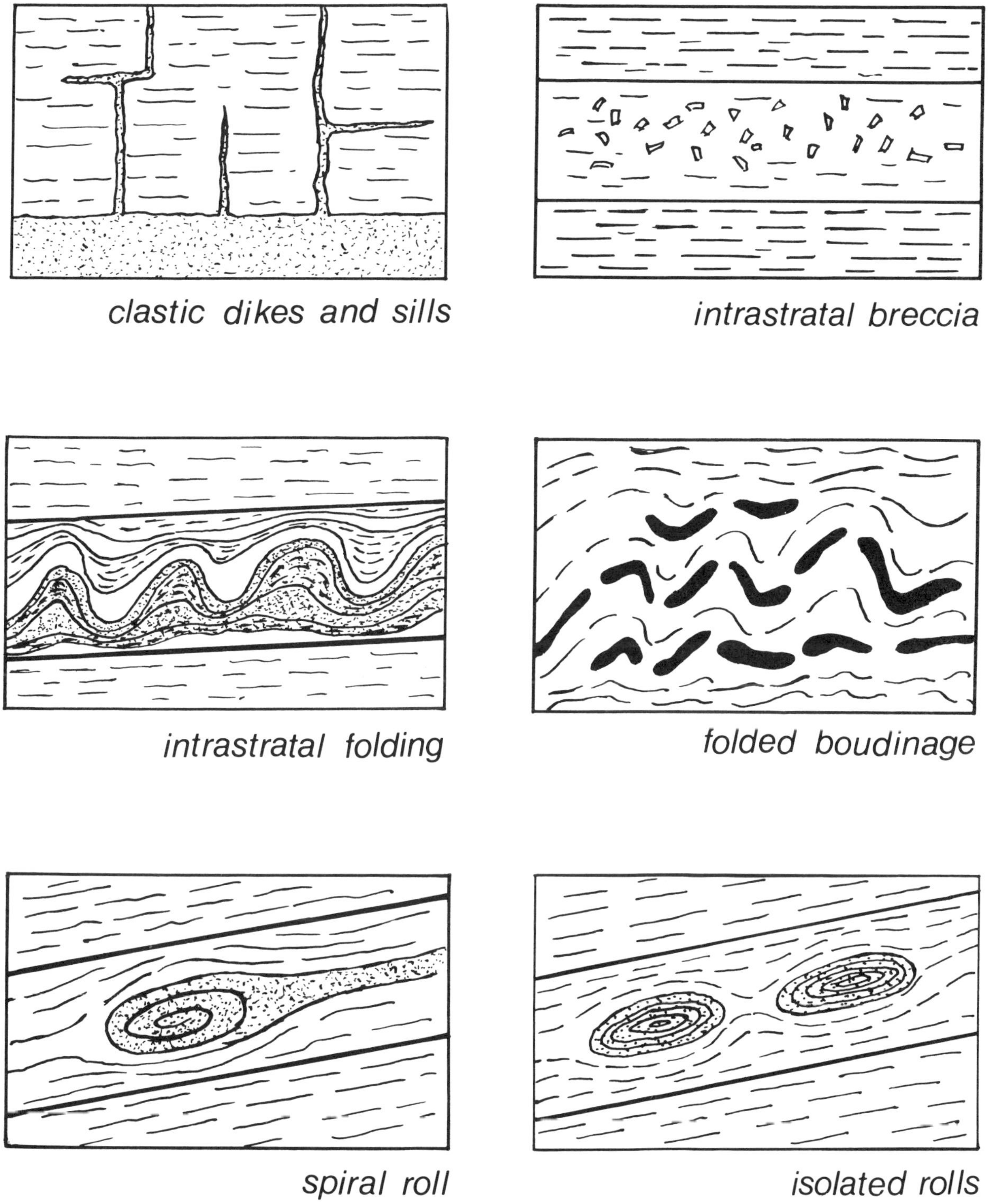

Fig. 1. Diagrammatic illustrations of selected dilation structures.

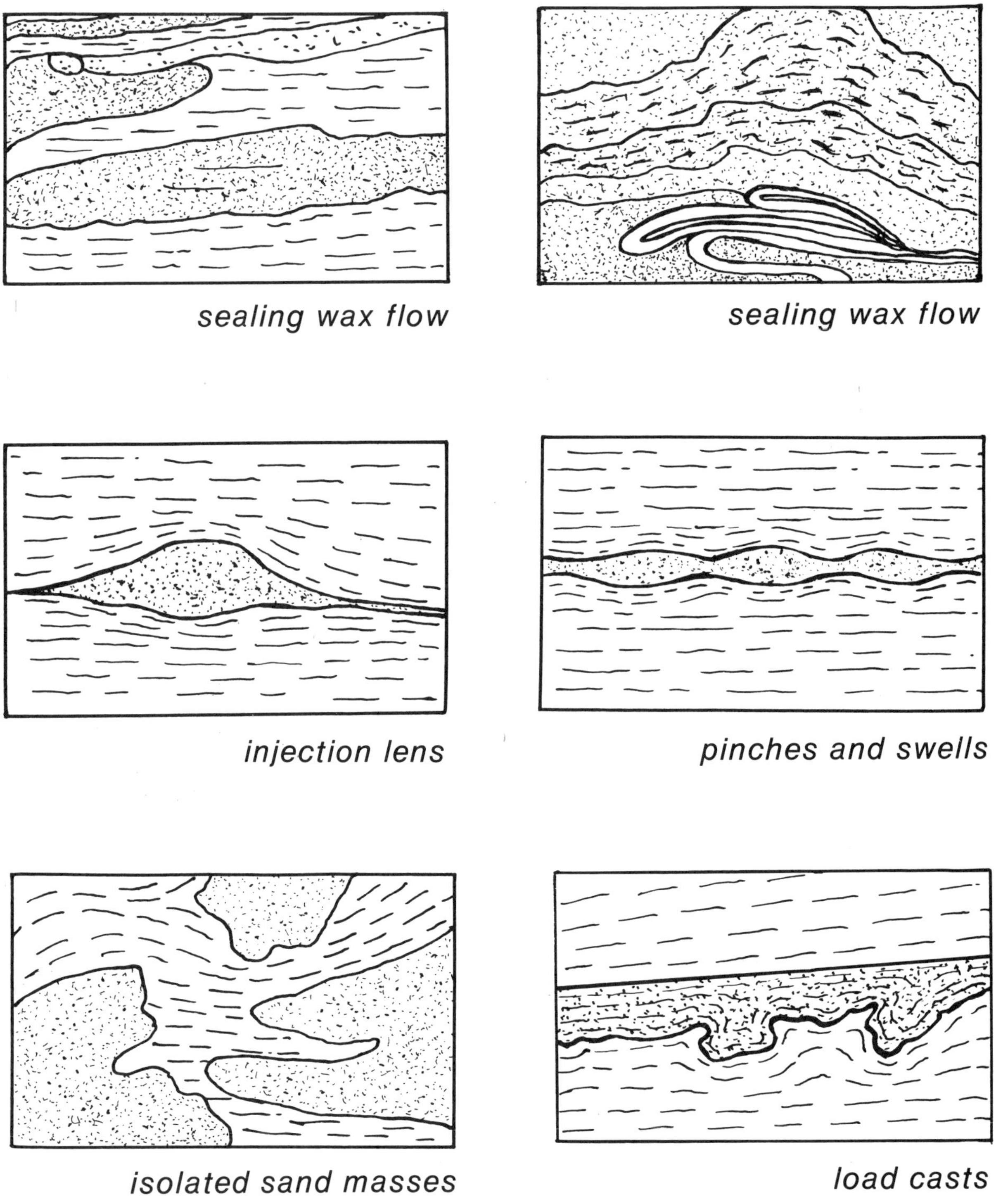

Fig. 2. Diagrammatic illustrations of selected dilation structures.

Fig. 3. A chalk dike of the Upper Taylor Formation of North Texas.

Fig. 4. A marl dike of the Upper Taylor Formation of North Texas.

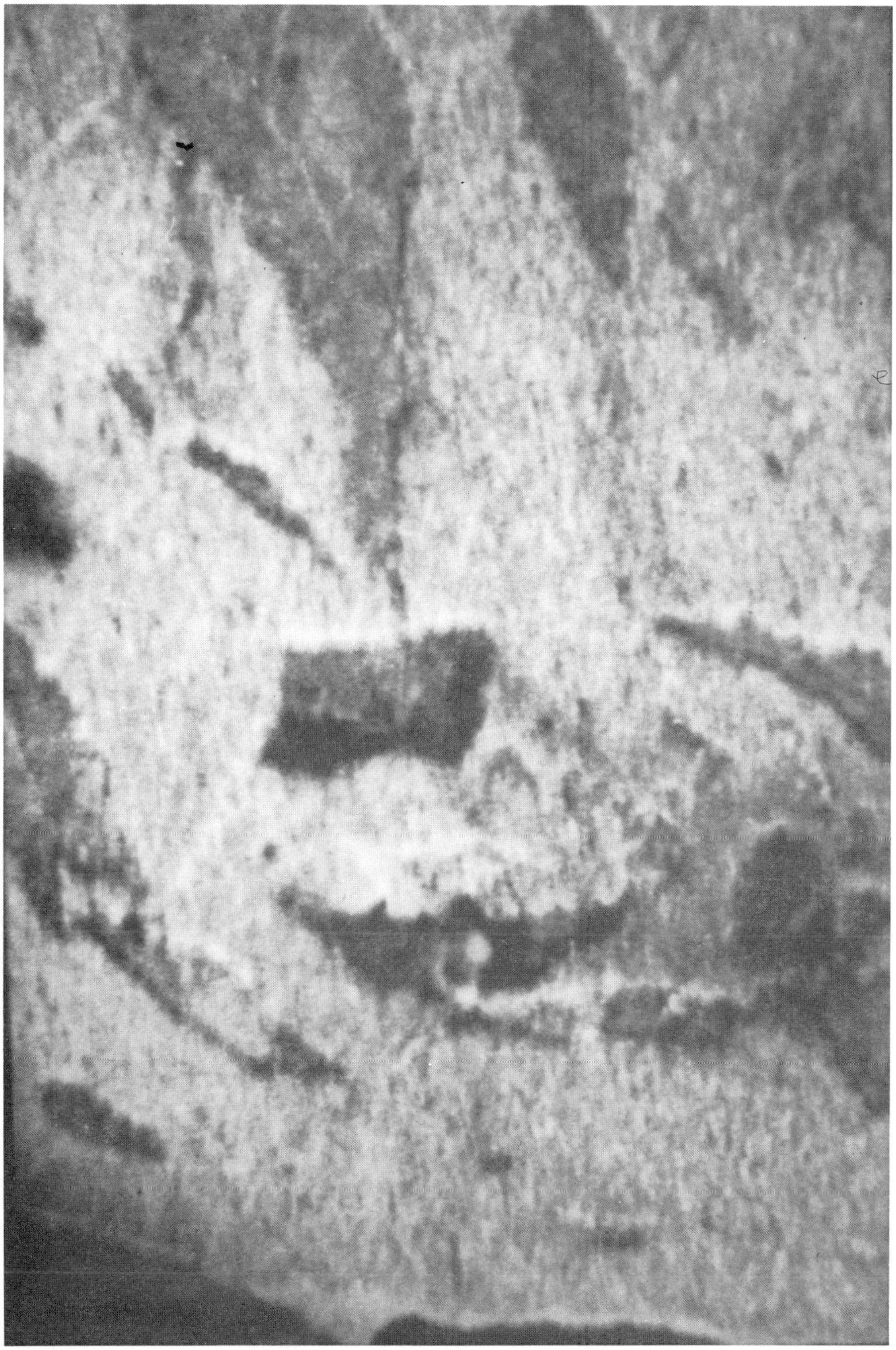

Fig. 5 Angular chips of shale which were carried along with dike sand during injection. The length of the rectangular fragment is ¼ in.

Fig. 6. Intrastratal folding from the California Pliocene, Santa Paula Creek, Ventura County.

Fig. 7. Folded Boudinage from the California Pliocene, Santa Paula Creek, Ventura County.

Fig. 8. A spiral roll formed within a sandstone dike of the Pepper Shale, Elm Creek, McLennan County, Texas.

Fig. 9. A sandstone dike which forces the turn of Elm Creek, McLennan County, Texas. Note the roll in the exposed end of the dike. This is the same roll pictured in Figure 8.

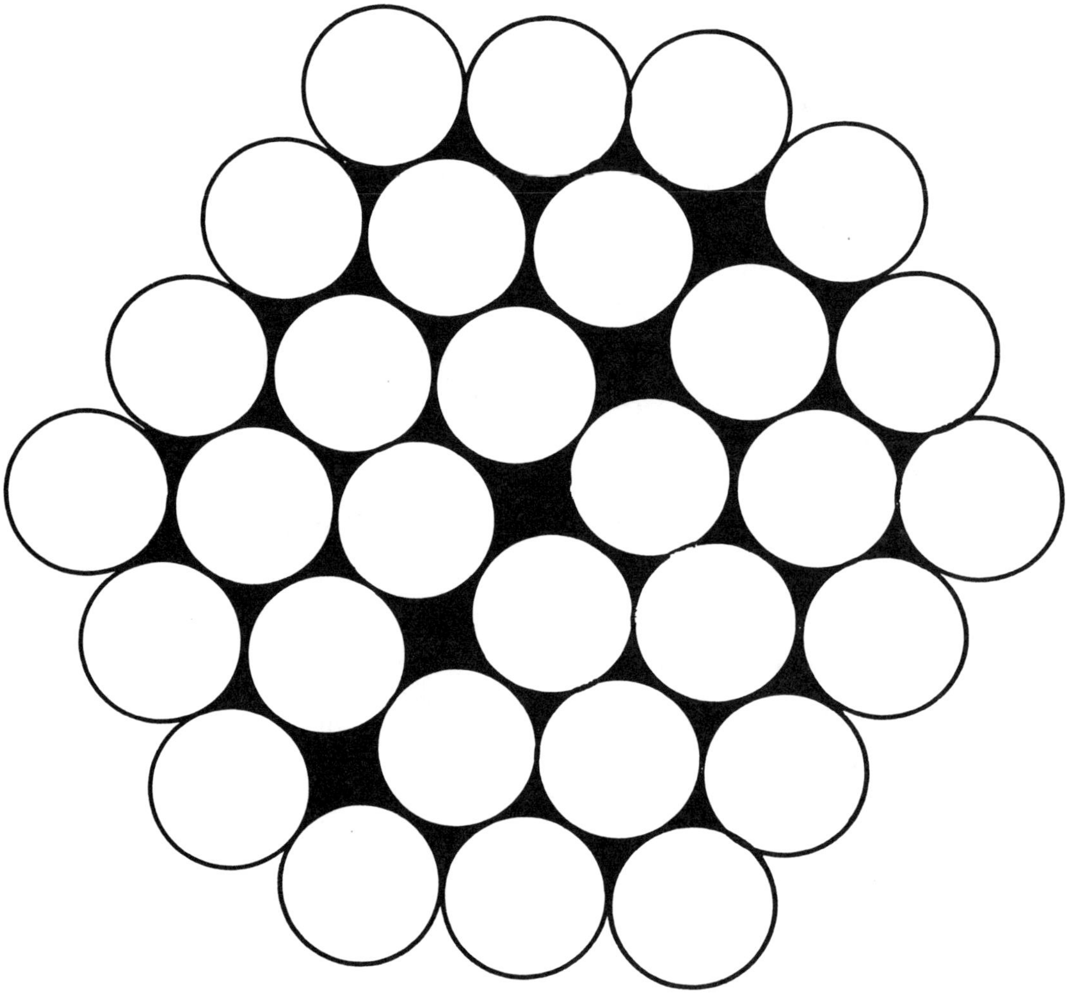

Fig. 10. A diagrammatic cross section of uniform spherical grains in rhombohedral packing undergoing shear. Note the larger pore space (in black) within the shear plane. This increased pore space represents the minimal expansion of the mass required for deformation.

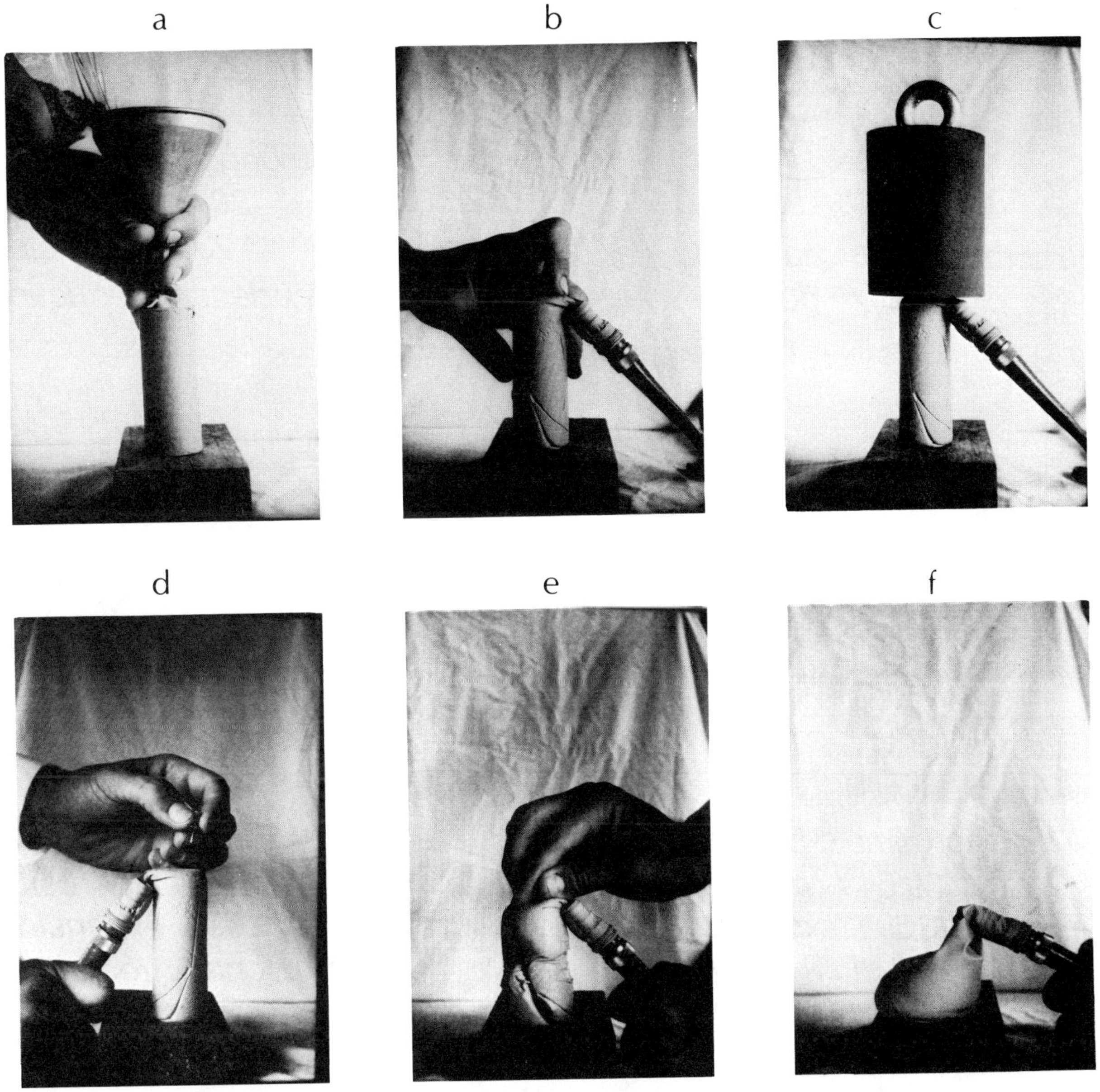

Fig. 11. This series of photographs illustrates the behavior of sand in a closed container and under confining pressure. (a) Sand is poured into a balloon which was previously placed within a paper tube. (b) After air is exhausted from the balloon of sand, the tube is peeled away. The cylinder of sand is firm. (c) A solid case iron-weight is balanced on the cylinder of sand. The shear strength of the sand (in terms of differential pressure) is approximately equal to the confining pressure. The hose on the right is the sealed-off vacuum hose. (d) The balloon is punctured. (e) A firm push results in shear of the mass. (f) When confining pressure disappears completely, the balloon of sand collapses.

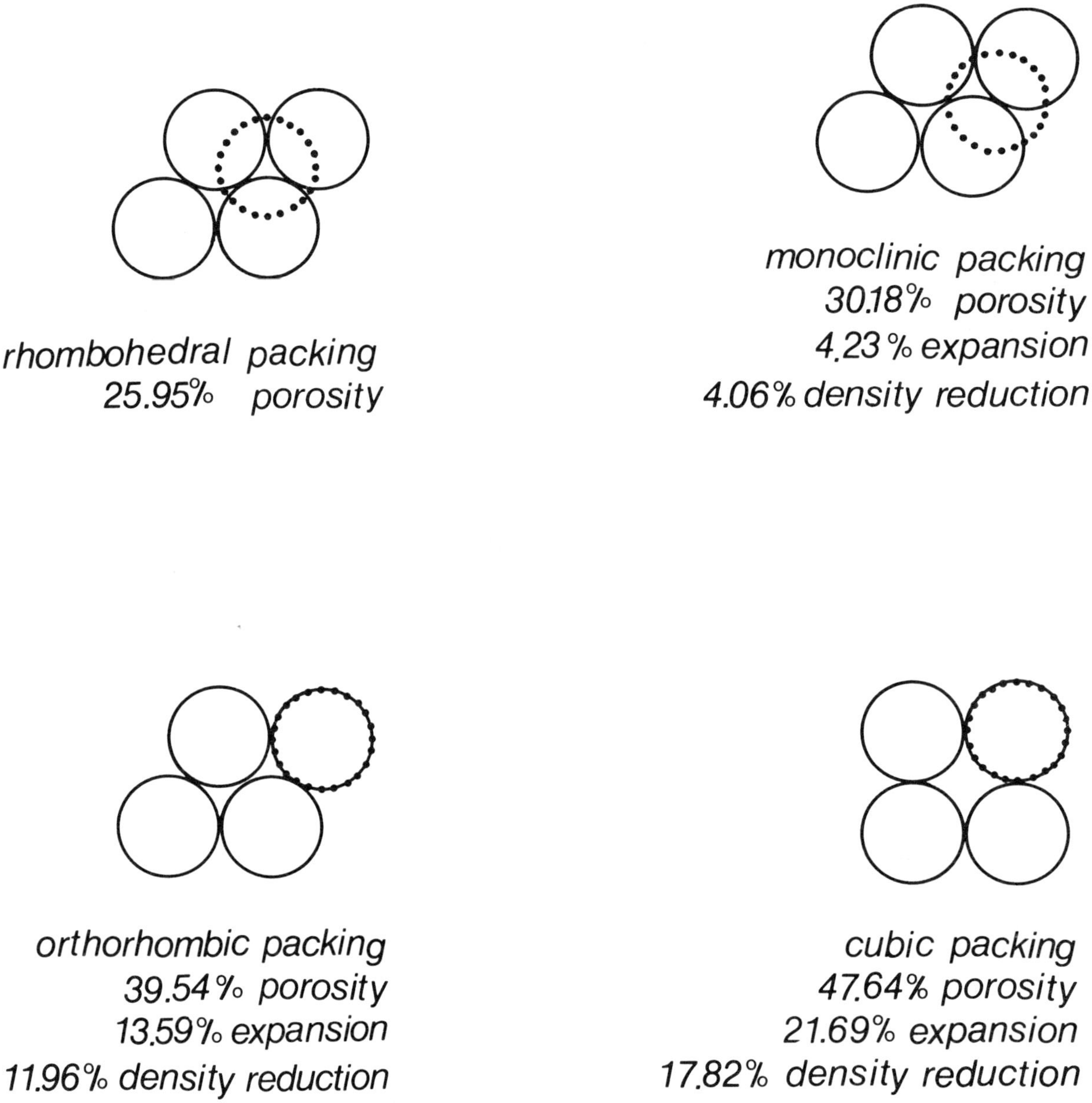

Fig. 12. Diagrams and some dilation parameters for uniform spheres. The solid lines for each case represent a lower level of grains. The dotted lines represent, for each case, the location of a grain stacked on top of the four grains below. In this way, the three-dimensional arrangement is illustrated.

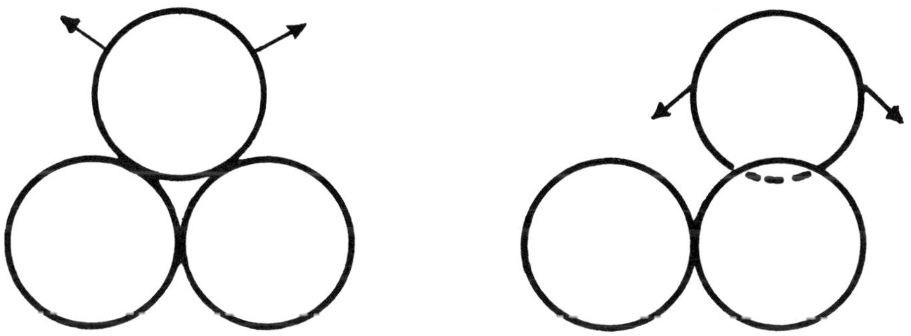

Fig. 13. Two cross-sectional views of the stacking of grains arranged in the monoclinic mode of packing. The cross section on the left shows that the upper grain must move upwardly or into expansion in order to change its position. The cross section on the right, however, shows that this "balanced" grain is free to move in the dimension illustrated.

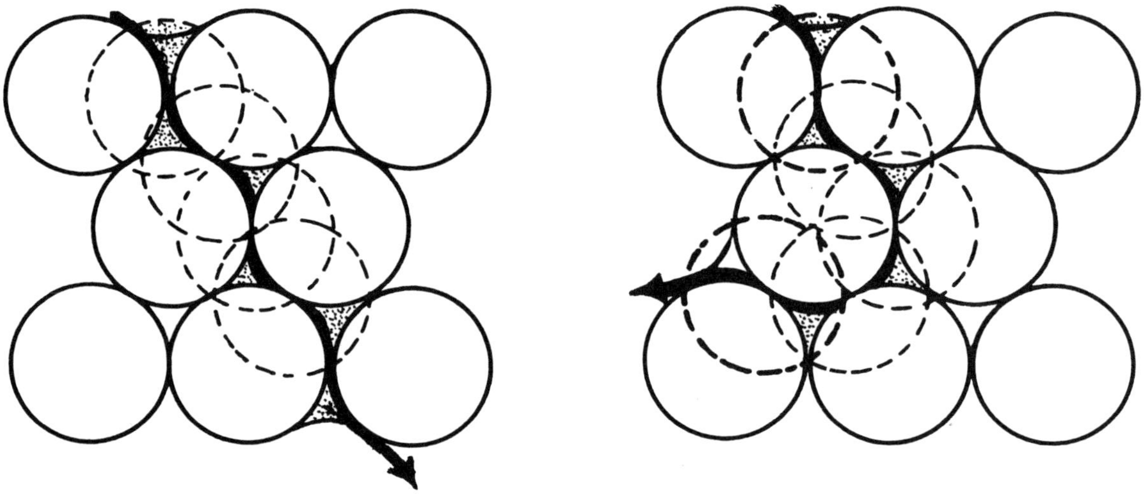

Fig. 14. Two of many routes of motion of a single unstable grain as illustrated in Figure 13.

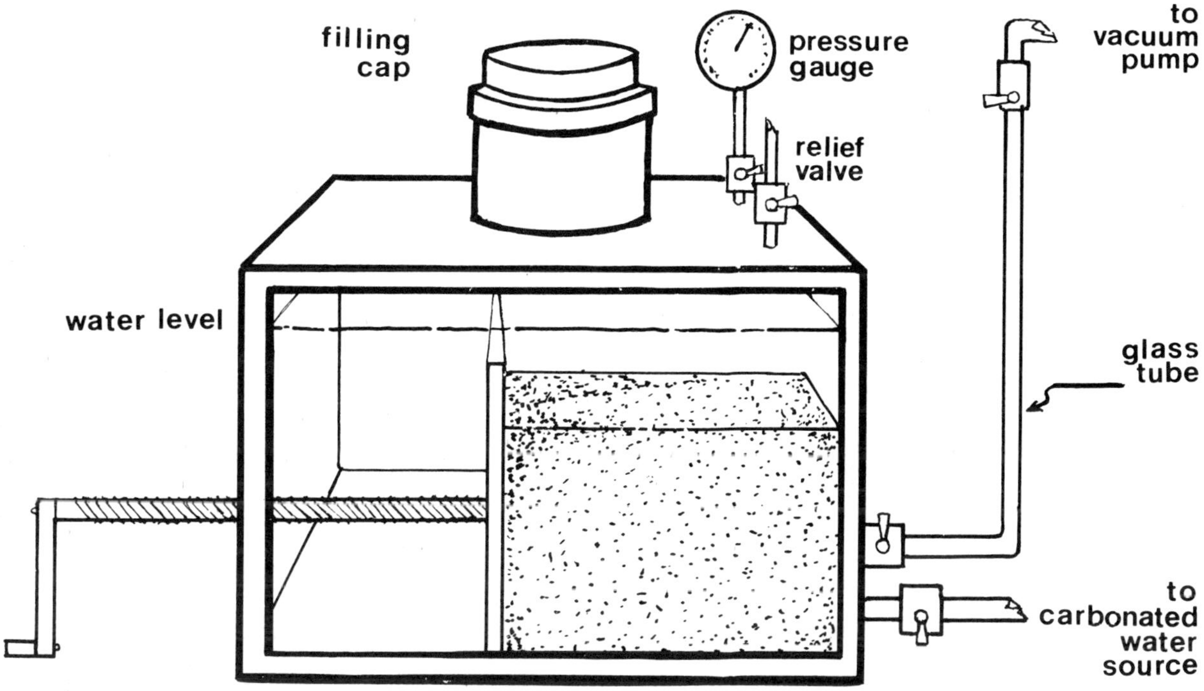

Fig. 15. Pressure box used for generation of dilation structures. The glass tube was used for simulated dike injection.

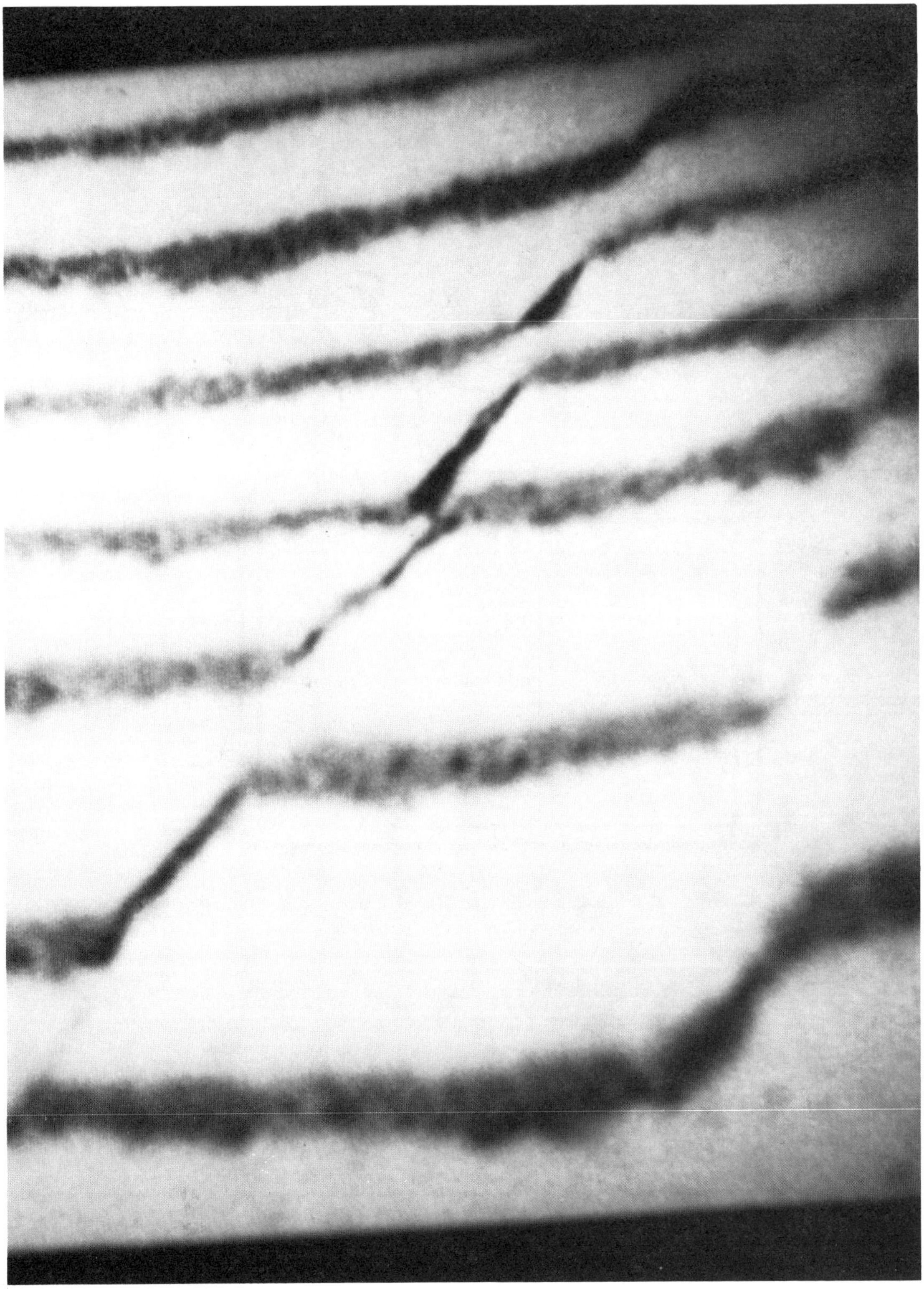

Fig. 16. Granular materials (the spherical beads in this case) fail by shear. This normal fault resulted from backing the piston away from the sand mass, allowing failure.

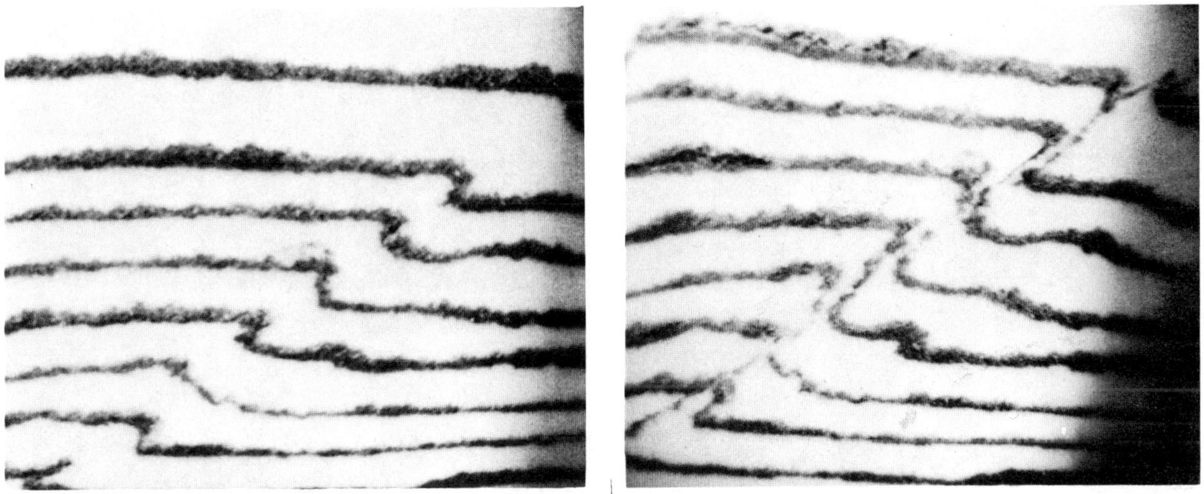

Fig. 17. When uniform spheres of granular material are dilated by about 5%, failure involves a combination of folding and faulting. The initial failure by flowage results in incipient shear folds which mark a location of later shearing. It is shown here that the term *Drag-Fold* is less than appropriate for describing the folding involved. It should be well known that cracks do not contribute to stress; rather they relieve stress.

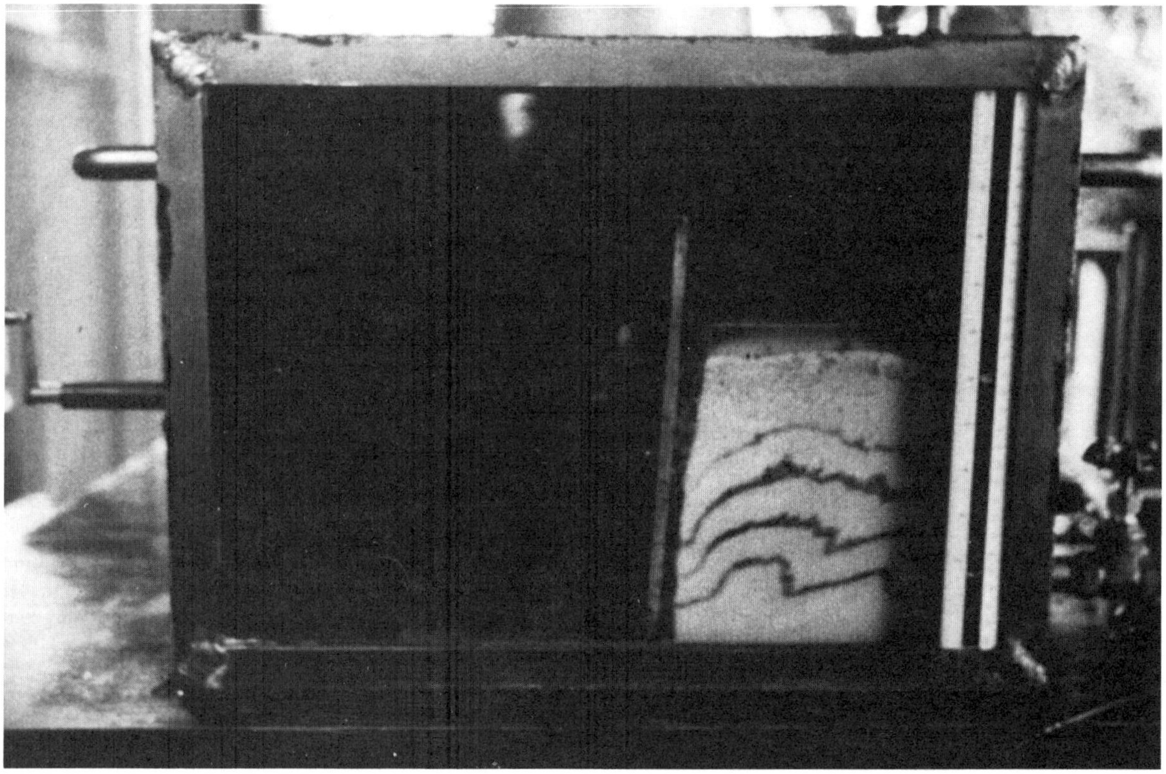

Fig. 18. Compression of a sand mass dilated by 8% results in parallel folding.

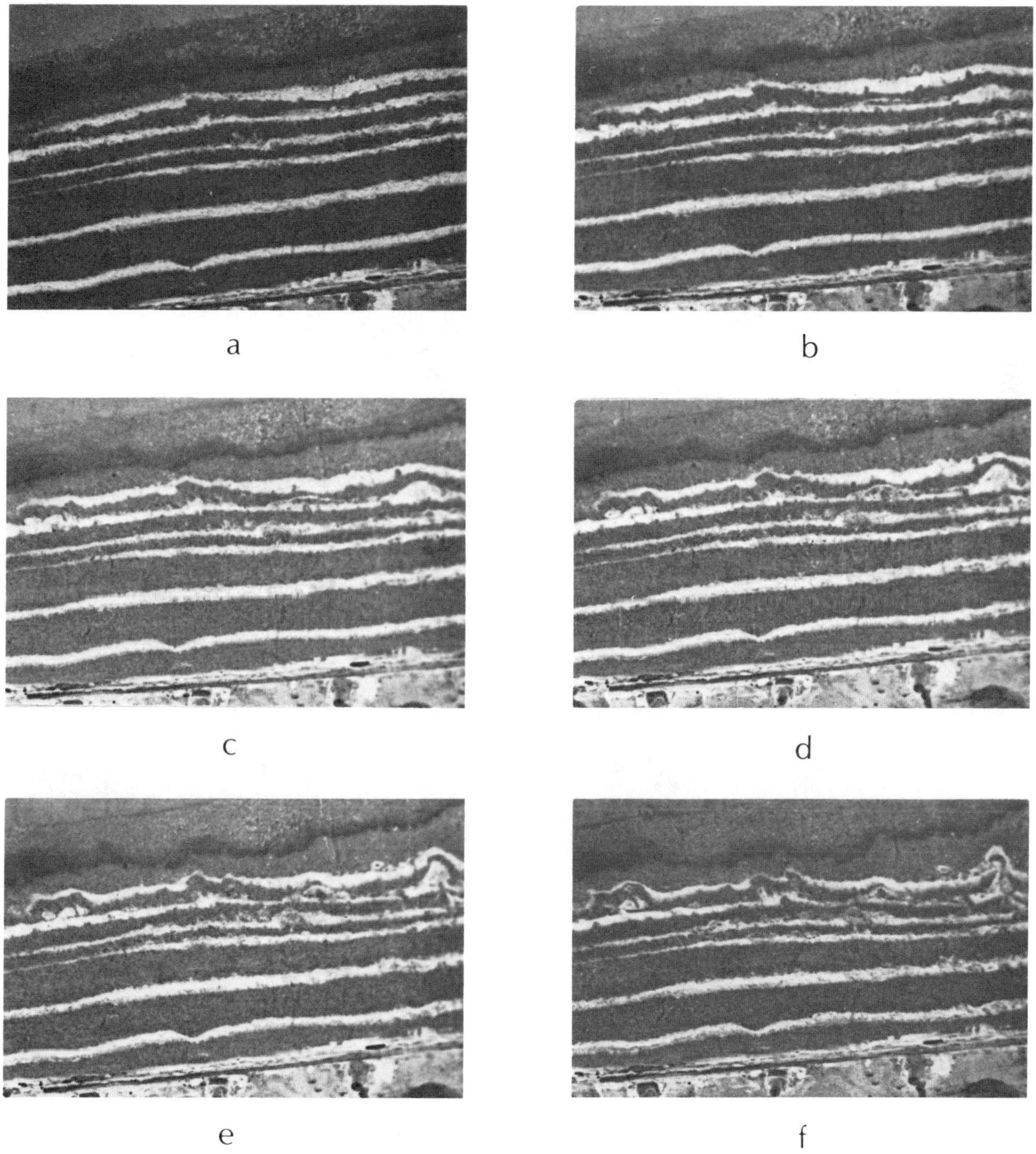

Fig. 19. A series of photographs showing the development of autogenous dilation structures. (a) Note the broad warp developing near the top of the slope on the right. This is a buoyancy structure. (b) Note the continuing development in the upper right. The underlying beds remain undisturbed. (c) A structure on the right takes the form of a dome. Still the lower beds are unaffected. (d) Here the dome shape suggests a diapir. (d) The diapir continues to develop. Note the white structure in the layer second from the top near the lower part of the slope. (f) The diapir shows real piercement characteristics. An incipient spiral roll develops near the left end of the box; *still* the lower strata are not disturbed.

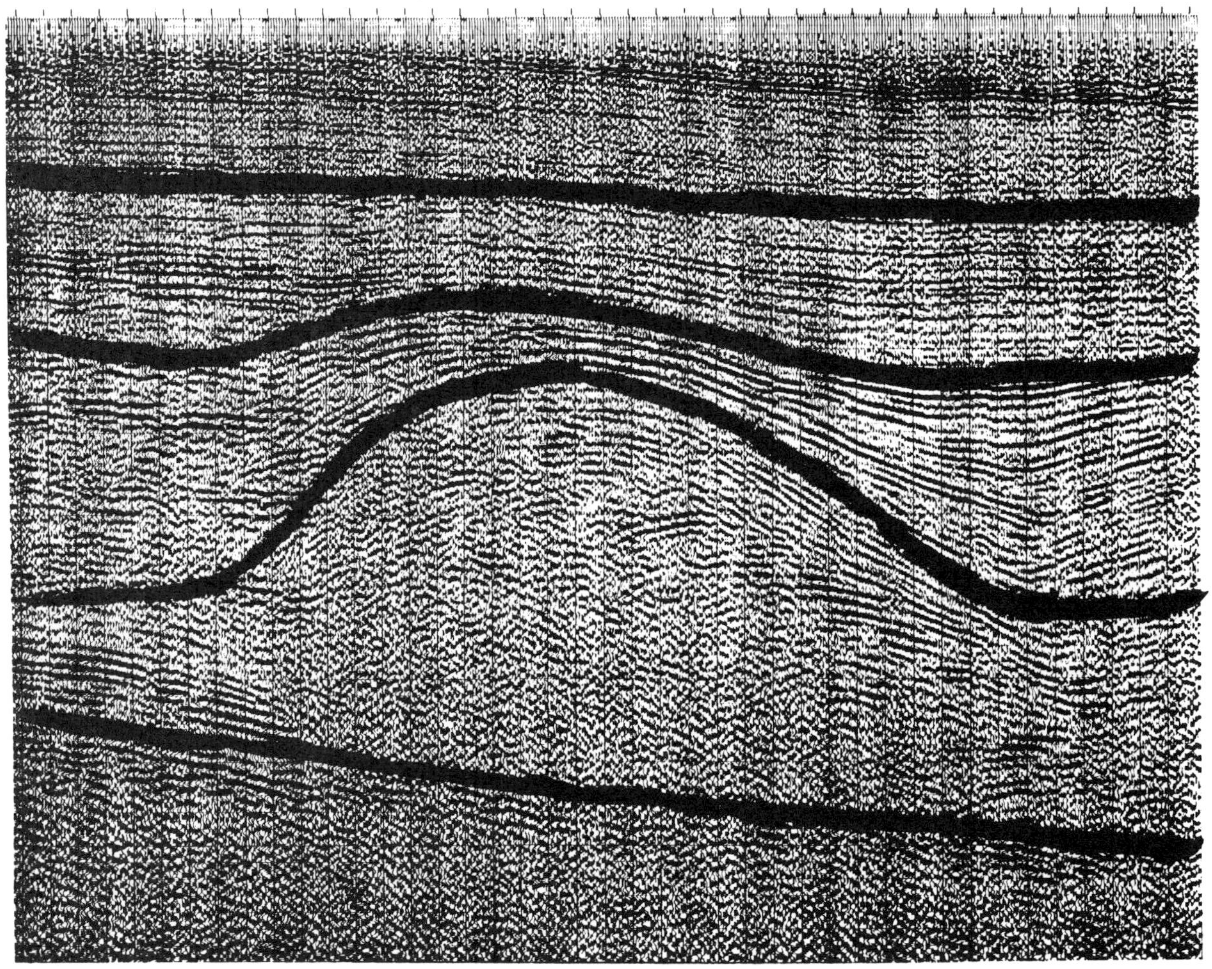

Fig. 20. A seismic section of the Fandango Field, Zapata County, Texas. The black lines were added by the writer for purposes of clarification. Note the similarity between this structure and that of Figure 19 (c) and (d). In both cases underlying strata are undisturbed. These seismic-section data were furnished through the courtesy of Dibler Seismic Service.

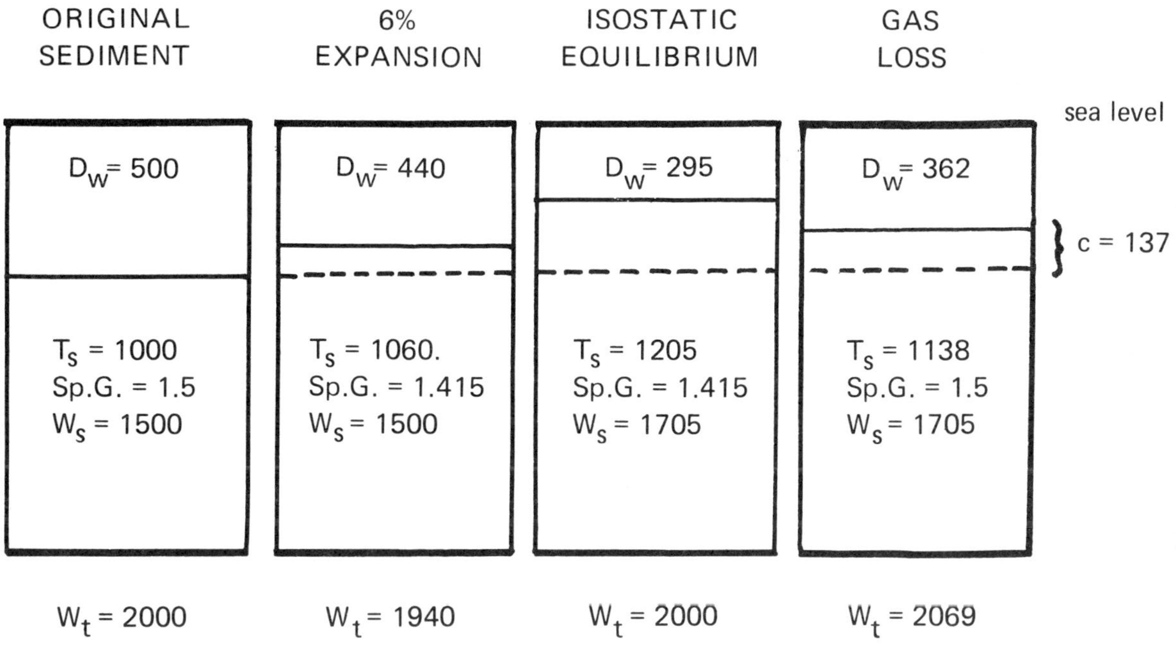

Fig. 21. Isostatic diagram showing the effects of dilation and ultimate contraction of sediment. The original sediment is under-water depth (D_W) of 500 units. The original thickness of sediment (T_S) is 1,000 units. Since the specific gravity of the sediment is taken as 1.5, the original sediment weight (W_S) is 1,500 units. The total original weight (W_t) is 2,000 units. The effects of a 6% expansion, the reestablishment of isostatic equilibrium through upward translation, and the effect of gas loss for the achievement of final contraction are illustrated in the other columns. The final closure of the structure is 137 units, as shown.

ELECTRICAL METHODS FOR HYDROCARBON EXPLORATION I. INDUCED POLARIZATION (INDEPTH) METHOD

Ben K. Sternberg and Dorothy Z. Oehler

CONOCO, INC.

ABSTRACT

Induced polarization (IP) measurements can be used to detect shallow electrical anomalies that occur over some deeper hydrocarbon reservoirs. These anomalies can be caused by the presence of shallow pyrite that forms as a result of upward seepage of hydrocarbons. Conoco's technique for using these IP measurements is known as the INDEPTH (Indirect Electrical Prospecting Tool for Hydrocarbons) method. In order to apply IP surveys to hydrocarbon exploration effectively, several procedures should be used, including removal of electromagnetic (EM) coupling interference (decoupling), use of multiple array spacings, precautions and tests for electrical leakage, avoidance of interference from grounded structures (cultural interference), and geological/geochemical evaluation of mapped anomalies to determine whether or not they are related to seepage. In order for seepage-related electrical anomalies to form, a potential for hydrocarbon seepage must exist, and appropriate near-surface host rocks must be present. Several years of testing at Conoco suggest that the necessary combination of seepage potential and near-surface geology is fairly rare. Despite its limited applicability, the induced polarization method could be useful for exploration in selected areas.

In this paper, we describe the procedures we have used for acquiring and interpreting induced polarization measurements, and we present some of the results of our field tests.

Once electrically valid data have been acquired and valid anomalies are located, it is important that the shallow geology and geochemistry be studied to determine whether or not the anomalies are related to hydrocarbon seepage. This geochemical/geological evaluation is discussed in the section on evaluation of anomalies.

When the conditions are appropriate for these induced polarization effects to occur, the resulting anomalies may provide a fairly large-amplitude and distinctive target. Seepage from a deep reservoir, even in extremely small quantities, may have occurred for millions of years (ever since the reservoir formed). With the IP measurements, the *integrated* effects of seepage over all this time are measured. This is one of the advantages of IP surveys. Another advantage is that with the IP arrays used, large volumes of earth, rather than single localities, are investigated. Methods aimed at looking directly for geochemical anomalies in soils or shallow sediment at selected sampling sites might miss localized anomalies.

Results of our geological/geochemical study have suggested a model that relates the occurrence of seepage-related pyrite (and therefore IP anomalies) to the presence of a deeper hydrocarbon reservoir. The model requires that hydrocarbons seep to the near surface and that they accumulate in an iron-bearing, near-surface host rock. Therefore, the INDEPTH method (and similar IP techniques) will be applicable to hydrocarbon exploration only in geologic settings where the seepage potential is adequate and where there are suitable near-surface host rocks in the shallow geologic section. Indeed, our field results indicate that the necessary combination of near-surface geology and seepage potential is fairly rare.

The limited applicability of induced polarization methods for hydrocarbon exploration is, in some respects, similar to that of bright-spot technology. Seismic-amplitude anomalies (bright spots) reliably indicate gas accumulations only in certain areas and are not associated with every known gas deposit. Thus, like the use of bright spots, we expect that, in selected areas, the INDEPTH method could significantly affect exploration strategy and success.

INTRODUCTION

Figure 1 shows a schematic illustration of how the INDEPTH method can be used to explore for hydrocarbons. The basic hypothesis is that oil and gas reservoirs can leak and that some hydrocarbons will seep upward to the surface. These hydrocarbons can lead to a reduced zone above the reservoir at a depth that would otherwise be oxidized. This reducing zone can in turn result in the formation of altered near-surface materials, which can have distinctive electrical properties. The seepage may be vertical or it may occur primarily along faults or carrier beds. Thus, the resultant electrical anomalies may be directly over a reservoir or they may be offset somewhat from a vertical projection of a reservoir.

In order to probe these shallow electrical properties, a transmitter is used to inject a current waveform into the earth via a relatively short (a few hundred to a few thousand feet) wire grounded to the earth at each end. The resulting potentials set up in the earth are detected at the center of this array with a receiver. The receiver is connected to a short wire which detects the potential difference (both magnitude and phase) between two grounded points. This general type of electrical measurement is well known in the industry and is referred to as induced polarization, or IP measurement (e.g., Keller and Frischknecht, 1966).

The maximum depth of penetration of the current from this type of electrical array is of the order of one-half the overall array length (i.e., of the transmitter wire). However, the dominant effect on the received signals for any given array length is usually of the order of 10 to 20 percent of the overall array length. Since the longest arrays are usually a few thousand feet long (with the optimum arrays in any given area often being just several hundred feet long), the anomalies that are detected with these electrical measurements are usually in the first few hundred feet of section.

Several papers have been published in recent years by authors who have used electrical measurements of this type and have claimed that they were directly detecting the presence of oil or gas by recording electromagnetic reflections from the high-resistivity hydrocarbons. Electrical arrays of this type do generate electromagnetic (EM) fields, and the effects of induction in the earth on the EM fields can be calculated in a fairly straightforward manner through the application of Maxwell's equations and boundary conditions (e.g., Anderson, 1974). These calculations show that for arrays of this type and size and for the techniques that are in use at the present time, the effect of a high-resistivity hydrocarbon layer on the observed EM fields is below the limits of detectability and well below the contribution from the near-surface IP effects.

DATA ACQUISITION

Figure 2 shows the typical field setup that is used in the INDEPTH method. The series of electrodes labeled C1 and C2 are the current electrodes from which the current from the generator is injected into the earth (typically 3 to 5 amps at 300 to 500 vs). The generator is an Inland Motors rotary amplifier and is basically a DC generator powered by a diesel engine. The excitation to the rotary amplifier is sinusoidally varied to produce a low-frequency sine-wave output. It is standard procedure in routine production surveys to use a single frequency (typically 0.1 Hz). A suite of frequencies may also be transmitted for wide-band spectral IP measurements. The received signals are detected at potential electrodes P1 and P2 and then are fed to the receiver. The receiver consists of a microprocessor-controlled data acquisition system. The potential signal and a sample of the transmitted current waveform are amplified, filtered, and digitized; and then successive waveforms are stacked or averaged to improve signal to noise. The signals may be analyzed either in the time domain (i.e., the characteristics of the decay of the waveform as a function of time) or in the frequency domain (magnitude and phase). The results from either domain are comparable, and both generally respond about the same to anomalous electrical properties. We have found some advantage to narrow-band recording in the frequency domain for these measurements, and the data in this paper are therefore expressed in the frequency domain.

The particular electrical array that was used for the measurements is known as the Schlumberger array (Keller and Frischknecht, 1966). Many other arrays can be used (such as parallel dipoles, in-line dipoles, etc.). One advantage of the Schlumberger array is that field experience has indicated that it is less sensitive to the normal variability in near-surface resistivities (geologic noise) and can provide a more reliable picture of the average electrical properties surrounding the recording site. The Schlumberger array typically has an unusually large amount of EM coupling interference, but this interference can be very effectively removed by the decoupling procedure described in the following section on EM field coupling. Controlled experiments with both Schlumberger and dipole arrays have shown that both types of arrays have a similar response to subsurface IP anomalies.

The magnitude and phase (at the fundamental frequency of the waveform) are calculated by the receiver microprocessor for both the received potential signal and the transmitted current signal. The final measured phase angle is the phase of potential signal minus the phase of current signal. The apparent resistivity of the earth is also calculated as a geometric factor (determined for the particular array used) times the ratio of the potential signal magnitude to the current signal magnitude.

FIELD PROCEDURES

Several field procedures must be followed in order to obtain valid data. First, since the transmitter and receiver wires are so close together in this type of array, considerable care must be exercised to avoid leakage of the high-voltage current signal directly into the very low-voltage receiving circuit. Various tests must be routinely run to verify that there is no leakage. Second, it is necessary to prevent any distortion of the received signal caused by interference from grounded structures (cultural interference), such as pipelines, power lines, fences, etc. Cultural interference either can cause false electrical anomalies, or can reduce the effect of a valid anomaly. This consideration is especially important when running test surveys in producing oil or gas fields, both of which usually have flow lines scattered throughout the field. Experience has shown that there will be no significant cultural interference when the array is one-half any array length (i.e., the total length, 5a, of the transmitting wire in Fig. 2) away from a grounding line that is parallel to the array and one-quarter of an array length away from a grounding line that is perpendicular to the array.

ELECTROMAGNETIC (EM) FIELD DECOUPLING

Figure 3 illustrates a third important technique which must be used. Since the transmitter and receiver wires (Fig. 2) are close together, there is substantial wire-to-wire EM coupling which distorts the signals. These fields may be calculated in a straightforward manner, however, and subtracted from the data. The quantities that must be known for this calculation are (1) the array configuration, (2) the spacing between the transmitter and receiver wires (this is kept accurately known and constant by using a parallel lead cable for the overlapping part of the array), and (3) the magnitude of the received signal (or, equivalently, the observed apparent resistivity). Figure 3 shows the measured phase angles (phase delay between transmitted and received signals) recorded over a seismically defined prospect. Also shown are the apparent resistivities that were derived from the magnitude of the received signals. The measured phase angles are positive and are virtually a mirror image of the apparent resistivity, as one would expect for wire-to-wire EM coupling. When a theoretical phase angle calculated for a half space, which has a resistivity equal to the measured apparent resistivity, is subtracted from the measured phase angles, the curve labeled ''decoupled phase angle'' is produced. In this example, there is no *valid* anomaly over the prospect, which is in agreement with the drilling results: no producible hydrocarbons.

VARYING ARRAY LENGTH

Figure 4 illustrates a fourth essential field technique. The electrical anomalies that we have found in the reducing zones over hydrocarbon reservoirs are often concentrated in relatively narrow zones. Because of the varying depth of penetration of each array length, some arrays will be ''focused'' on a particular anomalous layer and some will not. Figure 4 is a typical example. The optimum array length to detect the anomalous layer in this area is about 600 ft; arrays with lengths less than 100 ft or greater than 2,000 ft do not ''see'' the anomalous layer at all. Thus, the field procedures in a new area should include the use of multiple array spacings so that valid anomalies that may be present are not missed.

CASE HISTORIES

Figure 5 shows an example of an INDEPTH survey at the Ashland Field in southeastern Oklahoma. This field produces gas from the Hartshorne sandstone at a depth of about 3,000 ft. The contours on these two maps are net-pay isopachs and show the location of the producing field. INDEPTH recordings were made at this field with array lengths (i.e., the length of the transmitting wire) of 78 to 2,500 ft in increments of a factor of 2. The strongest anomalies were found with the 625-ft-long array. The phase angles corresponding to this array are indicated by the triangular symbols plotted at the center of each recording array. The large phase angles represent anomalous induced polarization in the earth, and they correlate closely with the outline of the field. The right-hand map shows that there is a corresponding resistivity high over the field. (In the following section on the evaluation of anomalies we discuss the geochemical model that explains how these IP and resistivity anomalies were formed.) Although there is a resistivity high over this field and in others, the common variations in near-surface resistivity which are unrelated to hydrocarbon seepage preclude the use of resistivity anomalies alone as reliable indicators of hydrocarbons at depth.

Figure 6 shows another example of an INDEPTH survey. These data were collected by a licensee for the INDEPTH method (Petty-Ray, Geosource, Inc.) for a survey in North Africa. On this map, the decoupled phase angles at 679 sites for 1,000 ft-long arrays have been contoured. The darkest gray areas are the regions of largest phase angle shifts (having the most anomalous electrical properties). The well symbols indicate the location of the producing field. Again, there is a very close correlation between the electrical anomalies and the producing field.

We feel that these two case histories illustrate that distinctive electrical anomalies can form over deeper hydrocarbon reservoirs. We wish to emphasize, however, that in other areas surveyed, we have not found valid and significant anomalies over known hydrocarbon reservoirs. By valid, we mean anomalies that are not caused by cultural interference (e.g., flow lines or well casings) in the field. By significant, we mean anomalies that are clearly distinguishable from normal background variation.

EVALUATION OF ANOMALIES

Extensive geochemical and geological study has shown that some electrically *valid* INDEPTH anomalies are related to hydrocarbon seepage, while others are unrelated to the presence of any deeper hydrocarbons and are thus "false" anomalies. "False" IP anomalies can be caused by shallow clay-bearing formations, pyritic shales, mudstones, or coals, zeolites and certain carbonaceous sediments (Nelson et al., 1982), hydrothermal metallic sulfide deposits, etc. Pyritic shales, mudstones, and coals are extremely common, and in most cases, the presence of this type of pyrite will not be related to seepage (Berner, 1971; Krauskopf, 1979). Because anomalies may be "false," it is essential that measured anomalies be evaluated, particularly in new areas with no known production for comparison.

In an evaluation, it must first be determined if a measured anomaly is *significant* compared to the normal variation in background electrical response surveyed well beyond the prospect or field being investigated. The electrical response of the anomaly should be compared to that of surrounding areas in terms of magnitude, areal extent, and surface morphology. The surface distribution of anomalous trends should also be compared to the distribution of near-surface and surface geological formations. For instance, if a measured IP anomaly were simply a reflection of a clay-bearing formation, then the surface distribution of the "false" anomaly might overlie the horizontal distribution of the clay-bearing formation. At this stage, even anomalies thought to be *significant* should be geochemically evaluated to determine their source and to see if there is independent evidence to suggest that they are related to hydrocarbon seepage.

Detailed geochemical studies at Ashland Field have provided data from which a model has been developed that explains the relationship between shallow electrical anomalies and deeper hydrocarbon reservoirs. This model depends on hydrocarbon seepage from the reservoir to the near surface, and it provides criteria for evaluating detected, significant electrical anomalies. Basically, the model can be summarized as follows:

1. Reservoirs may leak; seeping hydrocarbons will migrate upward and accumulate in porous, shallow host rocks.
2. Within those host rocks, aerobic bacteria can consume the hydrocarbons, and in so doing, they also use up oxygen; this process can convert aerobic, oxidizing pore waters to anaerobic, reducing pore waters.
3. Under the new anaerobic conditions, sulfate-reducing bacteria can grow if there is sufficient sulfate ion and organic material. The seeping hydrocarbons may be metabolized in part, or cometabolized by these bacteria (Freely and Kulp, 1957; Davis and Yarbrough, 1966; Berner, 1971; Barnes and Goldberg, 1976; Reeburgh, 1976; Martens and Berner, 1977; Jobson et al., 1979; Postgate, 1979; Reeburgh, 1980 and 1981).
4. The result of the bacterial metabolism is that sulfate is reduced to sulfide, and seeping hydrocarbons are oxidized to bicarbonate (Berner, 1971). A generalized reaction summarizing this process is:

$$\underset{\text{(methane)}}{CH_4} + \underset{\text{(sulfate ion)}}{SO_4^{=}} \rightarrow \underset{\text{(bicarbonate ion)}}{HCO_3^{-}} + \underset{\text{(bisulfide ion or other sulfide species)}}{HS^{-}} + H_2O$$

5. The produced sulfide species (HS^-) are highly reactive and will react with oxidized iron in the host rock to form (eventually) pyrite (FeS_2); the bicarbonate (HCO_3^-) can react with anions, such as calcium, to form a carbonate cement ($CaCO_3$).

6. This type of pyrite will be distributed in the pore spaces of a porous host rock and is a source of large-amplitude IP anomalies.

7. The calcite cements can effectively plug porosity, resulting in a "tight" layer that is a source of high-resistivity anomalies.

This model implies that reservoired H_2S is not required for the formation of near-surface IP anomalies and that seepage-induced IP anomalies may be accompanied by high-resistivity anomalies. It also implies that seepage from a reservoir must reach a shallow zone containing otherwise oxidized pore waters; that within this shallow zone, porous host rocks must be present within which the seeping hydrocarbons can accumulate; and that these host rocks must contain iron if pyrite is to form.

Evidence for this model was derived primarily from geochemical evaluation of two 300-ft-deep cores from the Ashland Gas Field in southeastern Oklahoma: one on-field and "on-anomaly" (the Frease core well) and the other off-field and"off-anomaly" (the Shores core well). These wells were logged with Conoco's downhole IP tool to determine first the main vertical zone of electrical anomaly (Fig. 7). Then geological comparisons of the two cores were made to correlate stratigraphically the anomalous on-field zone with the equivalent off-field zone. Finally, detailed geochemical comparisons were made of the two equivalent zones.

These data show that methane is much more abundant in the Frease well; that it is concentrated in the very shallow, iron-rich sandstones and siltstones of the zone of major electrical anomalies (between about 20 and 60 ft deep); that this shallow gas, while about 94 percent methane, has an overall hydrocarbon composition very similar to that in gas samples taken from the Hartshorne gas reservoir (at a depth of 3,000 ft); that both the reservoir and shallow gas samples lack measurable H_2S; and that the $\delta^{13}_{PDB}C$ value of the shallow methane (-43 to -45 per mil) is similar to that of the reservoir methane (-49 per mil). These analyses also show that pyrite and calcite accumulations occur in the anomalous sandstone/siltstone zone and that, while these minerals are also present in the equivalent off-field zone, they are nearly twice as abundant in the on-field well. Finally, the calcite cements in the Frease well have $\delta^{13}C$ values of about -26 per mil, suggesting that they have incorporated, at least in part, some oxidized organic matter (in this case, probably the seeping hydrocarbons). Similar pyrite accumulations and/or low $\delta^{13}C$ value carbonate cements have been reported by others over several hydrocarbon reservoirs where seepage is presumed to have occurred (Donovan, 1974; Donovan et al., 1974; Donovan and Dalziel, 1977; Ferguson, 1977; Donovan et al., 1981).

Laboratory IP analyses of the pyritic sandstones and siltstones show large-amplitude phase angle shifts, while similar tests of nonpyritic, nearby sandstones and siltstones show no phase angle shifts. Tests of deeper, pyritic shales (containing more pyrite than the shallower sandstones and siltstones) show only low- to medium- amplitude IP response. Laboratory analyses also show that the samples with highest concentrations of calcite cement have the lowest porosities and the highest resistivities.

These data illustrate the features of one example of seepage-related IP and resistivity anomalies. These characteristics can be used to formulate a set of criteria that can be used to evaluate other, electrically valid anomalies that are detected. Such an evaluation should include a determination of (1) the specific downhole interval (depth range) of electrical anomalies, (2) the source of anomalies within that interval, and (3) the likelihood that the source of anomalies is related to hydrocarbon seepage. Independent checks for anomalous accumulations of seeping hydrocarbons or their oxidized remnants (the low $\delta^{13}C$ value carbonates) should be made, as should "on-anomaly/off-anomaly" comparisons of lithological, hydrocarbon, and mineralogical data.

SURVEY STATISTICS

We have compiled statistics for 41 INDEPTH surveys run by Conoco during the period 1976 to 1980. Some of these surveys were run during the early stages of development of the INDEPTH method and were not adequate tests of the potential success of the method. Many more surveys were run over prospective areas where no anomalies were recorded but where either no well has been drilled or the well drilled was a dry hole—again not an adequate test of the method, since we cannot be certain that anomalies would have been detected had hydrocarbon reservoirs been present. Ten of the surveys did provide tests of indirect electrical detection. In four of these ten cases, there was a close correlation between recorded anomalies and the producing oil or gas field. A fifth example showed a correlation between anomalous electrical readings and a deep H_2S reservoir. A sixth example showed low IP readings over the field and high readings surrounding the field. In the remaining four cases, there were no anomalies over producing fields. In each of these latter four, however, the absence of an anomaly might be explained by either the presence of an impermeable seal which could have blocked seepage from reaching the surface, or the existence of inappropriate near-surface host rocks that could have been unsuitable for the accumulation of seeping hydrocarbons or the formation of pyrite. Test surveys are continuing at Conoco to determine which areas have appropriate geological conditions for use of the INDEPTH method in hydrocarbon exploration.

From this review of the forty-one surveys, it appears that large-amplitude IP anomalies (more negative than -10 mrad) are relatively rare. Where broad regions of large-amplitude IP anomalies do occur, these anomalies correlate with deeper reservoirs.

These results also suggest that not all reservoirs will be associated with shallow electrical anomalies. The INDEPTH method (and similar IP techniques), therefore, will be applicable to hydrocarbon exploration only in geologic settings where the seepage potential is adequate and where the shallow geologic section contains appropriate host rocks. While some regions can be reasonably eliminated because of inappropriate geology, the positive determination of the appropriateness of the subsurface conditions is extremely difficult in practice. Therefore, often the best approach from which to make this judgment is to template known producing fields. If a significant electrical anomaly correlates closely with the producing field, then IP methods can probably be applied to nearby prospective areas which have very similar near-surface geology and seepage potential. Since near-surface conditions often vary over relatively short distances, it is likely that one of the most effective uses of the IP technique will be in developmental exploration—that is, in extending the limits of known fields.

STRATEGY FOR USE OF SHALLOW-INVESTIGATION ELECTRICAL SURVEYS

Conclusions from the geochemical case studies and from review of forty-one INDEPTH surveys suggest the following strategy for use of shallow-investigation electrical surveys in hydrocarbon exploration:

1. Avoid areas with poor seepage potential or those which lack shallow, porous, iron-bearing host rocks.
2. For better interpretation of surveys over prospects, template nearby production.
3. Carefully avoid cultural interference (well casings and flow lines) and use decoupling and other acquisition procedures to obtain electrically *valid* data.
4. Determine which of the measured, valid anomalies are *significant* (this involves consideration of their magnitude, areal extent, and correlation with known fields, prospects, surface and subsurface geology).
5. *Evaluate* significant anomalies to determine if they are related to hydrocarbon seepage. This process involves looking for independent evidence of seepage and determining the source of the IP response in addition to the host rock type within which shallow methane, carbonate, pyrite, or other polarizable materials are distributed.

Using this strategy, shallow-investigation electrical prospecting methods can be applied more effectively to the search for hydrocarbons. It should be emphasized, though, that this strategy is based on just one model for electrical anomalies caused by hydrocarbon seepage, i.e., one developed primarily from studies at Ashland Field. There may well be other mechanisms by which shallow electrical anomalies can result from hydrocarbon seepage. As more information is obtained on the geochemical reactions caused by hydrocarbon seepage, this strategy can be refined and expanded.

ACKNOWLEDGEMENTS

We are grateful to Ron Cormick and Jerry Ware for support of this work in the Exploration Research Division at Conoco. We also are grateful for the contributions made by numerous others in the division during the development and testing of the INDEPTH method. These people include Dhari Bahjat, Alicia Boan, Paul Buller, Don Dunster, Juli Fahy, Dale Miller, Elizabeth Murphy, and Ray Skipping. Finally, we would like to thank the personnel in Conoco's Exploration offices, who provided information and helped plan surveys, and the crew and managers of Geophysics Operations Party 31, who collected much of the field data in this paper.

REFERENCES

Anderson, W.L., 1974, Electromagnetic fields about a finite electric wire source, program EMFIN, United States Geological Survey publication, available from National Technical Information Service, Report no. PB-238199.

Barnes, R.O., and Goldberg, E.D., 1976, Methane production and consumption in anoxic marine sediments, Geology, v. 4, p. 297-300.

Berner, R.A., 1971, Principles of Chemical Sedimentology, McGraw-Hill, New York, 240 p.

Davis, J.B., and Yarbrough, H.F., 1966, Anaerobicc oxidation of hydrocarbons by Desulfovibrio desulfuricans, Chemical Geology., 1, pp. 137-144.

Donovan, T.J., 1974, Petroleum microseepage at Cement, Oklahoma: Evidence and mechanism, American Association of Petroleum Geologists Bulletin, v. 58, no. 3. p. 429-446.

Donovan, T.J., and Dalziel, M.D., 1977, Late diagenetic indicators of buried oil and gas, United States Geological Survey Open-File Report 77-817, 44 p.

Donovan, T.J.; Friedman, I.; and Gleason, J.D., 1974, Recognition of petroleum-bearing traps by unusual carbon isotopic compositions of carbonate-cemented surface rocks, Geology (July), p. 351-354.

Donovan, T.J.; Roberts, A.A.; and Dalziel, M.D., 1981, Epigenetic zoning in surface and near-surface rocks resulting from seepage-induced redox gradients, Velma Oil Field, Oklahoma, American Association of Petroleum Geologists Bulletin, v. 65, no. 5, p. 919.

Ferguson, J.D., 1977, The subsurface alteration and mineralization of Permian red beds overlying several oil fields in southern Oklahoma, Master's thesis, Oklahoma State University, Stillwater, 95 p.

Freely, H.W., and Kulp, J.L., 1957, Origin of Gulf Coast salt-dome sulfur deposits, American Association of Petroleum Geologists Bulletin, v. 41, no. 8, p. 1802-1853.

Jobson, A.M.; Cook, F.D.; and Westlake, D.W.S., 1979, Interaction of aerobic and anaerobic bacteria in petroleum biodegradation, Chemical Geology, v. 24, p. 355-365.

Keller, G.V., and Frischknecht, F.C. 1966, Electrical Methods in Geophysical Prospecting, Pergamon, Oxford.

Krauskopf, K.B., 1979, Introduction to Geochemistry, 2d ed., McGraw-Hill, New York, 617 p.

Martens, C., and Berner, R.A., 1977, Interstitial water chemistry of anoxic Long Island Sound sediments: I. Dissolved gases, Limnology and Oceanography, v. 22, p. 10-25.

Nelson, P.H.; Hansen, W.H.; and Sweeney, M.J., 1982, Induced-polarization response of zeolitic conglomerate and carbonaceous siltstone, Geophysics, v. 47, no. 1, p. 71-88.

Postgate, J.R., 1979, The Sulphate-Reducing Bacteria, Cambridge University Press, New York, 151 p.

Reeburgh, W.S., 1976, Methane consumption in Cariaco Trench waters and sediments, Earth and Planetary Science Letters, v. 28, p. 337-344.

Reeburgh, W.S., 1980, Anaerobic methane oxidation: Rate depth distributions in Skan Bay sediments, Earth and Planetary Science Letters, v. 47, p. 345-352.

Reeburgh, W.S., 1981, Methane oxidation in anoxic sediments, American Association of Petroleum Geologists Bulletin, v. 65, p. 977-981.

Sternberg, B.K.; Miller, D.E.; and Bahjat, D.S., 1981, Electrode prospecting method providing calculable electromagnetic coupling for the indirect detection of hydrocarbon reservoirs, U.S. Patent No. 4,295,096.

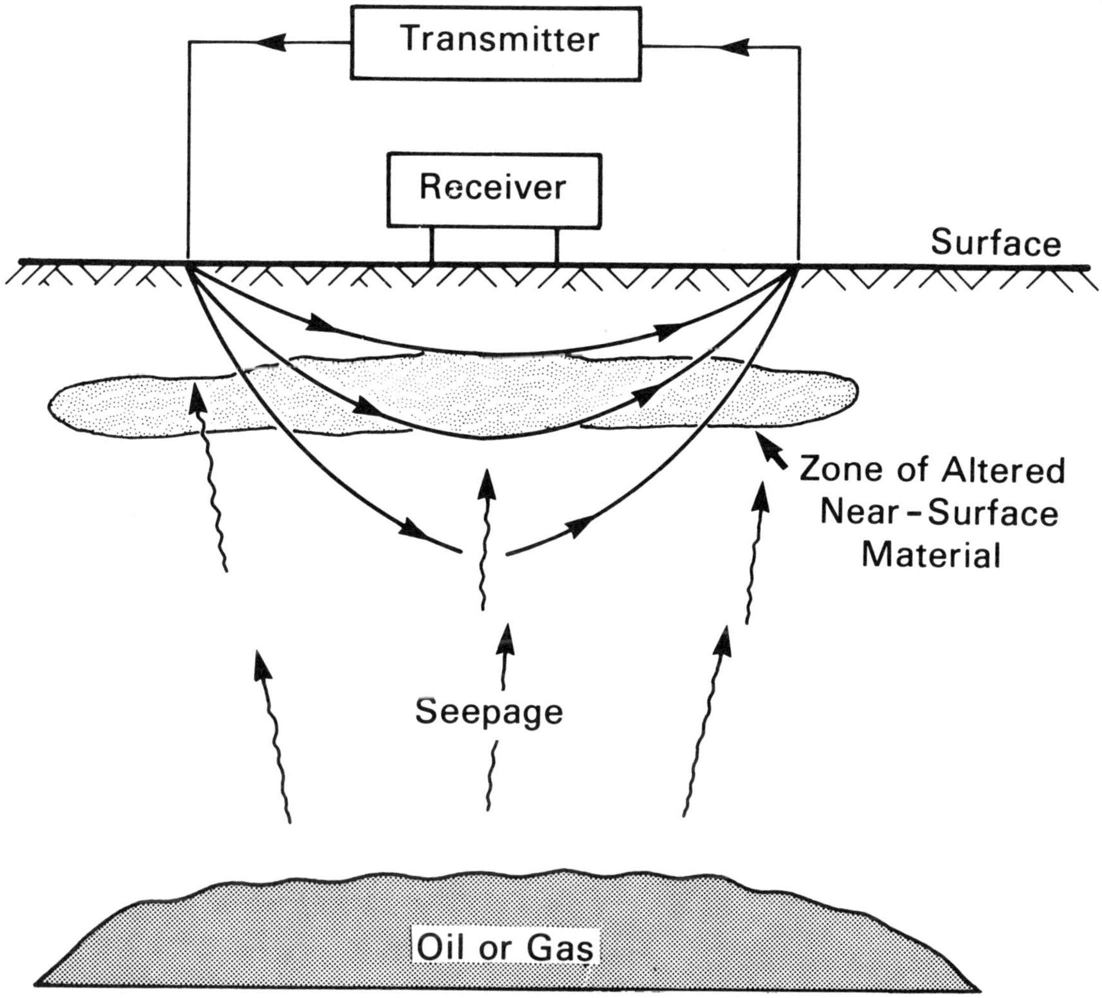

Fig. 1. Schematic illustration of hydrocarbon seepage and the INDEPTH electrical array that is used to detect zones of altered near-surface material that have been caused by the hydrocarbon seepage.

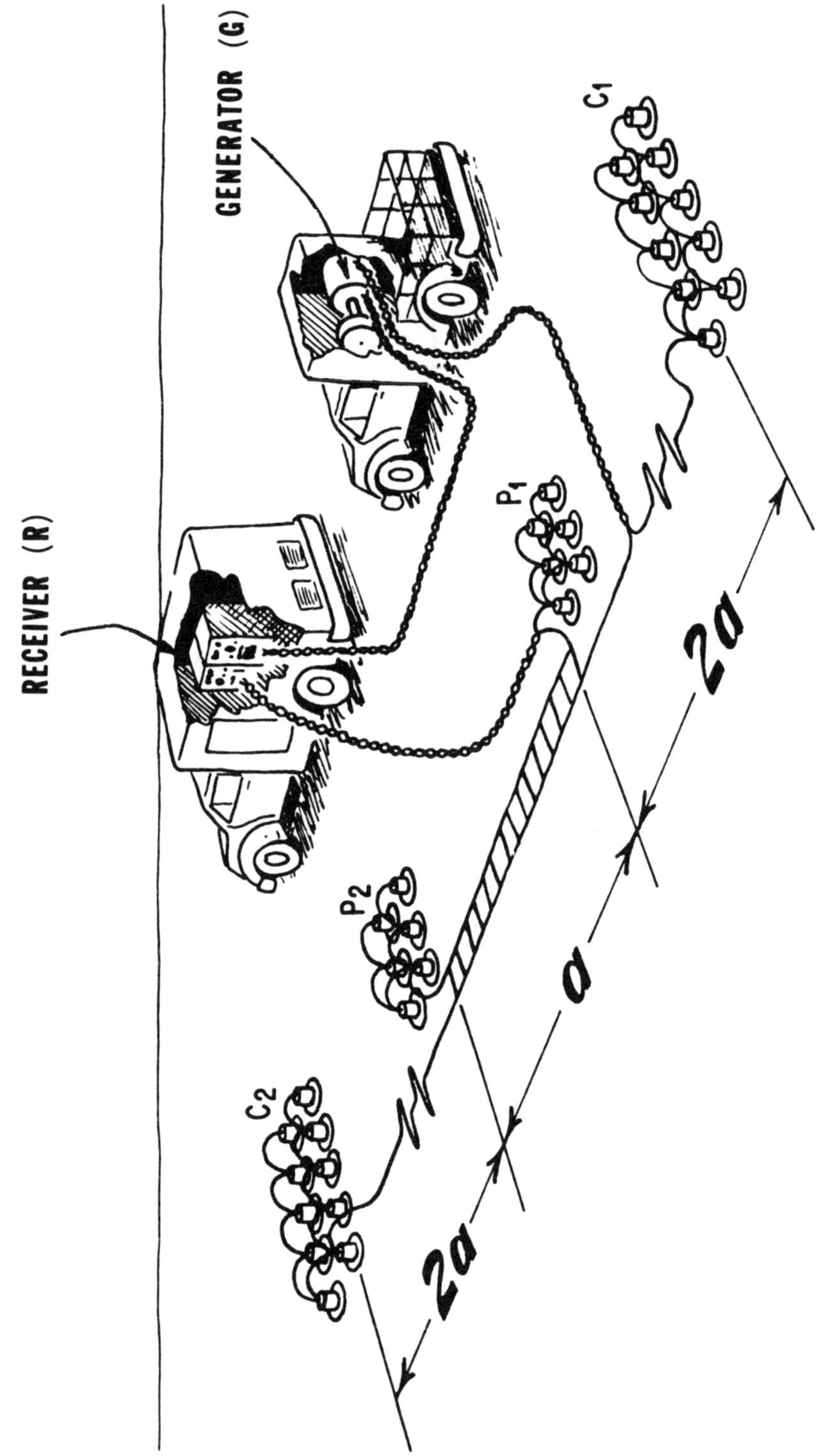

Fig. 2. Typical INDEPTH method field setup.

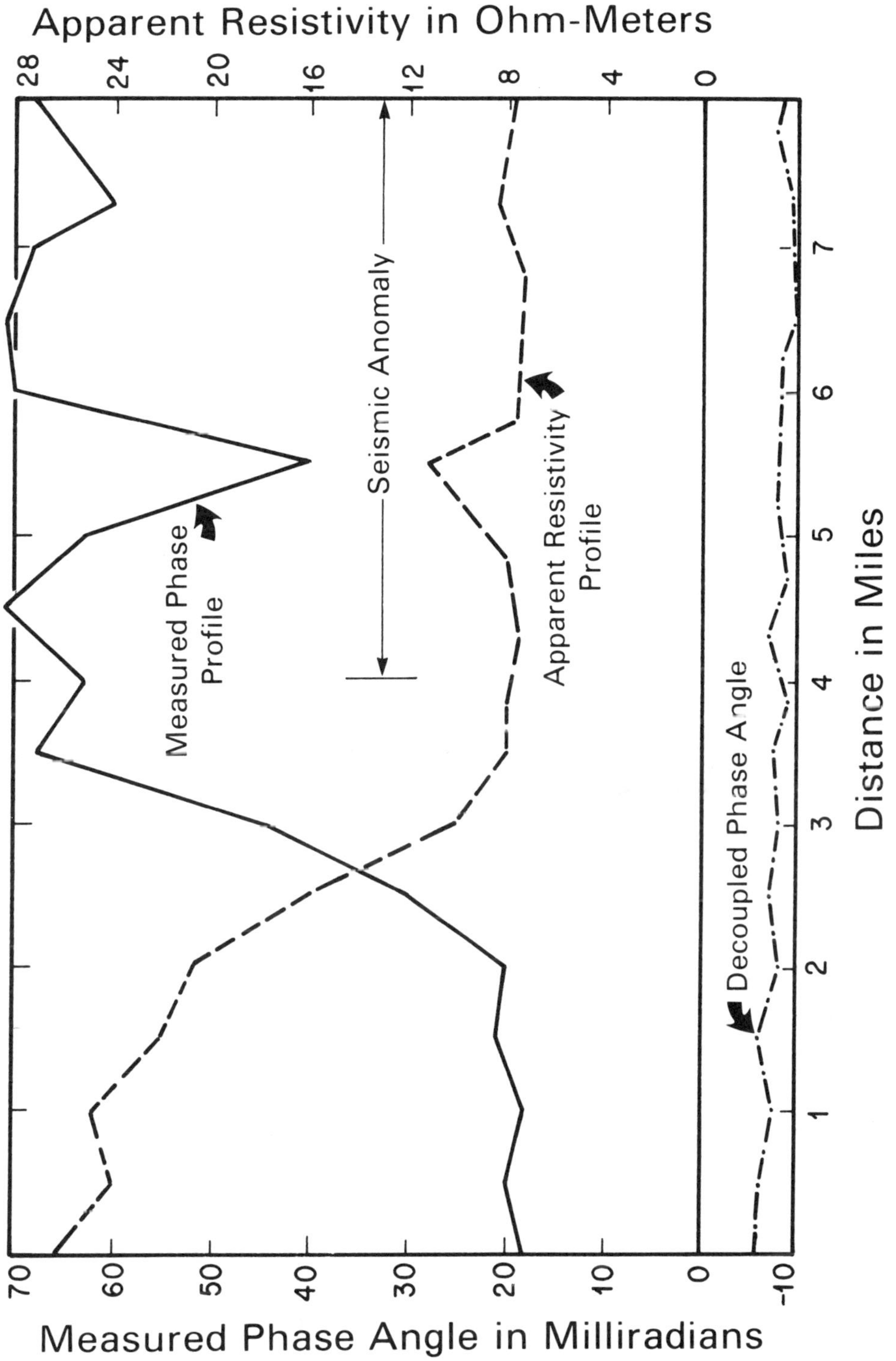

Fig. 3. Electromagnetic (EM) field decoupling.

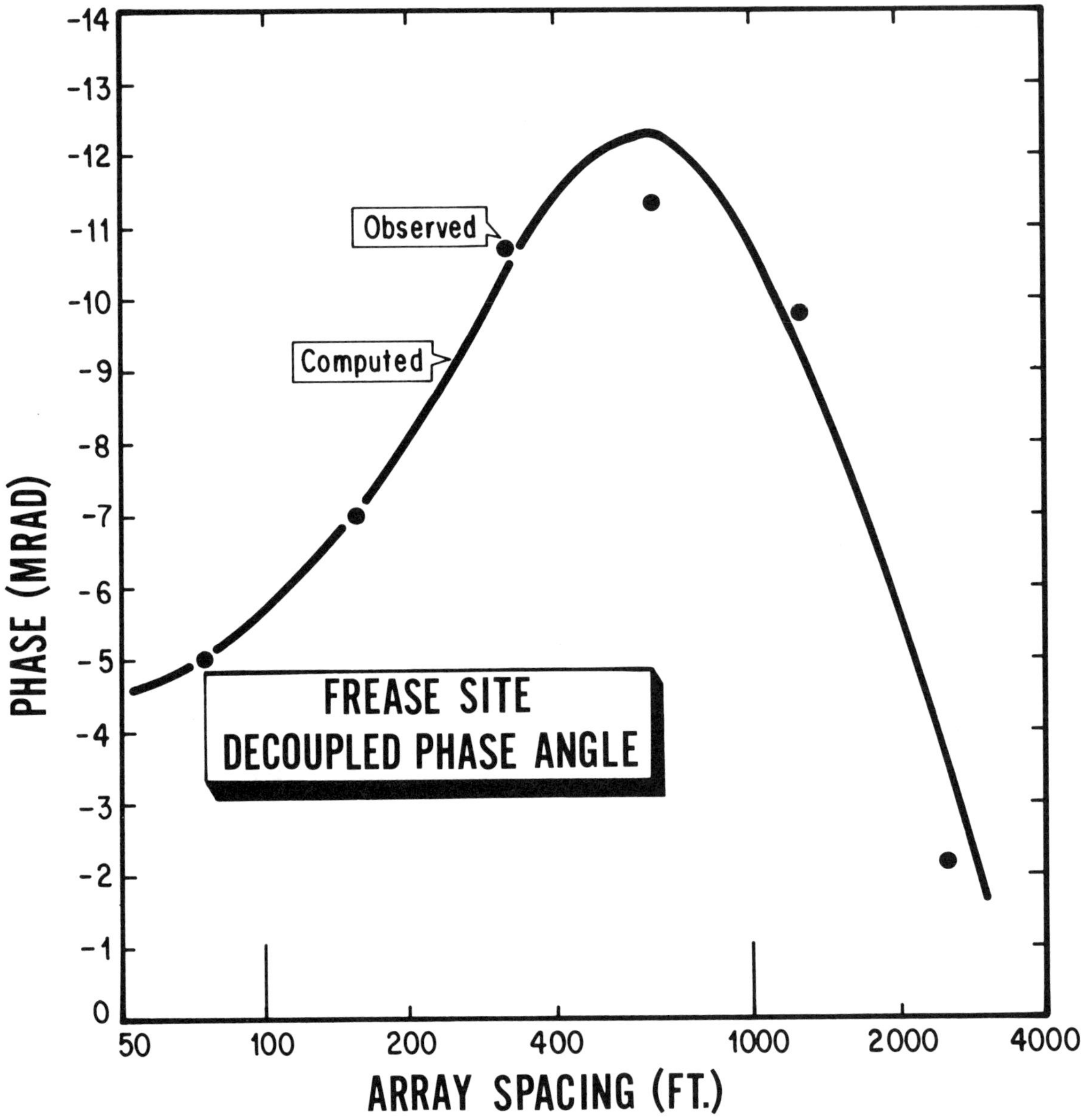

Fig. 4. Effect of variable array spacing.

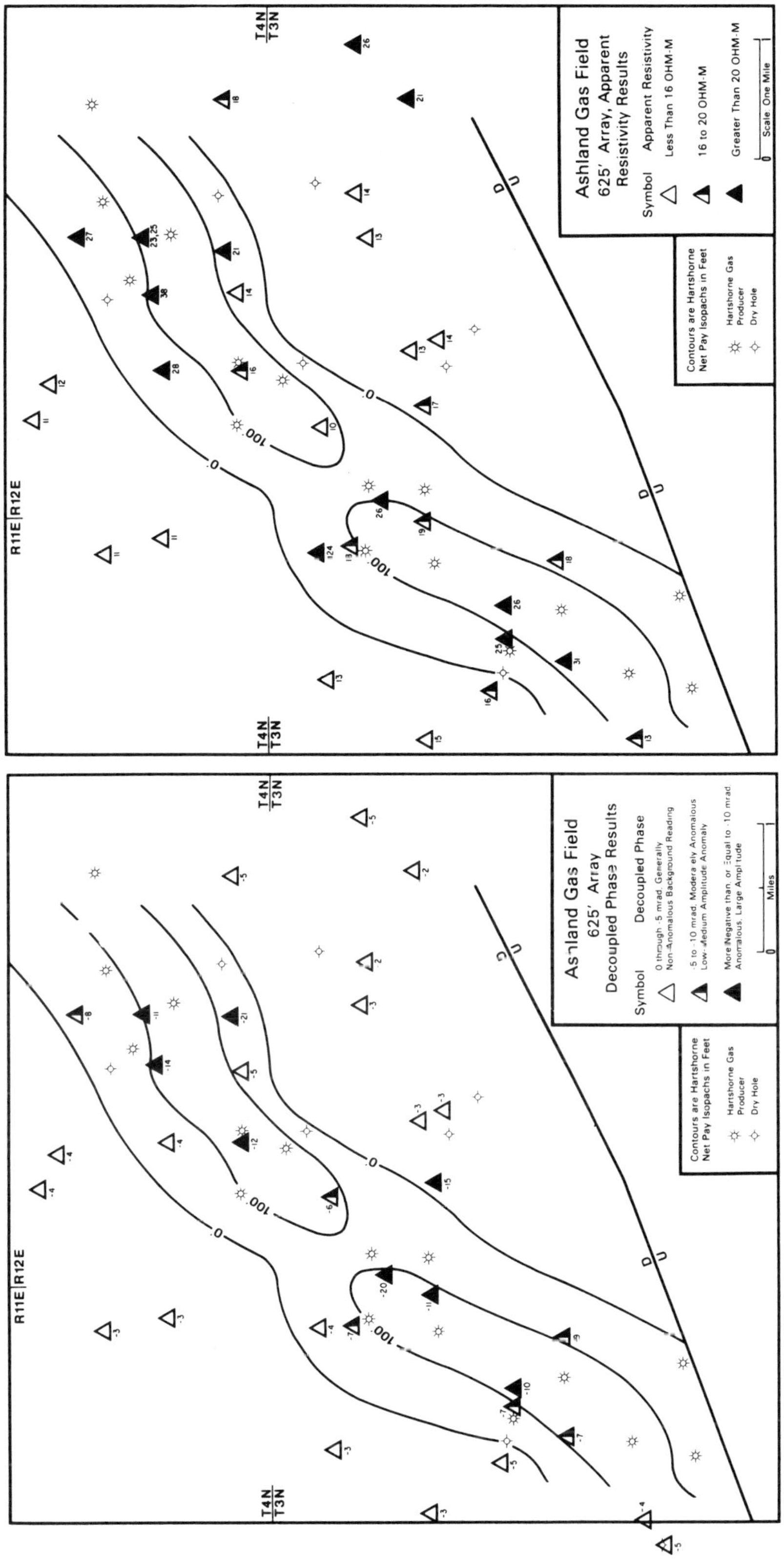

Fig. 5. INDEPTH survey results. Ashland Gas Field, Oklahoma.

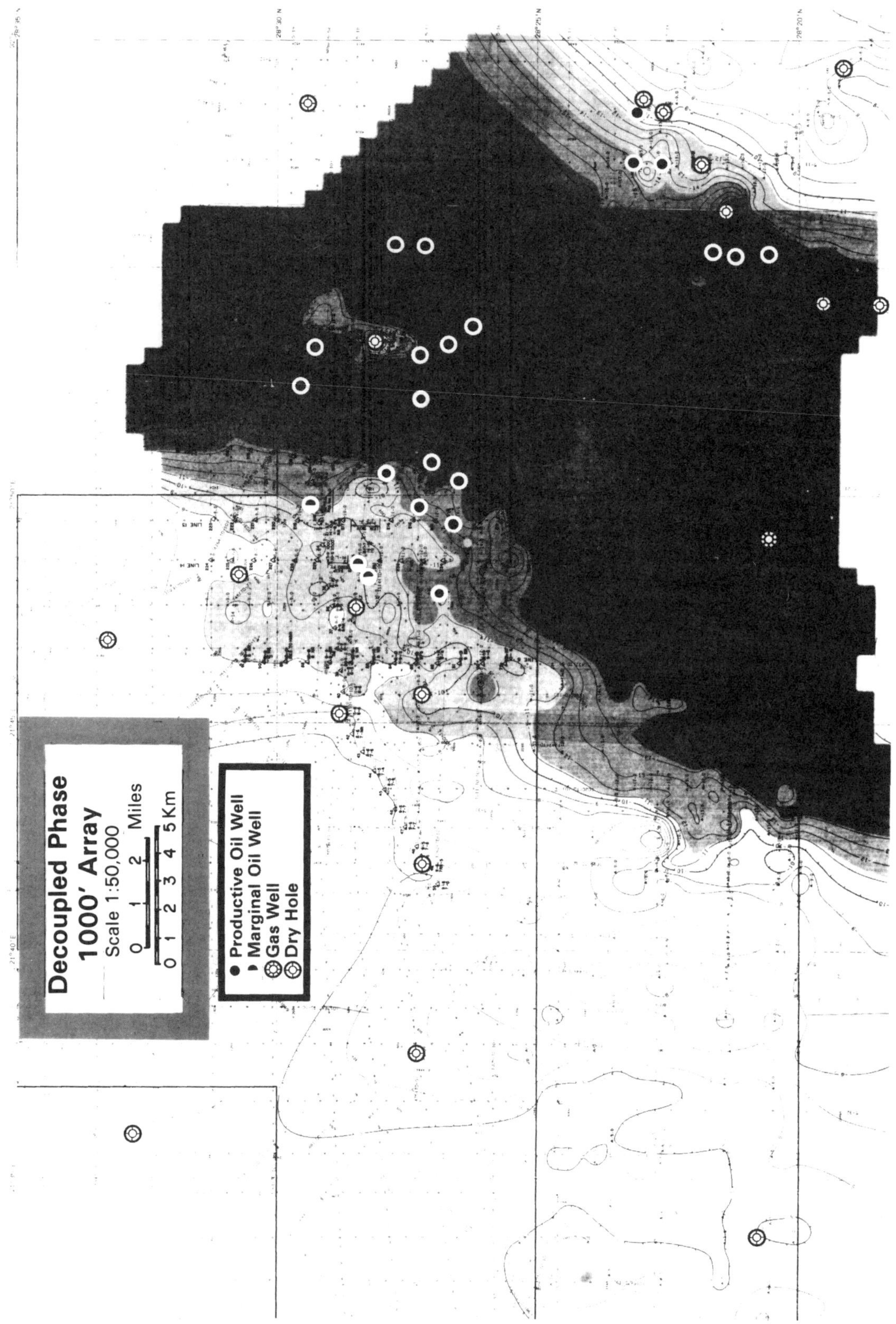

Fig. 6. INDEPTH survey results, North Africa.

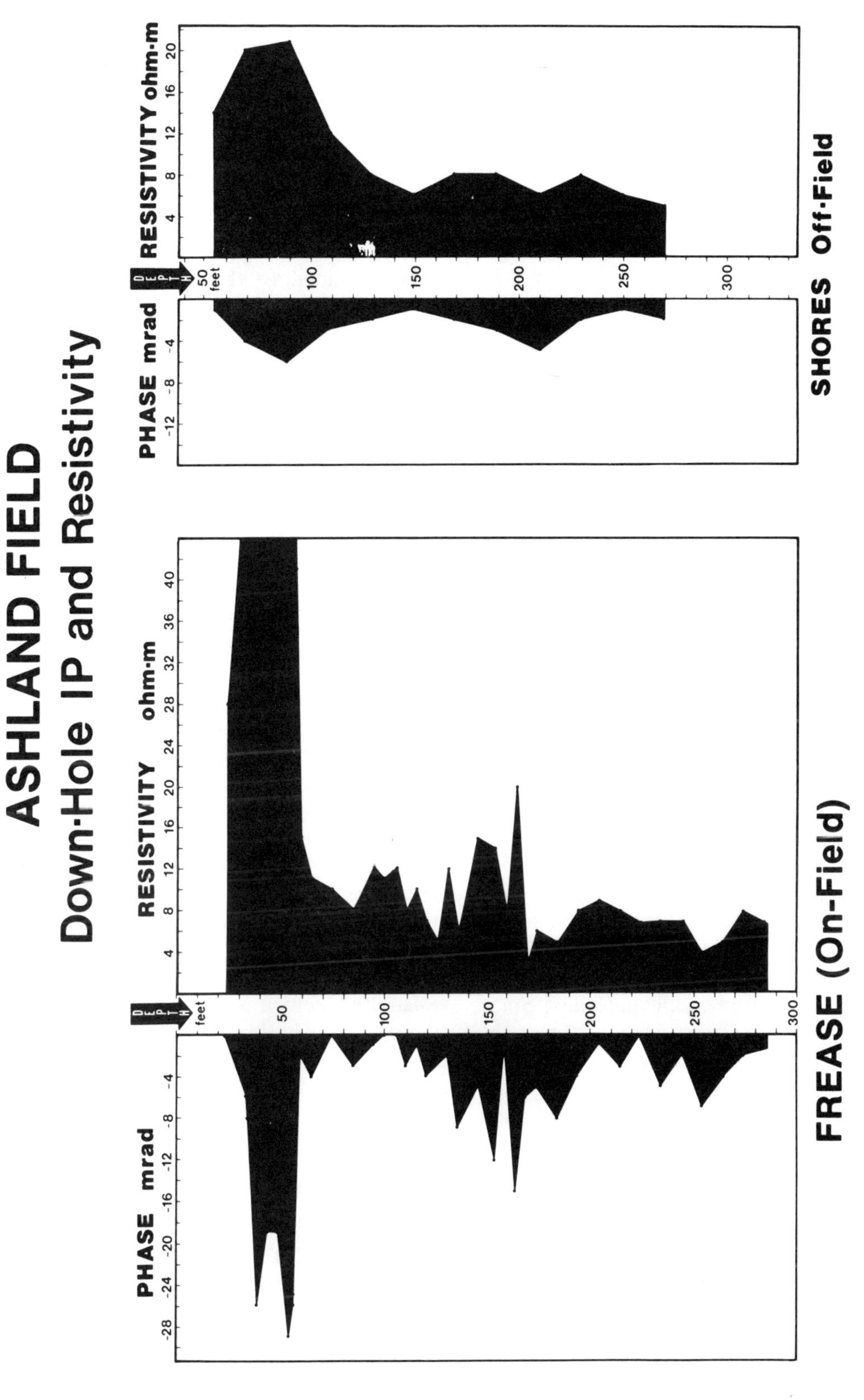

Fig. 7. Downhole IP (Phase) and resistivity at Ashland Field.

ELECTRICAL METHODS FOR HYDROCARBON EXPLORATION
II. MAGNETOTELLURIC (MT) METHOD

Ben K. Sternberg, Paul L. Buller, Jerry L. Kisabeth, and Efrem Mehreteab

CONOCO, INC.

ABSTRACT

The use of electrical methods has increased dramatically in recent years as the search for hydrocarbons has expanded from heavily developed, relatively well-known basins to areas that have not been effectively explored with seismic or other conventional techniques (e.g., seismic no-record (NR) areas, difficult-access areas, and stratigraphic trap prospects).

One such electrical method which is being used increasingly is magnetotelluric (MT) surveying. Recent advances in MT instrumentation and processing have led to greatly improved data quality. MT can be successfully applied to areas where there are large resistivity contrasts (e.g., between conductive clastic rocks and more resistive carbonate or basement rocks) and where there are large structures (i.e., where the structural relief is of the order of 10 percent or more of the depth to the structure). An example from Conoco's MT surveys illustrates the potential of MT for delineation of prospective structures in seismic NR areas.

INTRODUCTION

Figure 1 shows a schematic representation of the field layout for an MT sounding. The electromagnetic signals that are used for MT soundings originate predominantly in the ionosphere at low frequencies (less than about 1 Hz) and from worldwide thunderstorm activity at higher frequencies. This electromagnetic energy impinges on the surface of the earth and is refracted vertically into the earth because of the large contrast in conductivity between air and earth materials. Since the ionosphere is far from the surface of the earth, we may assume that the wavefronts are planar (at least for frequencies of exploration interest). Thus, in the absence of electromagnetic scattering from lateral inhomogeneities, we are dealing with vertically propagating plane waves, a fairly straightforward mathematical problem. The electromagnetic fields observed at the surface of the earth are comprised of both the incident fields and fields induced at depth within the earth.

The surface electric (E) fields are detected with sensors, as shown in Figure 1. These sensors consist of a length of wire (generally 100 to 1,000 ft long) grounded to the earth at each end via a stable, low-noise electrode (e.g., a cadmium rod immersed in $CdCl_2$ solution in contact with the earth through a porous membrane). The potential difference between each pair of electrodes is detected and recorded in the field truck. The associated magnetic (H) fields are detected with sensors such as induction coils or cryogenic (squid) magnetometers.

In the absence of any inhomogeneities (i.e., a homogeneous or layered earth), the ratio of any orthogonal pair of electric and magnetic fields in any horizontal direction (e.g., E_x/H_y or E_y/H_x) will be the same, and the magnetic field in the vertical direction will be zero. In practice, the electric fields are always measured in the x and y directions and the magnetic fields in the x, y, and z directions so that the effect of fields scattered from inhomogeneities can be accounted for.

The ratio of observed orthogonal electric and magnetic fields (for the case of a one-dimensional, layered earth) is known as the impedance. This ratio is independent of the strength of the primary signal and is dependent only on the electrical properties of the earth. It is common to convert these observed ratios to a quantity called "apparent resistivity," which is defined as the resistivity of a homogeneous half space that would give the same observed electric and magnetic field ratio as was actually observed. The fields over a half space are

$$\frac{E_x}{H_y} = (1 - j) \left(\frac{\omega\mu\varrho}{2}\right)^{1/2} \qquad (1)$$

where $\omega = 2\pi f$; f = frequency (hertz); μ = magnetic permeability, usually $\mu = \mu_o = 4\pi \times 10^{-7}$ (henries per meter); ϱ = resistivity (ohm-meters); and $j = \sqrt{-1}$. Note also that the magnetic field strength (B) equals the magnetizing force (H) times the magnetic permeability (μ). If E is in mv/km and B is in gammas (or nT), then solving for ϱ in the above equation, the apparent resistivity is defined as:

$$\varrho a = \frac{0.2}{f} \left(\frac{|E_X|}{|B_Y|}\right)^2 \quad (2)$$

An MT depth sounding is made by recording signals over a wide frequency range. Because of the skin effect, high-frequency EM energy will penetrate only a short distance into a conductor and lower frequencies more deeply. The skin depth is defined as the depth at which the EM fields have decayed in amplitude to 1/e (or 37 percent) of their values at the surface and is given by:

$$\delta = \sqrt{\frac{2\varrho}{\omega\mu}} \quad \text{(meters)} \quad (3)$$

$$\approx 1/2 \sqrt{\frac{\varrho}{f}} \quad \text{(kilometers)} \quad (3a)$$

High-frequency apparent resistivities, therefore, indicate the approximate resistivities of shallow layers. Low-frequency apparent resistivities indicate the approximate resistivities of deeper formations.

INSTRUMENTATION

Figure 2 shows a generalized diagram of one approach to MT data acquisition and analysis—Conoco's research MT system. This system has just recently been built, and the testing and refinement of this system are continuing. The system consists of two data acquisition (or DA) vehicles, each being equipped with the necessary electronics to record three components of the magnetic field with either squid magnetometers or induction coils along with four electric field sensors to record two pairs of orthogonal electric fields. The extra satellite electric field sensor is typically deployed at a distance of a few hundred to a few thousand feet from the main-base electric field sensor. The two DA vehicles may be separated by any distance but are generally 1 to 10 miles apart. Time synchronization between the two vehicles is obtained by signals from WWVB.

A signal analysis (SA) truck is located in the vicinity of both DA trucks and is used to process the MT data. Magnetic tapes are transported from the DA trucks to the SA truck. We have found that the use of telemetry links to transmit the data from one vehicle to another can severely restrict one's flexibility in choosing site locations. Furthermore, transportation of the tapes usually leads to an insignificant delay in comparison with the site occupation time, particularly for a research system. The SA truck is capable of performing all of the MT processing, including remote reference processing (described in a later section), modeling, and generation of final resistivity cross sections.

Figure 3 shows the specifications for the DA unit. We have emphasized very high accuracy in the development of this research system. The channel-to-channel match (which is the most important quantity for E/H measurements) is better than 1 percent and 1°. The stability of the system is better than 0.1 percent and 1 milliradian for at least three hours, i.e., the length of a single data-recording period. This stability is insured, since all the critical analog components are in a temperature-stabilized oven. Although these high accuracies are not normally required for routine MT surveys, they do allow the determination of the resolution that is possible in a given geologic setting. Research is under way to determine what accuracy in the measurements can be effectively exploited. This research includes the use of a 200,000-watt signal source for controlled source measurements with this recording system.

A detailed block diagram of the DA system is shown in Figure 4. An LSI 11/23 microcomputer is used to control all aspects of the data acquisition, and an HP 85 calculator terminal is used to input parameters and print out the status of the system. The DA units are synchronized to each other to within 1 msec by the use of WWVB receivers. Alternatively, very precise timing can be achieved for certain experiments (e.g., controlled source experiments) by an external timing signal from a rubidium clock, which has an accuracy of 1 part in 10^{11}. Data can be stored on either standard 9-track tapes or on small cartridge tapes. A header recorder is placed at the beginning of each data record which contains information on all the equipment and field setup parameters.

Figure 5 shows the details of the analog section of the DA unit. Some parts of this section (particularly the preamp and the separation into a high and low band) were adopted from the University of Texas MT system (Bostick, personal communications, 1981, and Becker, 1980).

The simultaneous acquisition of the high bands with the low band is an important feature of this system. These bands are defined in Figure 5. The minimum expected time to acquire band 1 data is 3 hours. The minimum expected time to acquire band 2 data is 1.7 hours; for band 3, 1 hour; for band 4, 5 minutes; and for band 5, 1 minute, for a total of about 3 hours of high-band data. Since the high-band data and the low-band data are acquired simultaneously, the time required to occupy a site is effectively reduced by a factor of two.

The 11/23 has been programmed in the FORTH language to control all aspects of data acquisition, including all the filter and gain settings. The microprocessor reads the level detectors at each stage before a data run, so it can adjust the gains to optimize the dynamic range and still avoid any chance of saturating an amplifier during the run. It can also adjust the post amp during a run.

Considerable attention has been paid to the dynamic range requirements (e.g., a 16-bit digitizer). The MT power spectral amplitudes frequently span 6 decades (over the frequency range 0.001 to 250 Hz) so that even with a 16-bit digitizer, it is desirable to break up the data acquisition into several bands to insure high-accuracy recordings in the low signal strength parts of the spectrum. A preemphasis filter can also be included prior to the midamp to prewhiten the spectrum (see Fig. 7).

The power for the DA system is the 12-V DC to 110-V AC, 400-Hz inverter. The 400-Hz power source was chosen to avoid beating against the 60 Hz from the power line grid and to remove any direct interference within the band of interest.

A block diagram of the signal analysis system is shown in Figure 6. The basic computer used in this system is the LSI 11/23 microprocessor, but all the time-consuming calculations, (like Fourier transforms) are done in the array processor. This array processor can perform a 2048-point Fourier transform in 10.2 msecs. Data and intermediate calculations can be stored on either floppy disks or an 80-megabyte Winchester disk.

A TI Silent 700 printing terminal is used to enter all the processing parameters and to output status information. The final calculated quantities (such as apparent resistivity versus frequency plots) are displayed on the HP 9845 computer. In addition to the built-in graphics printer, plots can be made on a four-color pen plotter. A WWVB clock, 12-bit digitizer, and 12-bit digital-to-analog converter are also included in the system for special projects.

This computer system is capable of processing all the MT data in the field. In addition, all of the modeling and interpretation can be done on a computer of this size. Having the computer in the field allows one to look at data quality before leaving the site and, furthermore, allows experimentation with various field parameters right on the site without the costly delay of transporting the data back to a central computer facility.

We wish to emphasize that the MT instrumentation described here was intended to comprise a research system. As such, it has more features than would normally be required on a production crew.

DATA PROCESSING

Figure 7 gives an overview of the steps involved in MT data processing. In the upper left is a strip chart record of representative MT signals. These signals are Fourier-transformed to produce the power spectra shown in the upper center of Figure 7. Ratios of orthogonal electric and magnetic fields are then calculated to produce the apparent resistivity versus frequency curves shown on the upper right. The phase differences between electric and magnetic fields are also calculated. Since two sounding curves are obtained, the curve which is based on the electric field component that is parallel to the general geologic (or geoelectric) strike direction is selected for further analysis. The apparent resistivity curve for E parallel component then shows, in a qualitative manner, the variations in resistivity beneath the site (lower right). This apparent resistivity curve is then modeled, or inverted, to produce a resistivity versus depth estimate (lower center). The final step is to produce a geologic interpretation (lower left) from this gross estimate of the resistivities at each depth interval. The next nine subsections describe these steps in greater detail.

MT Signals

The signals shown on the strip chart record in the upper left of Figure 7 are representative of good MT signals. There is a strong correlation between orthogonal pairs of electric and magnetic fields (e.g., E_x and H_y or E_y and H_x). Where an event or burst appears on E_x, it also appears in a similar form on H_y but not necessarily on E_y or H_x. It is important to look occasionally at the incoming data to see if these rough correlations between orthogonal components do exist. If one of the channels appears to be completely uncorrelated with the other channels, an open channel should be suspected. Likewise, if events appear on one channel and not the others, a noisy channel should be suspected. It is possible for all four channels to be essentially uncorrelated in the case of an extremely anisotropic earth. We have found this to be very rare, though, and usually a visual inspection of the correlation between orthogonal channels is a good indication of the data quality.

The magnitude of H_z on this record is small, as one would expect over a predominantly homogeneous or layered earth. If the magnitude of H_z is very large, this may be an indication that the site is set up very close to some inhomogeneity. Such an inhomogeneity may severely complicate the interpretation, and it might be desirable to move the site.

MT Power Spectra

The MT signals are Fourier-transformed to produce the power spectra shown in the upper middle of Figure 7. These spectra are quite similar to MT spectra we have recorded at numerous survey areas all over the United States. The actual level of the signals may vary one to two decades above or below this curve, but we have always found detectable MT signals present, and they have roughly this spectral shape.

Below about 0.1 Hz, the power spectra increase steeply. It is this steep increase and the variability in the amplitudes that impose such stringent requirements on the dynamic range of the recording system. The midband range (from about 0.1 Hz to 10 Hz) has quite low signal levels; thus MT data in this band are often relatively poor. The Conoco research MT system has been designed with very low instrument noise levels so that high-quality data can be obtained in the low-signal midband.

MT Impedance, Apparent Resistivity, and Phase

The details of the derivation of the MT impedance calculations are given in a number of references such as Sims and Bostick (1969); Swift (1962); Vozoff (1972); and Word, Smith, and Bostick (1970). The following is a brief summary of these results.

Since, in general, the measured apparent resistivity will vary with direction, the simple expression (1) relating the E and H fields must be extended. In the general case, E_x is related not only to H_y, but also to H_x. Likewise, E_y is related to both H_x and H_y. Therefore:

$$E_x = Z_{xx} H_x + Z_{xy} H_y \quad (4a)$$

$$E_y = Z_{yx} H_x + Z_{yy} H_y \quad (4b)$$

The impedances Z_{xx}, Z_{xy}, Z_{yx}, and Z_{yy} are complex. Like the simple impedance in the homogeneous case ($Z = E/H$), these impedances are independent of the strength of the incident field and are dependent only upon the earth's electrical properties.

Since the desired information is in the Z's, we need to solve for each of the Z_{ij}. At first, it would appear that we have two equations with four unknowns. Since we have more observed field quantities than unknowns, we can use this information by multiplying Equations 4a and 4b by the complex conjugate of each of the fields

$$\langle E_x A^* \rangle = Z_{xx} \langle H_x A^* \rangle + Z_{xy} \langle H_y A^* \rangle \quad (5a)$$

$$\langle E_x B^* \rangle = Z_{xx} \langle H_x B^* \rangle + Z_{xy} \langle H_y B^* \rangle \quad (5b)$$

$$\langle E_y A^* \rangle = Z_{yx} \langle H_x A^* \rangle + Z_{yy} \langle H_y A^* \rangle \quad (5c)$$

$$\langle E_y B^* \rangle = Z_{yx} \langle H_x B^* \rangle + Z_{yy} \langle H_y B^* \rangle \quad (5d)$$

where A^* and B^* are the complex conjugates of any two of H_x, H_y, E_x, or E_y, and the brackets denote frequency band averages. This procedure yields more possible equations than unknowns. Since each of the solutions to these equations responds differently to noise on any one of the field components, it is customary to discard some of the solutions and average others to obtain the best estimate.

Solving for the Z_{ij}'s (e.g., Vozoff, 1972), we obtain:

$$Z_{xx} = \frac{\langle E_x A^* \rangle \langle H_y B^* \rangle - \langle E_x B^* \rangle \langle H_y A^* \rangle}{\langle H_x A^* \rangle \langle H_y B^* \rangle - \langle H_x B^* \rangle \langle H_y A^* \rangle}; \quad (6a)$$

$$Z_{xy} = \frac{\langle E_x A^* \rangle \langle H_x B^* \rangle - \langle E_x B^* \rangle \langle H_x A^* \rangle}{\langle H_y A^* \rangle \langle H_x B^* \rangle - \langle H_y B^* \rangle \langle H_x A^* \rangle}; \quad (6b)$$

$$Z_{yx} = \frac{\langle E_y A z^* \rangle \langle H_y B^* \rangle - \langle E_y B^* \rangle \langle H_y A^* \rangle}{\langle H_x A^* \rangle \langle H_y B^* \rangle - \langle H_x B^* \rangle \langle H_y A^* \rangle}; \quad (6c)$$

$$Z_{yy} = \frac{\langle E_y A \rangle \langle H_x B^* \rangle - \langle E_y B^* \rangle \langle H_x A^* \rangle}{\langle H_y A^* \rangle \langle H_x B^* \rangle - \langle H_y B^* \rangle \langle H_x A^* \rangle}. \quad (6d)$$

Normally, the measurement axes are laid out in some predetermined manner (e.g., x = north-south, y = east-west) so that the measurement coordinate axes will be at some arbitrary angle to the regional geoelectric strike. In general, then, it will be necessary to rotate mathematically the impedances found from Expression 6 to a new coordinate system. The rotated impedances are

$$2Z'_{xx}(\Theta) = (Z_{xx} + Z_{yy}) + (Z_{xx} - Z_{yy}) \cos 2\Theta + (Z_{xy} + Z_{yx}) \sin 2\Theta) \quad (7a)$$

$$2Z'_{xy}(\Theta) = (Z_{xy} - Z_{yx}) + (Z_{xy} + Z_{yx}) \cos 2\Theta - (Z_{xx} - Z_{y}) \sin 2\Theta \quad (7b)$$

$$2Z'_{yx}(\Theta) = (Z_{xy} - Z_{yx}) + (Z_{xy} + Z_{yx}) \cos 2\Theta - (Z_{xx} - Z_{yy}) \sin 2\Theta \quad (7c)$$

$$2Z'_{yy}(\Theta) = (Z_{xx} + Z_{yy}) - (Z_{xx} - Z_{yy}) \cos 2\Theta - (Z_{xy} + Z_{yx}) \sin 2\Theta$$

where Θ is the rotation angle.

The usual procedure is to rotate the impedances to the principal directions, defined as the directions that give Z'_{xy} and Z'_{yx} their largest and smallest values, respectively, and where Z'_{xx} and Z'_{yy} are minimized. In the case of a perfectly one- or two-dimensional earth, Z_{xx} and Z_{yy} will be zero. Z_{xx} and Z_{yy} are typically very small after rotation; that is, the earth is often close to locally one- or two-dimensional.

Now that the impedances in the principal directions have been determined, it is customary to convert these to apparent resistivities, as was done in Equation 1.

$$\rho_{xy} = \frac{0.2}{f} \left| Z'_{xy} \right|^2 \quad (8a)$$

$$\rho_{yx} = \frac{0.2}{f} \left| Z'_{yx} \right|^2 \quad (8b)$$

Several other important quantities are calculated from this analysis. First, the phase is used for modeling in combination with the apparent resistivity. This phase is simply the phase component of the complex Z's that were calculated in Equations 5, 6, and 7. The rotation angle (Θ) in Equation 7 gives information about the regional geologic strike. Coherencies between the measured E and H fields or coherencies between the measured E field and the E field predicted from Equation 4 using the measured H fields and the calculated Z's can also be calculated. The coherency is a more rigorous and quantitative measure of data quality, but it is similar in kind to the discussion of data quality in the preceding section on MT signals. Namely, the E and H signals will be correlated if we are seeing genuine MT signals from a distant source and not some local noise.

Remote Reference Processing

A significant enhancement to MT soundings involves the use of a remote reference (e.g., Gamble et al., 1979). The expressions in Equation 5 contain auto and cross powers of the various field components. In the presence of local noise sources (e.g., noise in the recording system instruments or power lines and motion of the magnetometer caused by winds or industrial equipment), these auto and cross powers will be contaminated by noise.

In turn, this contamination may bias the impedance estimates up or down, depending on which set of equations from Equations 5a through 5d is used.

If, instead of multiplying Equation 4 by a local field, we multiply by the magnetic fields at a distant site (H_{xr} and H_{yr}), we obtain, instead of Equation 6, the following expressions for the Z_{ij}

$$Z_{xx} = \frac{(\overline{E_x H_{xr}^*}\;\overline{H_y H_{yr}^*} - \overline{E_x H_{yr}^*}\;\overline{H_y H_{xr}^*})}{(\overline{H_x H_{xr}^*}\;\overline{H_y H_{yr}^*} - \overline{H_x H_{yr}^*}\;\overline{H_y H_{xr}^*})}, \quad (9a)$$

$$Z_{xy} = \frac{(\overline{E_x H_{yr}^*}\;\overline{H_x H_{xr}^*} - \overline{E_x H_{xr}^*}\;\overline{H_x H_{yr}^*})}{(\overline{H_x H_{xr}^*}\;\overline{H_y H_{yr}^*} - \overline{H_x H_{yr}^*}\;\overline{H_y H_{xr}^*})}, \quad (9b)$$

$$Z_{yx} = \frac{(\overline{E_y H_{xr}^*}\;\overline{H_y H_{yr}^*} - \overline{E_y H_{yr}^*}\;\overline{H_y H_{xr}^*})}{(\overline{H_x H_{xr}^*}\;\overline{H_y H_{yr}^*} - \overline{H_x H_{yr}^*}\;\overline{H_y H_{xr}^*})}, \quad (9c)$$

$$Z_{yy} = \frac{(\overline{E_y H_{yr}^*}\;\overline{H_x H_{xr}^*} - \overline{E_y H_{xr}^*}\;\overline{H_x H_{yr}^*})}{(\overline{H_x H_{xr}^*}\;\overline{H_y H_{yr}^*} - \overline{H_x H_{yr}^*}\;\overline{H_y H_{xr}^*})}, \quad (9d)$$

where the bar denotes an average over a frequency window as well as over all data sets.

The expressions in Equations 9a through 9d involve only cross powers between the base and the remote stations. If the noise at the base station is not correlated with the noise at the remote reference station, and if a sufficient number of data sets are averaged, these impedance estimates will be unbiased by noise. Furthermore, since Equations 4a and 4b were multiplied in turn by a single reference field, the values of the impedance elements are independent of the magnitudes and phases of the reference fields and of the resistivity structure at the reference site. Further details on the remote reference processing method are given in Goubau et al. (1978); Gamble (1978); Gamble et al. (1979a); and Gamble et al. (1979b).

EFFECT OF INHOMOGENEITIES

As can be seen from the upper right plot in Figure 7 and from the discussion on MT impedance, apparent resistivity, and phase, we end up with two MT sounding curves at each site. If the MT sounding sees only two-dimensional inhomogeneities (i.e., lateral changes in resistivity but with long strike length), and if the impedances have been rotated to the principal directions, then one of the apparent resistivity curves will be parallel to the strike direction and one perpendicular to it.

Determination of which of these two apparent resistivity curves is the parallel curve and which one is the perpendicular curve is made by using the vertical component of the magnetic field (H_z). The vertical magnetic field can be written as a linear combination of H_x and H_y (in analogy with Equation 4).

$$H_z = AH_x + BH_y \quad (10)$$

We may solve for A and B in a manner similar to the procedure used for solving for the Z's (Equations 5 and 6). The strike direction Φ (in degrees from the x axis) is then given by

$$\Phi = \frac{(a_r^2 + b_r^2)\arctan(b_r/a_r(+ (a_i^2 + b_i^2)\arctan(b_i/a_i)}{(a_r^2 + a_i^2 + b_r^2 + b_i^2)} \quad (11)$$

where a_r, a_i, b_r, and b_i are the real and imaginary parts of A and B (Jupp and Vozoff, 1976).

If the rotation angles for the impedance tensor and the angle Φ in Equation 11 (which is generally known as the tipper angle) are approximately the same, the parallel-to-strike component is ϱ_{xy}. If they differ by approximately 90°, ϱ_{yx} is parallel to strike.

The behavior of the parallel and perpendicular components of apparent resistivity is shown in Figure 8. This is a two-dimensional model (i.e., infinite strike direction in and out of the page) and is adopted from Lee et al. (1978). This model is representative of a typical lateral inhomogeneity. The resulting apparent resistivity curves parallel and perpendicular to strike are shown in the upper left for a site (Location A) which is 3 km to the left of the right edge of the inhomogeneity and in the upper right of the figure for a site (Location B) 3 km to the right of the inhomogeneity. Also plotted on these figures is the apparent resistivity curve for a one-dimensional model based on the resistivity structure directly beneath the site. At both of these locations, the parallel-to-strike curve follows quite closely the one-dimensional apparent resistivity curve.

The procedure is then as follows: (1) determine the E parallel-to-strike component from the rotation and tipper angles, and (2) perform a layered (one-dimensional) interpretation on the parallel component.

Two-dimensional model calculations are generally made using finite-difference or finite-element mesh methods (e.g., Jones and Pascoe, 1971; Pascoe and Jones, 1972; and Stodt, 1978). In general, it is difficult to use two-dimensional model calculations routinely for production surveys, since the calculations tend to be quite time-consuming. Furthermore, there is so much flexibility in defining the resistivity distribution of the cells in the mesh that unless there is some additional control, it may be difficult to derive a realistic model of the geology. We have found that applying one-dimensional models to just the parallel component is useful to show the resistivity structure and can be automated for efficient use in routine production surveys. An important research area in MT should involve further study of two- and three-dimensional effects.

ONE-DIMENSIONAL (LAYERED EARTH) MODELS

Now that we have selected the parallel-to-strike apparent resistivity component, we apply a one-dimensional analysis to the data to derive the earth resistivity structure beneath the site. Figure 9 shows a resistivity model (left side) which is representative of the case history area discussed below, Carta Primitive.

The middle plot in Figure 9 is a schematic illustration of the skin effect (Equation 3). The 100-Hz signal has a skin depth of only ~900 feet. The ratio of electric to magnetic fields at 100 Hz will therefore be most sensitive to the resistivity of only the first layer in this model. At 1 Hz, the skin depth is ~9,000 feet, indicating the resistivity of the third layer. The result of this depth sounding is illustrated on the right-hand plot in Figure 9. At high frequencies (around 100 Hz), the apparent resistivity curve asymptotically approaches the true resistivity of the first layer. At intermediate frequencies (around 1 Hz), the apparent resistivity approaches the true resistivity of the second layer. Finally, at frequencies below 0.01 Hz, the apparent resistivity curve asymptotically approaches the third-layer resistivity.

MODELING (INVERSION) PROCEDURE

At present, there is no reliable and exact direct inverse solution for the MT sounding problem. We therefore make successive-forward calculations (i.e., given a model, what are the calculated apparent resistivity and phase curves?) and iterate on this process until the calculated and observed apparent resistivity and phase curves agree. The "best-fit" model is then used as an estimate of the resistivity structure for that site.

The equation for the forward calculation of the MT fields can be derived by a fairly straightforward application of Maxwell's equations and the boundary conditions at each layer. This application leads to the following set of recursive equations (e.g., Ward, 1967):

We first compute the impedance at the top of the first layer above the homogeneous half-space at the bottom of the model (note that the parameters in this equation are defined below).

$$\hat{Z}^n = Z^n \frac{Z^{n+1} + Z^n \tanh(jk^n h^n)}{Z^n + Z^{n+1} \tanh(jk^n h^n)} \quad . \text{(12a)}$$

Once this impedance is computed, we then use it as the terminating impedance of an equivalent homogeneous half-space and write for the impedance at the top of the (n − 1)th layer:

$$\hat{Z}^{n-1} = Z^{n-1} \frac{\hat{Z}^n + Z^{n-1} \tanh(jk^{n-1}h^{n-1})}{Z^{n-1} + \hat{Z}^n \tanh(jk^{n-1}h^{n-1})} \quad \text{(12b)}$$

and so on for each of the layers up to the surface, where:

$$\hat{Z}^1 = Z^1 \frac{\hat{Z}^2 + Z^1 \tanh(jk^1h^1)}{Z^1 + \hat{Z}^2 \tanh(jk^1h^1)} \quad \text{(12c)}$$

where $\hat{Z}^i$ is the impedance at the top of the ith layer; Z^i = characteristic impedance of the ith layer = $\omega\,\mu/k^i$; k^i = propagation constant of ith layer = $\sqrt{j\,\sigma^i\,\mu\omega}$;$\omega = 2\pi$ x frequency; $j = \sqrt{-1}$; σ^i = conductivity of ith layer = 1/resistivity; μ = magnetic permeability, usually taken equal to free space value; $\mu_o \times 4\pi \times 10^{-7}$; and h^i = thickness of ith layer.

Also, apparent resistivity = $\rho a = \frac{1}{\mu\omega} \left|\hat{Z}^1\right|^2$

We have found that it is convenient to use many layers in the model (far more than can actually be resolved from the data) to produce a quasi-continuous model. An example of the layers that were used for the case history that follows is shown in Figure 10. The thickness of each of these layers is fixed, but the resistivity is allowed to vary.

Marquardt's algorithm (Marquardt, 1963) was used to find the "best-fit" resistivities. Our adaptation of the Marquardt algorithm uses successive linear approximations to the nonlinear MT field equation (12), combined with successive projections along a gradient vector, to find models which minimize the sum of squares differences between the observed and calculated MT apparent resistivities and phases. Wu (1968) describes in further detail use of the Marquardt algorithm for finding resistivity models that fit MT data. Oldenburg (1979) uses a continuous resistivity distribution to model the resistivity structure using the Backus-Gilbert theory. Hobbs (1982) also used a continuous resistivity distribution and fitted it to the observed data in steps as we have done here.

An example of the use of our program is shown in Figure 11. The three-layer model shown by the dashed line in the bottom plot of Figure 11 was used to calculate the apparent resistivity and phase curves in the middle and top plots of Figure 11. These curves were then used as input to the inversion program. The resulting best-fit model is shown by the smooth curve at the bottom of Figure 11.

If we had perfect data at all frequencies, we could use this procedure to invert the apparent resistivity and phase curves to a model with an arbitrary number of layers and with perfect resolution. Unfortunately, with even a small amount of noise on the observed data, the range of models that will fit within the scatter of the data is very large. Indeed, even with noise-free synthetic data, the limited computer word length prevents a really precise inversion to the original model (see the example in Fig. 11).

Modeling experiments with a typical scatter in the data have led to the following rough rules of thumb:

1. It is usually possible to resolve only about three to five distinct layers from an MT sounding.
2. To be detectable, the thickness of a layer should be about 10 percent of its depth of burial or more.
3. The resistivity contrast of a layer should be at least a factor of 2 and preferably an order of magnitude for the layer to be detectable.
4. The combination of resistivity and thickness is crucial for the detection of a layer. A large resistivity contrast may lead to a relatively thin layer being detectable. Likewise, a very thick layer may be detectable even if it has a fairly small resistivity contrast.
5. When profiling the depth to a thick, deep layer (e.g., depth to basement), the accuracy of the depth estimates is on the order of 10 percent.
6. The resolution of MT varies roughly as the logarithm of the depth. As a result, the inversion curves are plotted on a logarithmic depth scale to display essentially equal resolution in depth.

In order to avoid letting this inversion procedure produce more resistivity variations than can really be justified by the data, a procedure similar to Hobbs's (1982) was used.

1. Begin with an initial model which has a uniform resistivity over the entire half space. Divide all the apparent resistivities and phases by 2 so that a fit is being made to curves with less severe variations.
2. Use the resulting model from Step 1 to fit the actual observed data.
3. Go back to the half-space model as an initial model, and fit it to the smooth apparent resistivity and phase curves that resulted from Step 2.
4. The convergence criteria at each of these steps must be no more stringent than can be justified by the scatter and uncertainty of the data.

Comparison of MT Contractors

Figure 12 shows a comparison of the MT measurements made by three different contractors at the same location. Contractor No. 1 used standard single-site processing. Contractors Nos. 2 and 3 used remote reference processing. Each contractor used different electric field line lengths, but the center of the electrode array was at approximately the same location.

The component of the apparent resistivity that is parallel to strike (shown by the solid line) has almost an identical shape in each of the three plots. However, the position of the apparent resistivity curve is different in each case. The apparent resistivity curves appear to have been displaced parallel to the apparent resistivity axis. Furthermore, the split between the two components varies in each of the three examples. The parallel shift of the apparent resistivity curves is an effect that appears, at least to some extent, in most of the MT surveys that we have looked at.

Figure 13 shows a further comparison of MT measurements at three locations. The same three contractors are compared at each of these sites. The smooth lines drawn on these plots are the best-fit curves to the original scattered data. At each of the three sites, the various recordings produced the same shaped curve, but there is a parallel shift between each recording. In some cases, one contractor will be higher than another contractor or the theoretically calculated curve based on the well log; in other cases, he will be lower. Therefore, there does not seem to be any evidence that this parallel shift is due to consistent errors in equipment calibration.

The parallel shift does not appear to be due to bias errors from noise, since at Site 12, curves referenced first to Site 13 and then to Site 10 are essentially the same. Furthermore, longer recording times did not shift the curves at these sites; they simply reduced the scatter. At Site 13, after the conventional run was made with the E field lines oriented north-south and east-west, the entire array (including magnetic sensors) was rotated 20°. The resulting curve shape was almost identical, but the curve was shifted parallel to the first one. At Site 12, the same contractor was sent back to occupy the site, but with the E field lines displaced slightly. Again, the same-shape curves result but hava a parallel shift. The MT curves appear to be very sensitive to the precise location of the electric field lines. The most likely explanation for the different results from the three contractors was the fact that each had different electrode line lengths and therefore slightly different locations.

A possible explanation for the parallel shifts of the apparent resistivity curves is related to the heterogeneity of the surface layer. The magnetic fields that are induced in the earth are induced over a radius of the order of a skin depth and will, in a sense, integrate all the minor resistivity variations over this radius. Likewise, the currents induced in the earth will be induced over an area something like a skin depth in radius. But the measurement of the electric fields is a very local measurement which depends on the resistivity in the immediate vicinity of the electrode array ($E = J\varrho$). The ϱ in this equation is the local surface resistivity, which we expect to vary rapidly. An electrode array set up at one location often sees a different surface ϱ than an array set up a short distance away. The observed electric fields at all frequencies will then be different by a factor of $\varrho 1/\varrho 2$ at the two locations, and the apparent resistivity curves will differ by $(\varrho 1/\varrho 2)^2$. When these two curves are plotted on a logarithmic scale, they will differ by a fixed parallel shift. Further study of the parallel shifts of the apparent resistivity curves will be made with our research MT system.

Area-Averaged Apparent Resistivities

The parallel shifts in the apparent resistivity curve can greatly distort the interpreted cross section. Not only will the modeled resistivities be off by a factor equal to the amount of parallel shift, but also the depths will be off by an amount related to the square root of the shift, since the skin depth will be relatively deep in a higher-resistivity material, but shallower in lower-resistivity material. It is necessary, therefore, to try to remove these parallel shifts.

Figure 14 shows a plot of the average apparent resistivities at each site along a profile line. This profile is from the case history that will be discussed in the next section. The averages were calculated as the arithmetic mean of the logarithm of the apparent resistivities (at all frequencies, in this particular case). Random jumps in the averages are seen all along this profile line. We know from well logs that the general resistivity of the sedimentary layers is relatively consistent in this area. Therefore, we have calculated an area average apparent resistivity for each site by weighting sites surrounding that site in an average with a Hanning window. The total length of the Hanning window in this case is two-thirds the profile length. The area average apparent resistivity is plotted at each site as the short horizontal bar. When the area average has been found, the logarithms of the apparent resistivities of each frequency for that site are shifted by the difference of (area average - site average).

It must be emphasized that in order to use this area average effectively, there must be relatively dense sampling of the electric field along the profile line. In areas where the structures of interest are changing rapidly, it will be necessary to have more sites per mile and a shorter window than in this example.

CASE HISTORY—CARTA PRIMITIVE

LOCATION AND GENERAL GEOLOGY

Figure 15 shows a location map for this survey area, which is known as Carta Primitive. It is located approximately 40 mi northeast of Del Rio, Texas. The topography is quite rugged, with somewhat limited access in parts of the survey area.

The survey area is in the Val Verde Basin. The surface formation is high-resistivity Edwards Limestone with highly variable thickness but averaging about 1,000 ft. This surface limestone has made most of the area historically seismic no-record (NR). Below the Edwards Limestone are low-resistivity sands and shales with thickness varying from a few thousand feet to 15,000 ft or more. Below the sand and shales is high-resistivity Pennsylvanian and older Paleozoic carbonate (including Ellenburger), which can be a potential reservoir where it has been fractured and where it has been upthrown so that it is in contact with good source rocks. There may be some relatively thin conductive sands below the Ellenburger, but they usually cannot be resolved with MT. Below these sands are high-resistivity, crystalline Precambrian rocks. The exploration objective is to locate block faulting and map the top of the Ellenburger carbonate. Sixteen sites were recorded by a contractor across this prospect with approximately 1-mi site spacing.

RESISTIVITY CROSS SECTIONS

The inversion procedure that was described in the section on modeling (inversion) prodecure (p. 222) was applied to each of the sites from the Carta Primitive survey. The local one-dimensional inversions were plotted beneath each site, and the resistivities were then contoured (Fig. 16).

The area-averaged apparent resistivities (see p. 222) are shown in Figure 17. Some of the artifacts that appeared in the contoured cross section of Figure 16 have been removed in the area-averaged section. In the example inversion that was shown in Figure 11, the step increase in resistivity occurred at the point where the quasi-continuous inversion had increased to about 100 ohm-m. Further model studies have shown that on the average, the 100-ohm-m point on the inversion curve indicates the approximate depth to the resistivity increase. Therefore, we have used the 100-ohm-m contour line on Figure 17 to indicate the top of the resistive carbonate.

The short horizontal line below the Gulf-Anderson-Dunham well indicates the top of the high-resistivity Ellenburger carbonate that was actually encountered in the well; however, the Conoco-Bishop never did encounter the carbonate. The MT-interpreted top of the carbonate, as shown by the 100-ohmmeter contour line, is in agreement with the well information.

On the basis of this MT survey, a prospective upthrown block was predicted to occur between Sites 20 and 18. A location at Site 13 was recommended, and a depth to the resistive carbonate was predicted, before the well was drilled, at about 7,500 ft ± 10 percent. The Doran well at Site 13 actually encountered the carbonate at a depth of 8,200 ft. The fractures in the well core were tightly cemented, so the well was a dry hole. Nevertheless, this survey does demonstrate that MT can map prospective structures in this seismic NR area.

COMPARISON WITH WELL LOGS

Figure 18 shows a comparison of the MT inversions with the three well logs. In each case, the agreement between the well logs and the inversions is very good.

It should be emphasized that MT can map only resistivity variations. These resistivity variations may not always correlate with lithologic boundaries. For instance, in the Bishop well the resistivity increases, even though the deep Ellenburger carbonate was not encountered. The high resistivity of these sands and shales may be due to carbonate cements. It is fairly common in this area to find medium-resistivity sands and shales filling the deeper parts of the downthrown areas but low-resistivity sands and shales overlying this and also covering the upthrown basement blocks. The total depth of the downthrown areas may be underestimated somewhat, but the prospective upthrown blocks will generally be correctly mapped, and the tops of the blocks can be correctly estimated.

ACKNOWLEDGEMENTS

Many people in the Exploration Research Division and in the Exploration division offices have contributed to the development of MT at Conoco. In particular, individuals who have been directly associated with the MT project during the early development discussed in this paper are Dhari Bahjat, Alicia Boan, Angie Chapman, Jan Sensintaffar, Bob Thomas, and Jerry Ware.

REFERENCES

Becker, J.D, 1980, A multi-site magnetotelluric measurement system with real time data analysis, Ph.D. thesis, University of Texas, Austin.

Gamble, T.D., 1978, Remote reference magnetotellurics with squids, Ph.D. thesis, University of California, Berkeley; also, Lawrence Berkeley Laboratory Report No. LBL-8062.

Gamble, T.D.; Goubau, W.M.; and Clarke, J., 1979a, Magnetotellurics with a remote magnetic reference, Geophysics, v. 44, no. 1.

Gamble, T.D.; Goubau, W.M.; and Clarke, J., 1979b, Error analysis for remote reference magnetotellurics, Geophysics, v. 44, no. 5, p. 959-968.

Goubau, W.M.; Gamble, T.D.; and Clarke, J., 1978, Magnototelluric data analysis, removal of bias, Geophysics, v. 43, no. 6, p. 1, 157-1, 166.

Hobbs, B.A., 1982, Automatic model for finding the one-dimensional magnetotelluric problem, Geophysical Journal of the Royal Astronomical Society, v. 68, p. 253-264.

Jones, F.W., and Pascoe, L.J., 1971, A general computer program to determine the perturbation of alternating electric currents in a two-dimensional model of a region of uniform conductivity with an embedded inhomogeneity, Geophysical Journal of the Royal Astronomical Society, v. 24, p. 3-30.

Jupp, D.L., and Vozoff, K., 1976, Discussion of "The magnetotelluric method in the exploration of sedimentary basins" by Keeva Vozoff, Geophysics, v. 41, no. 2, p. 325-328.

Lee, K.H.; Labson, V.; Wilt, M.; and Goldstein, N., 1978, Catalogue of magnetotelluric apparent resistivity pseudosections over two-dimensional models, Lawrence Berkeley Laboratory Report 7057.

Marquardt, D.L., 1963, An algorithm for least squares estimation of nonlinear parameters, Journal of the Society of Industrial and Applied Mathematics, v. 2, p. 431-444.

Oldenburg, D.W., 1979, One-dimensional inversion of natural source magnetotelluric observations, Geophysics, v. 44, no. 7, p. 1,218-1,244.

Pascoe, L.D., and Jones, F.W., 1972, Boundary conditions and calculation of surface values for the general two-dimensional electromagnetic induction problem, Geophysical Journal of the Royal Astronomical Society, v. 27, p. 179.

Sims, W.E., and Bostick, F.X., 1969, Methods of magnetotelluric analysis, EGRL Technical Report no. 58 and Ph.D. thesis (Sims), University of Texas at Austin.

Stodt, John A., 1978, Documentation of a finite element program for solution of geophysical problems governed by the inhomogeneous 2-D scalar Helmholtz equation, University of Utah, Department of Geology and Geophysics, Report on Contract AER 76-11155.

Swift, C.M., 1967, A magnetotelluric investigation of an electrical conductivity anomaly in the southwestern United States, Ph.D. thesis, MIT, Cambridge, Massachusetts.

Vozoff, K., 1972, The magnetotelluric method in the exploration of sedimentary basins, Geophysics, v. 37, no. 1.

Ward, S.H., 1967, Electromagnetic theory for geophysical applications in Mining Geophysics, v. 2, Theory, Society of Exploration Geophysicists, Tulsa, Oklahoma.

Word, D.R.; Smith, H.W.; and Bostick, F.X., 1970, An investigation of the magnetotelluric tensor impedance method, EGRL Technical Report no. 82 and Ph.D. thesis (Word), University of Texas at Austin.

Wu, F.T., 1968, The inverse problem of magnetotelluric sounding, Geophysics, v. 33, no. 6.

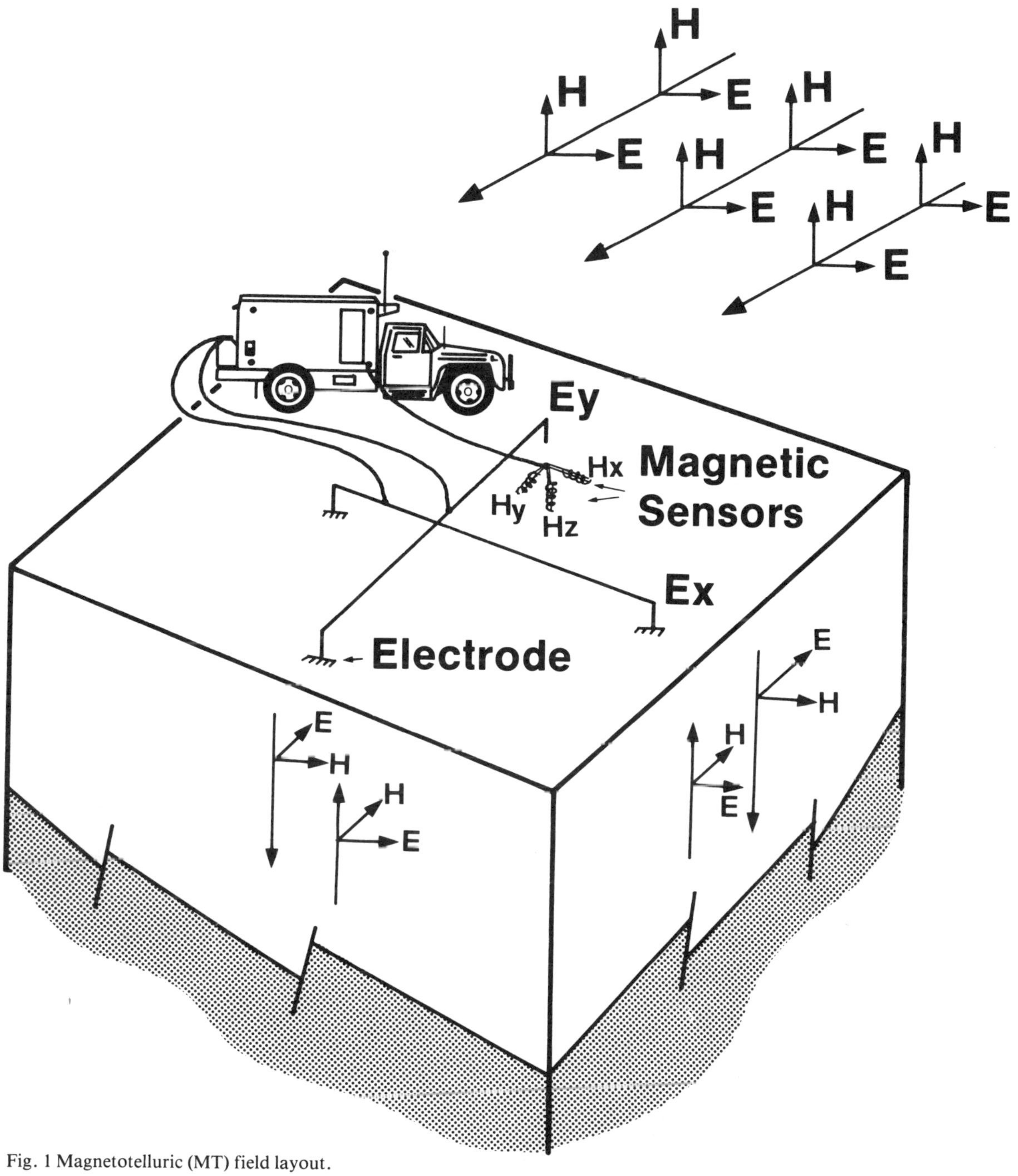

Fig. 1 Magnetotelluric (MT) field layout.

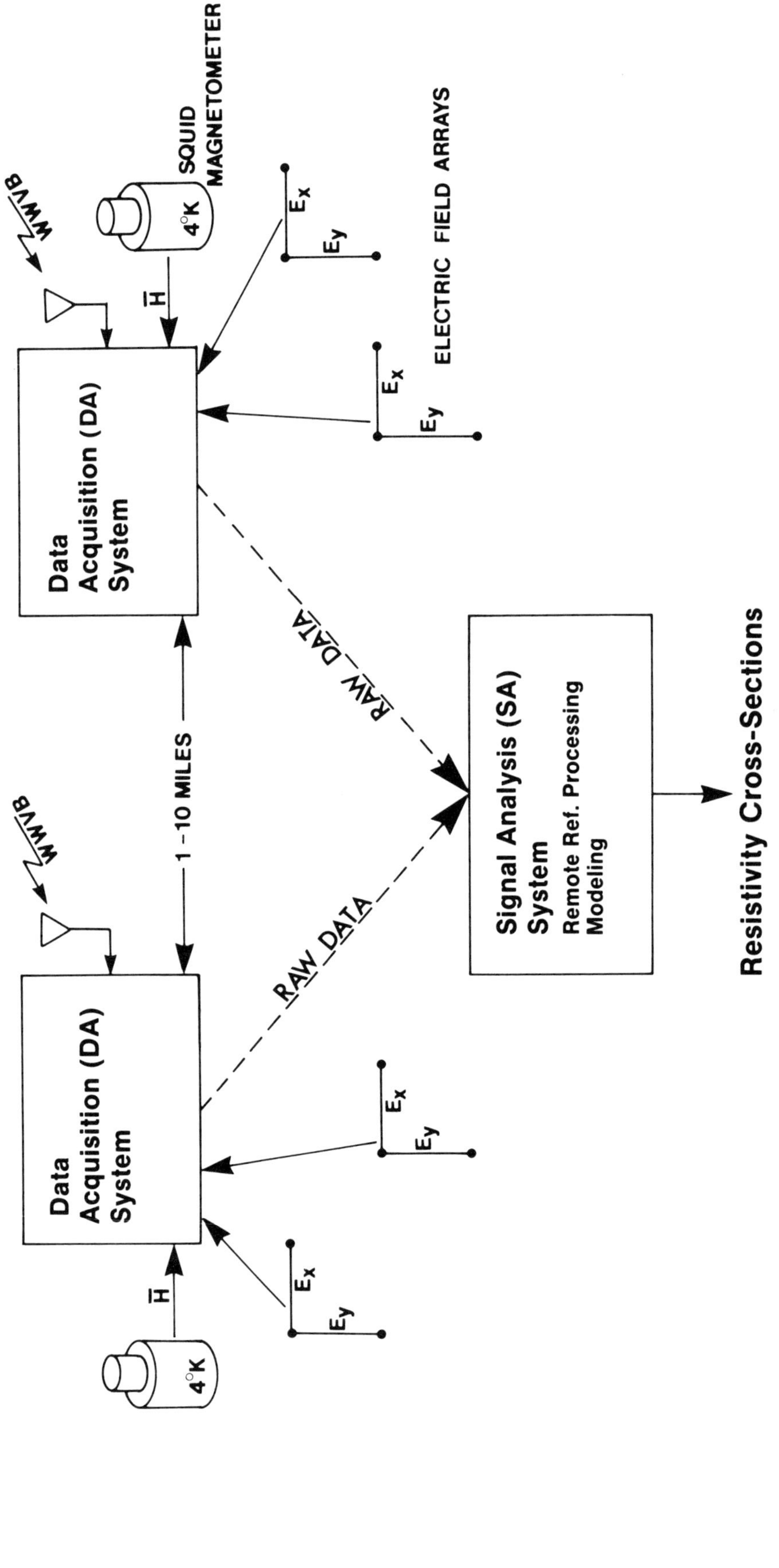

Fig. 2. Conoco Research MT system.

Seven Data Channels (3H,4E)

Bandwidth	0.001 Hz to 250 Hz
Noise (Referred to Input)	Less Than 0.2μV RMS
Channel to Channel Calibration	Better Than 1% Amplitude and 1° Phase (Over 0° to 50° C, and 3 Months)
Calibration Stability	Better Than 0.1% Amplitude and 1 Milliradian (0.057°) (Over 20° C to 30° C, and 3 Hours)

High Accuracy, Wide Dynamic Range Analog-to-Digital Converter (16 Bits)

Fig. 3. Data Acquisition (DA) specifications.

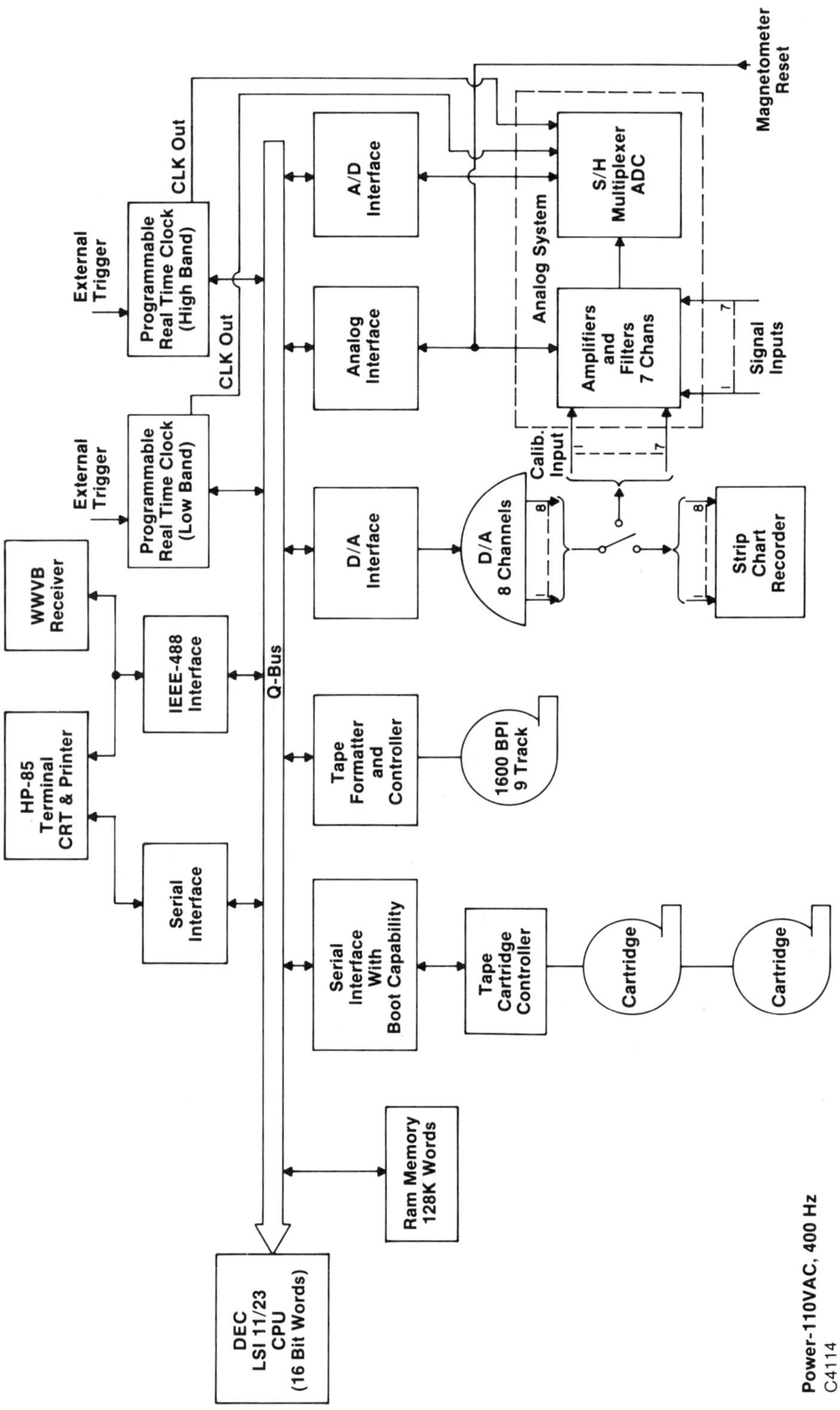

Fig. 4. Data Acquisition (DA) system.

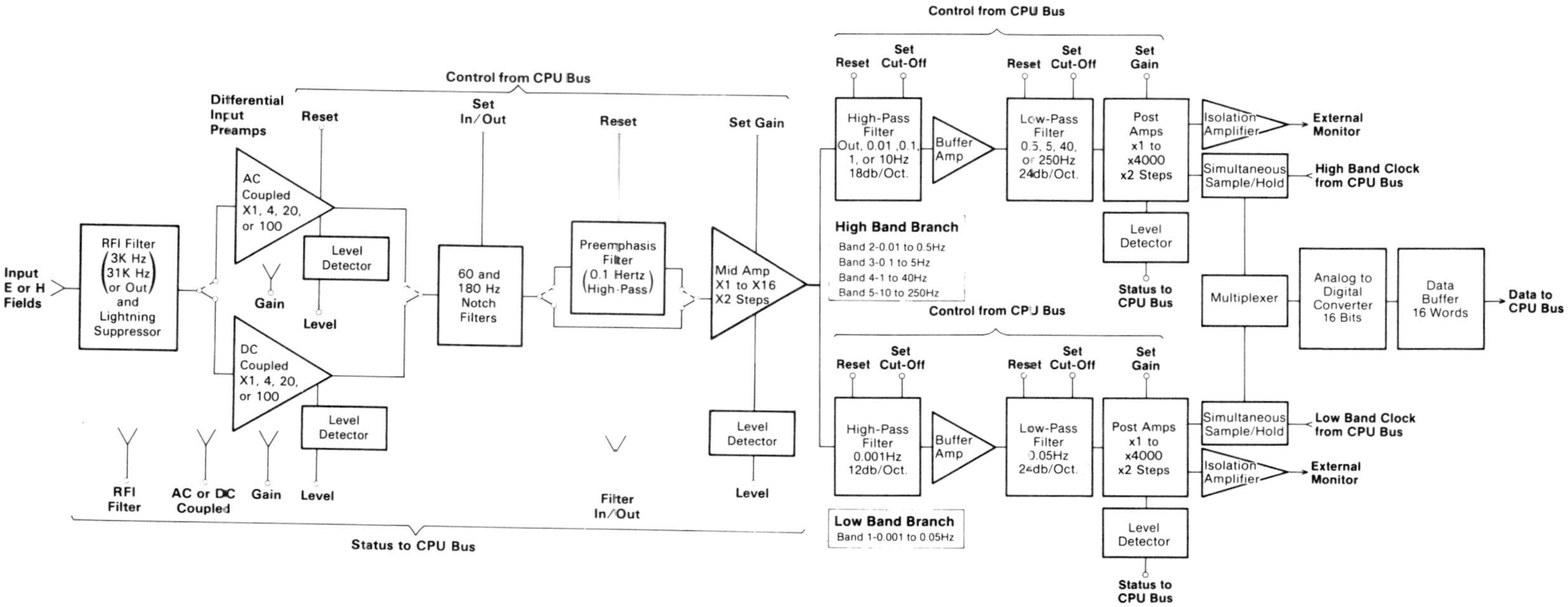

Fig. 5. Data Acquisition (DA) analog section (one of seven channels).

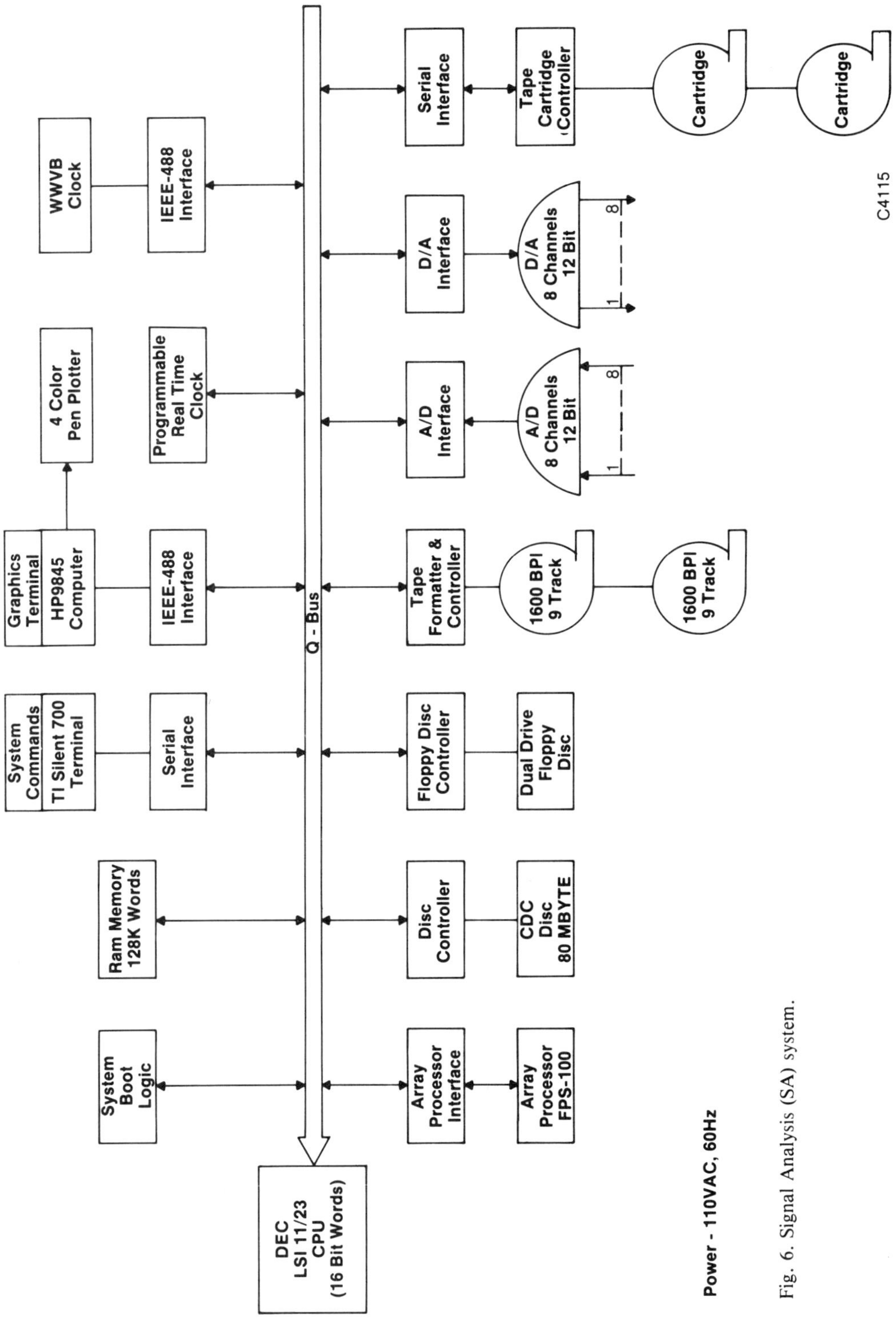

Fig. 6. Signal Analysis (SA) system.

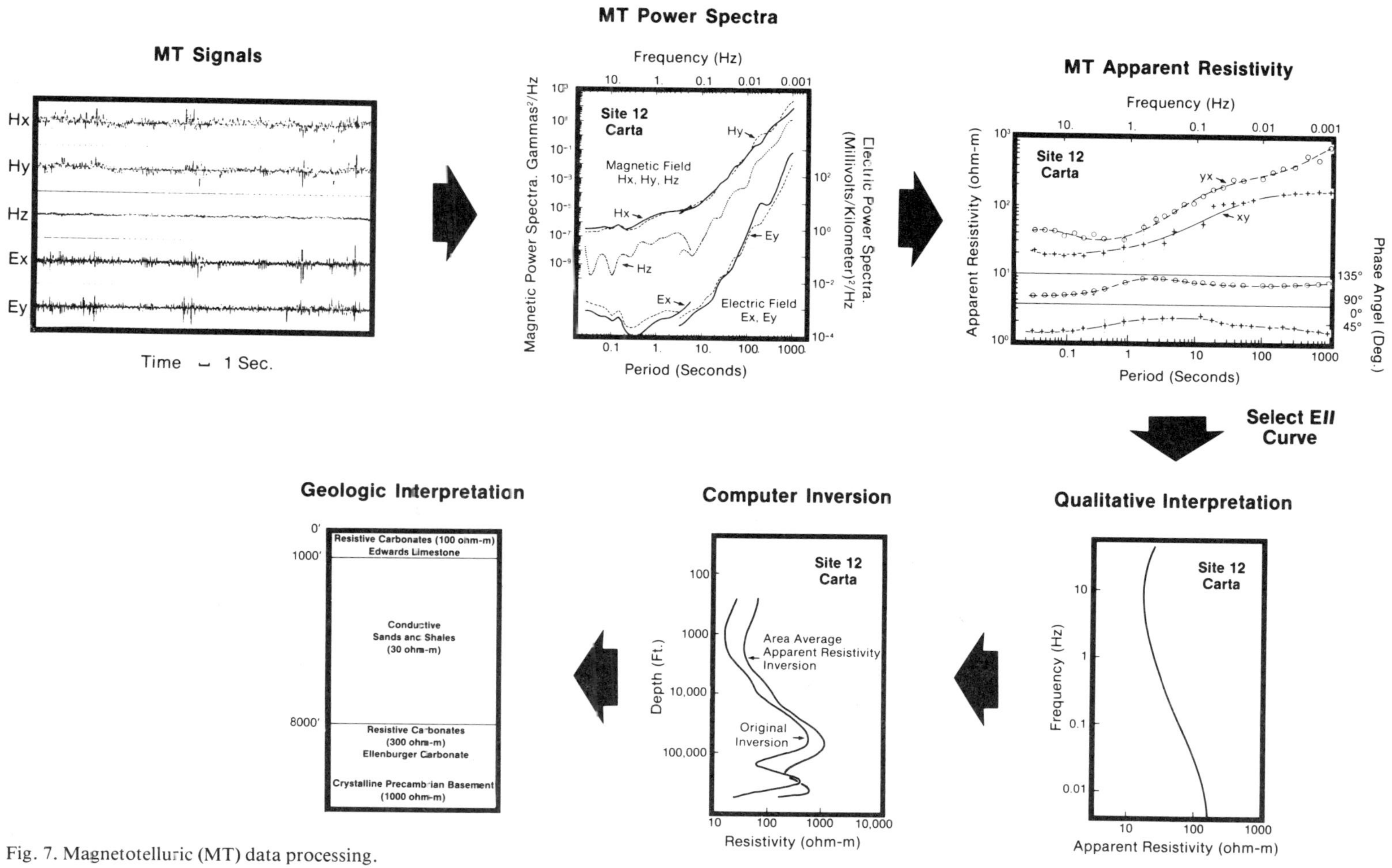

Fig. 7. Magnetotelluric (MT) data processing.

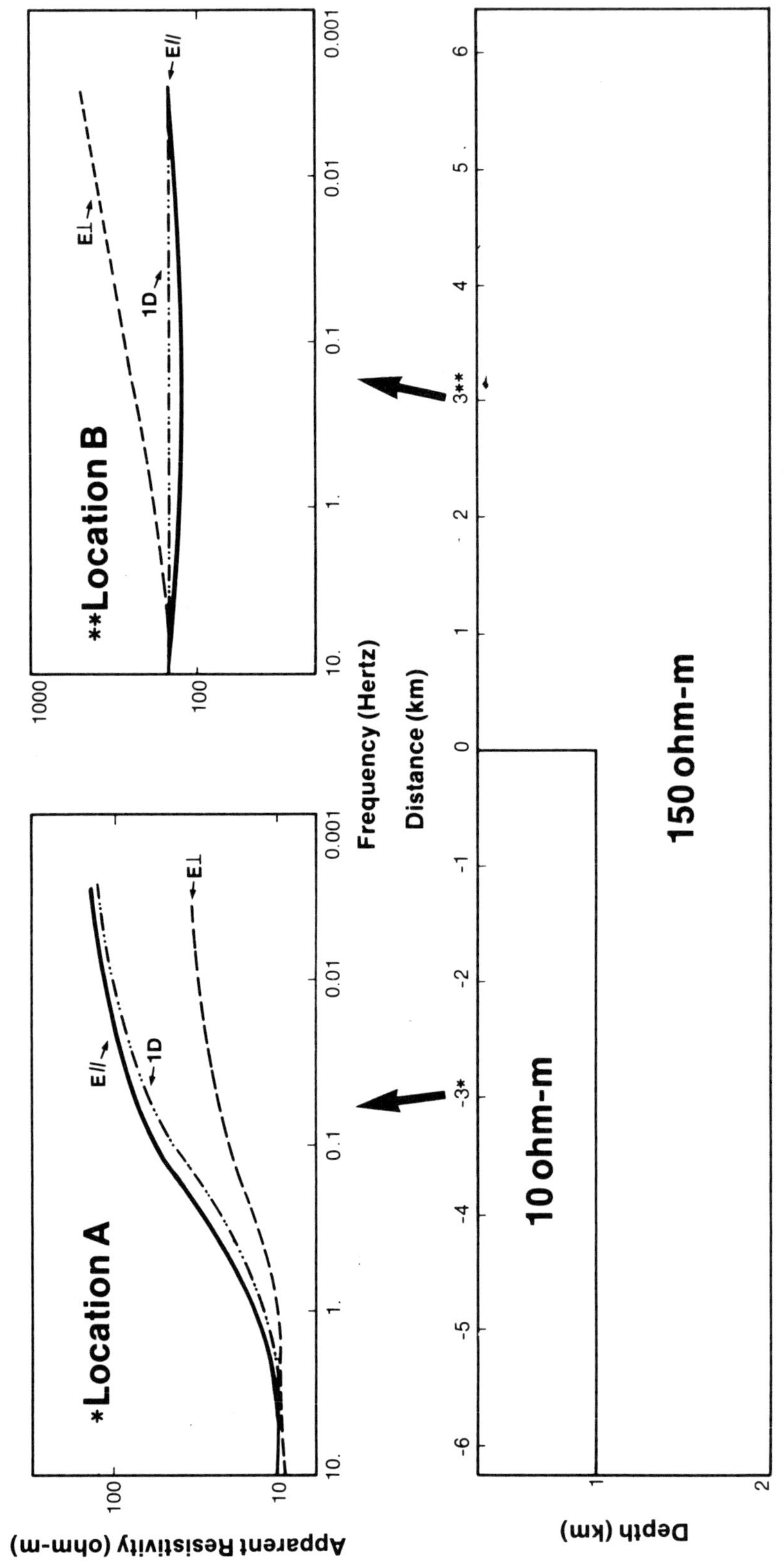

Fig. 8. Effect of inhomogeneities.

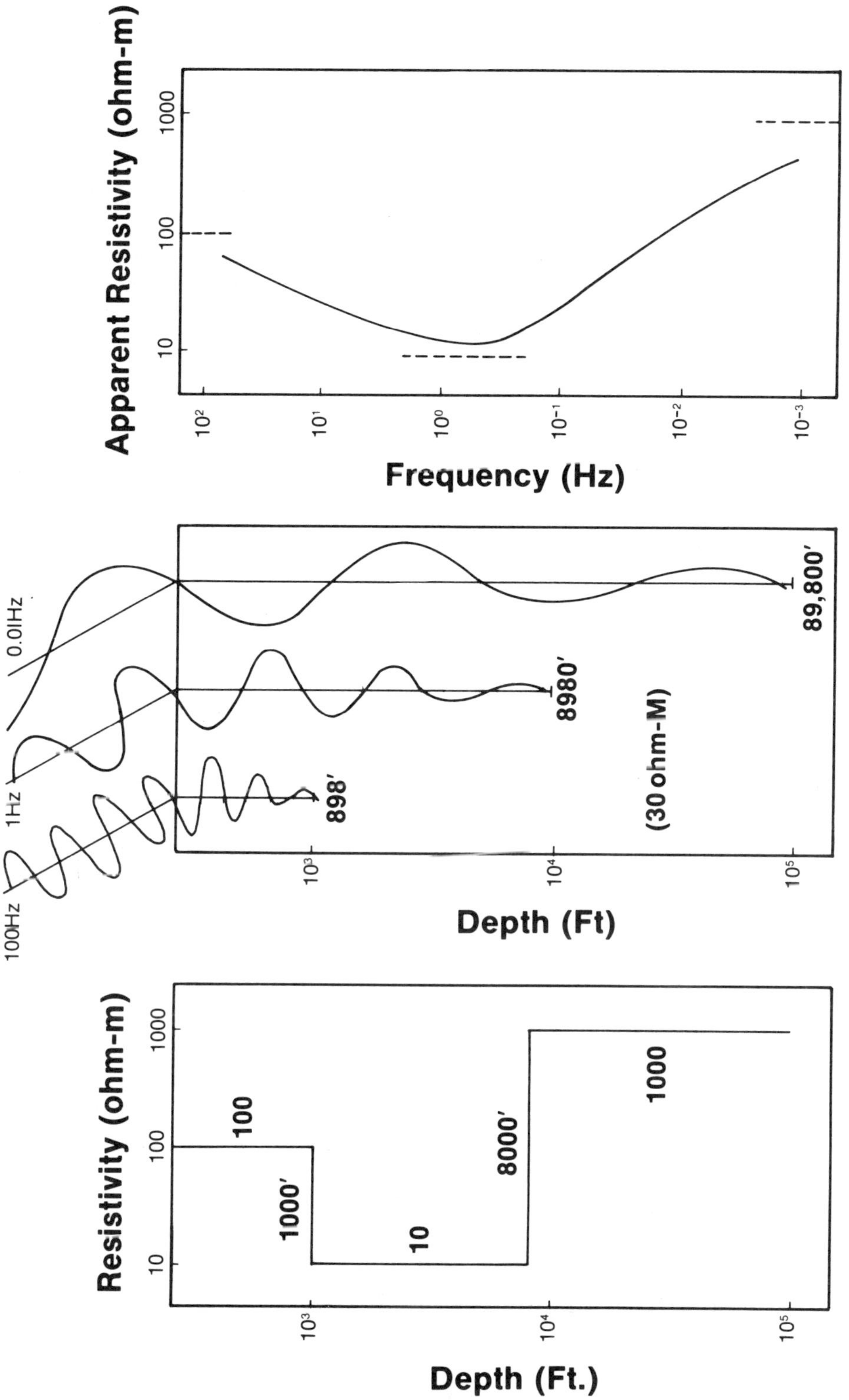

Fig. 9. MT layered model interpretation.

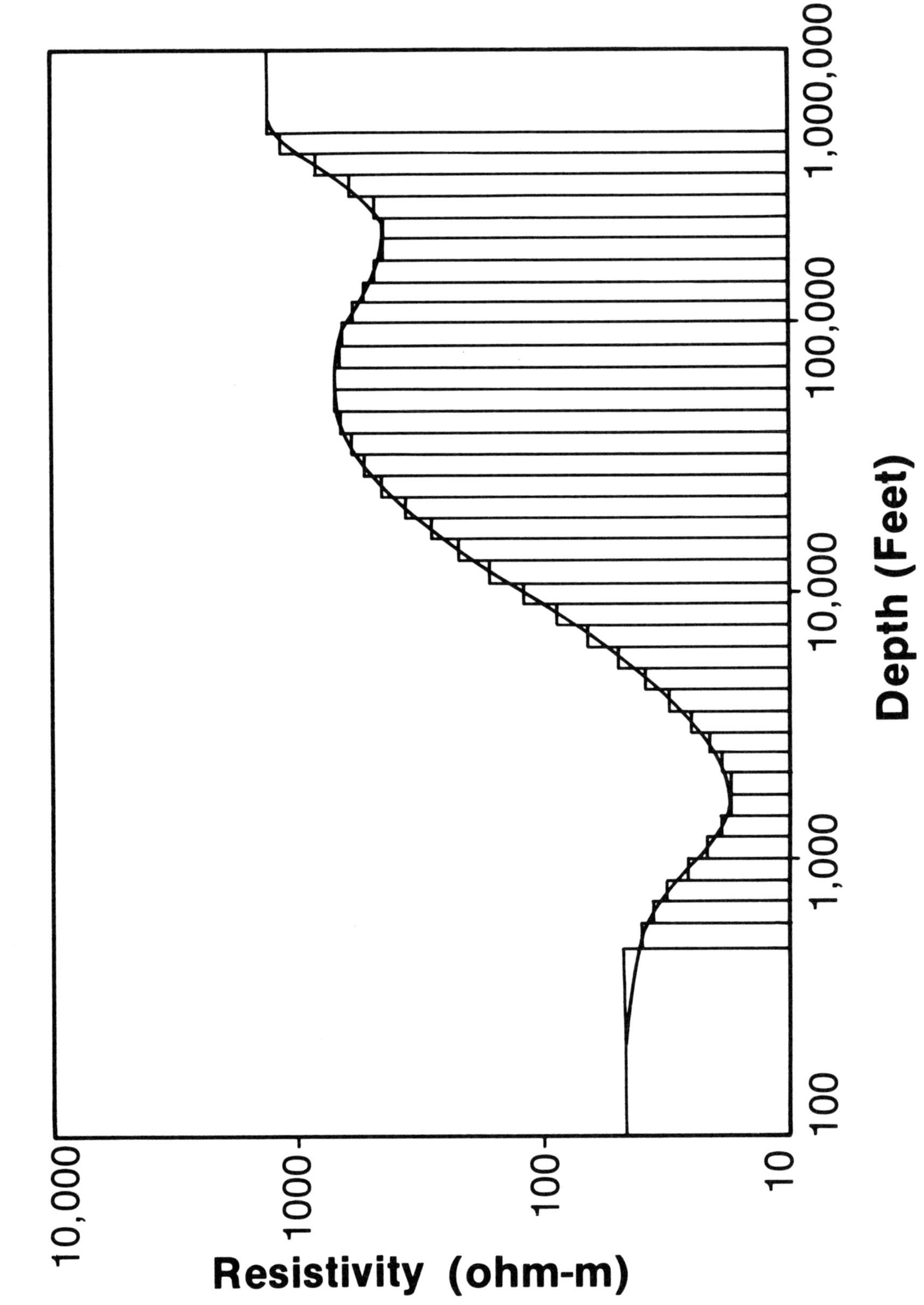

Fig. 10. Quasi-continuous model.

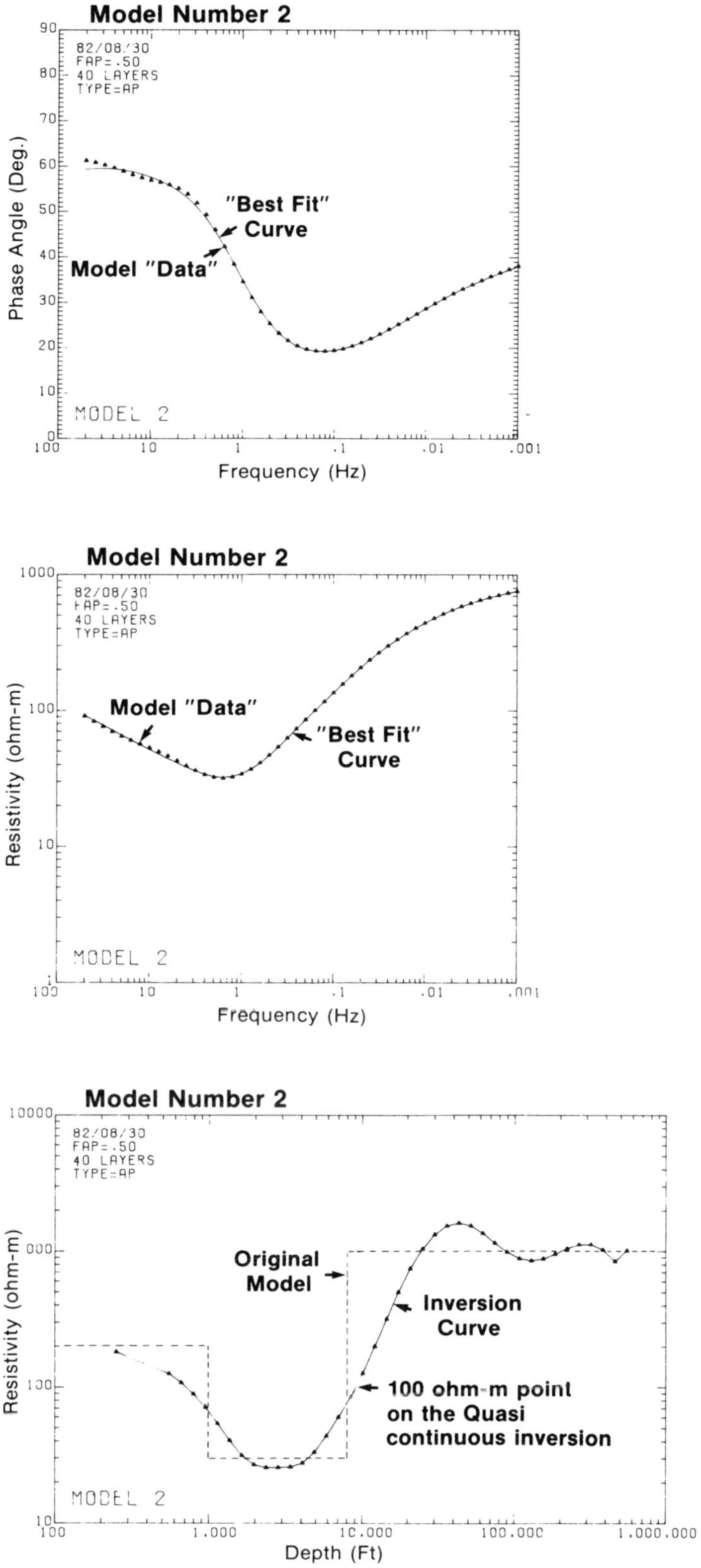

Fig. 11. Inversion example-synthetic data.

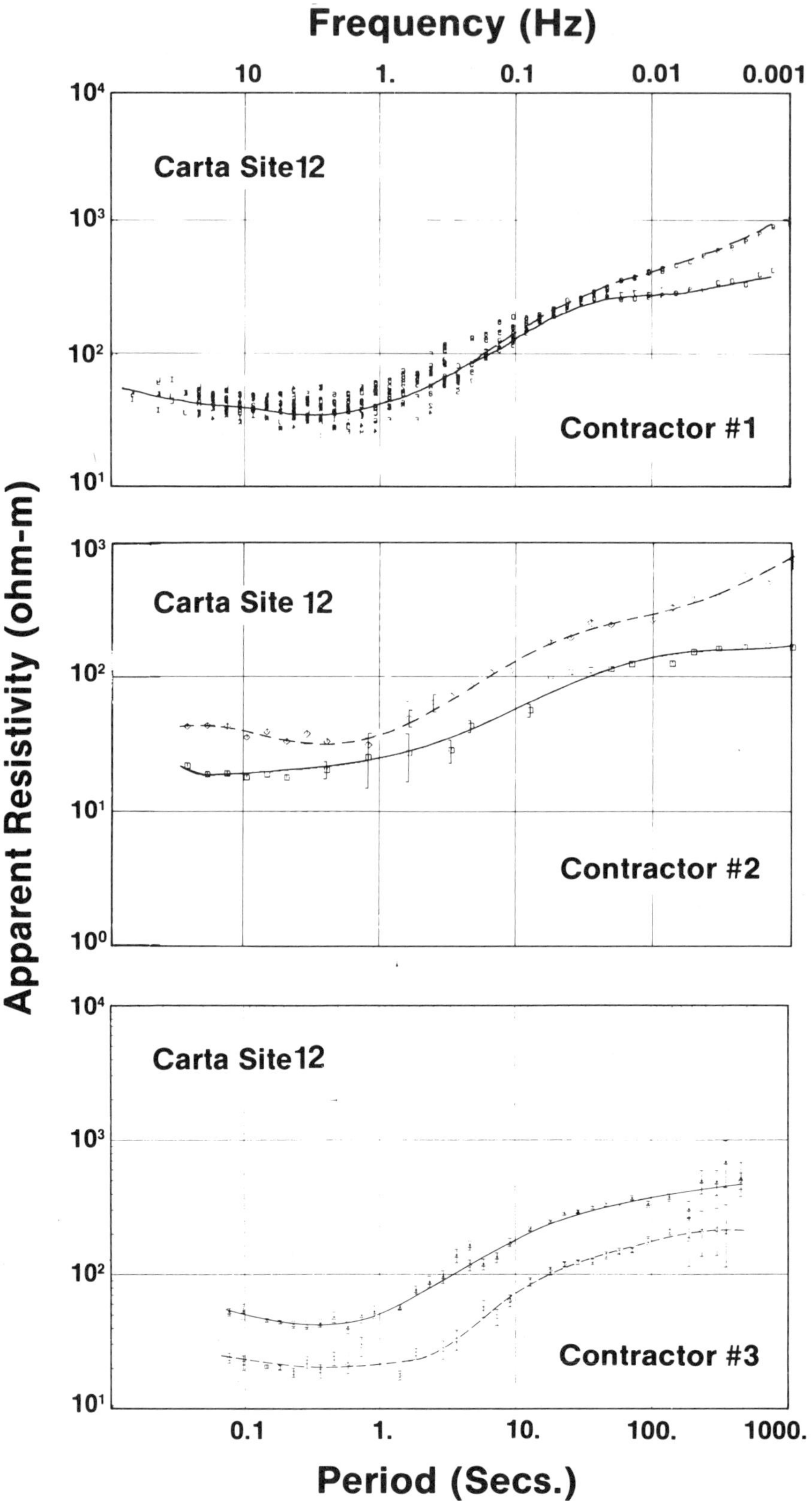

Fig. 12 Comparison of different contractors' MT measurements at the same site.

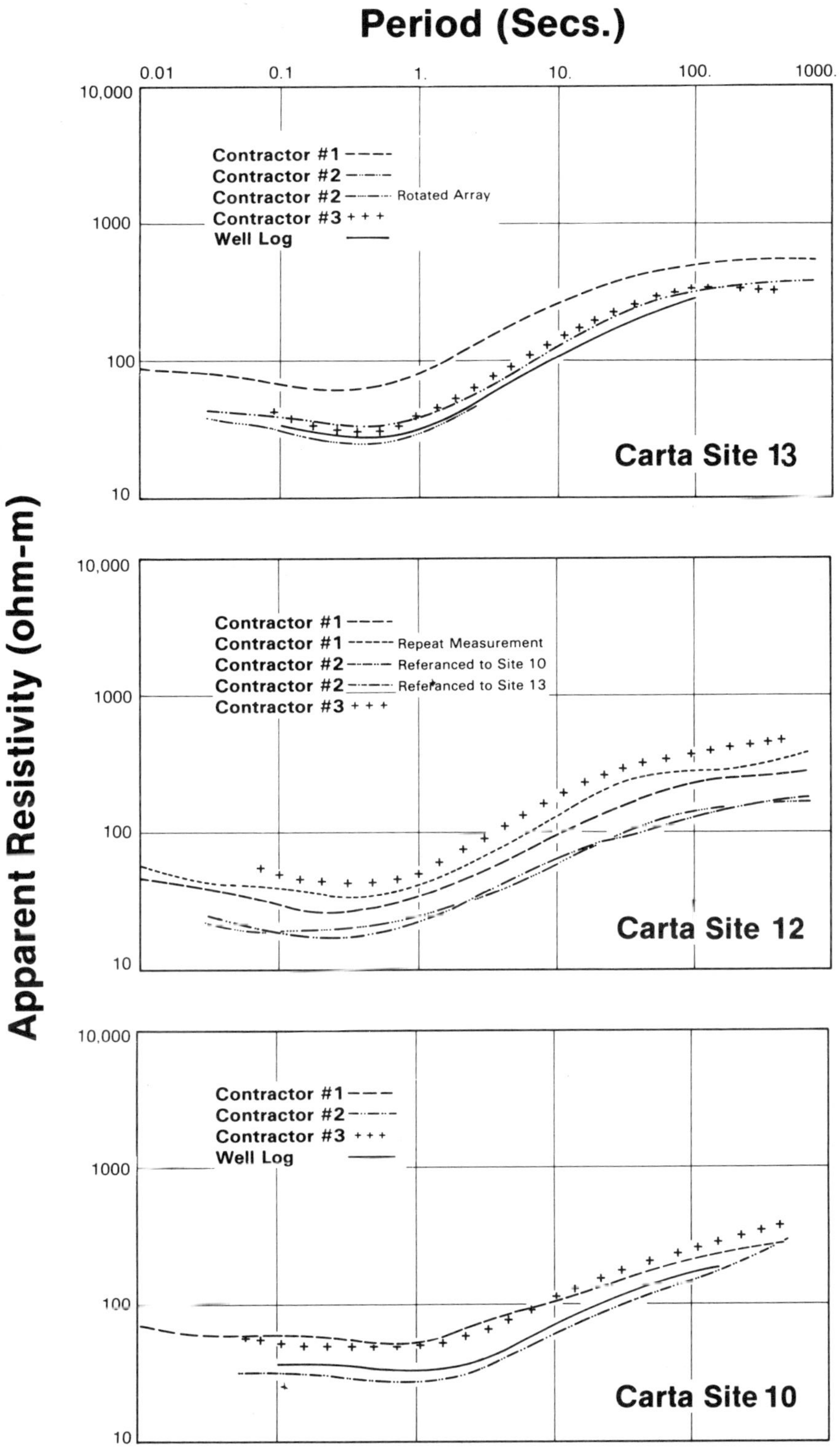

Fig. 13. Further comparisons of MT measurements at the same site.

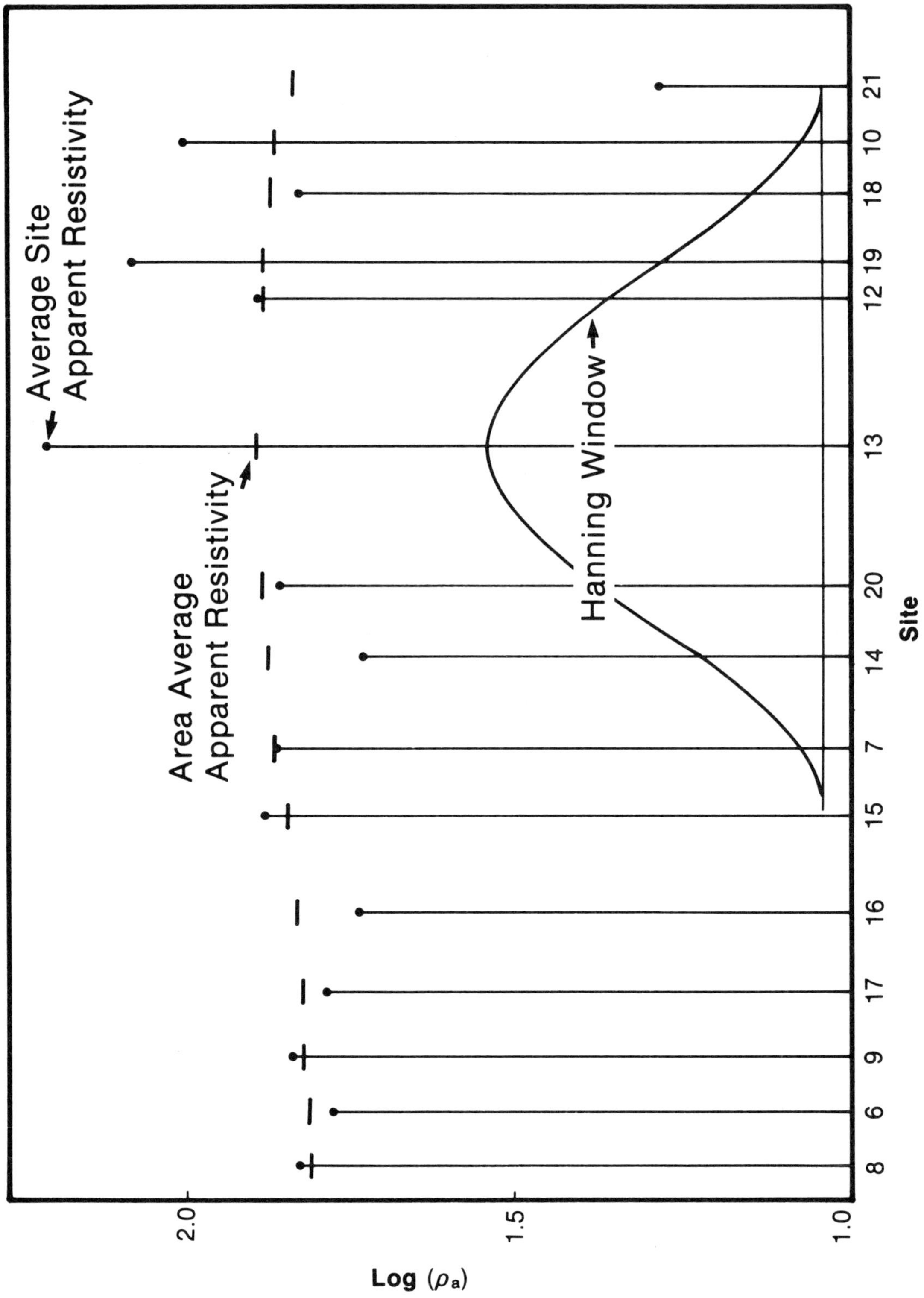

Fig. 14. Area average of the apparent resistivities.

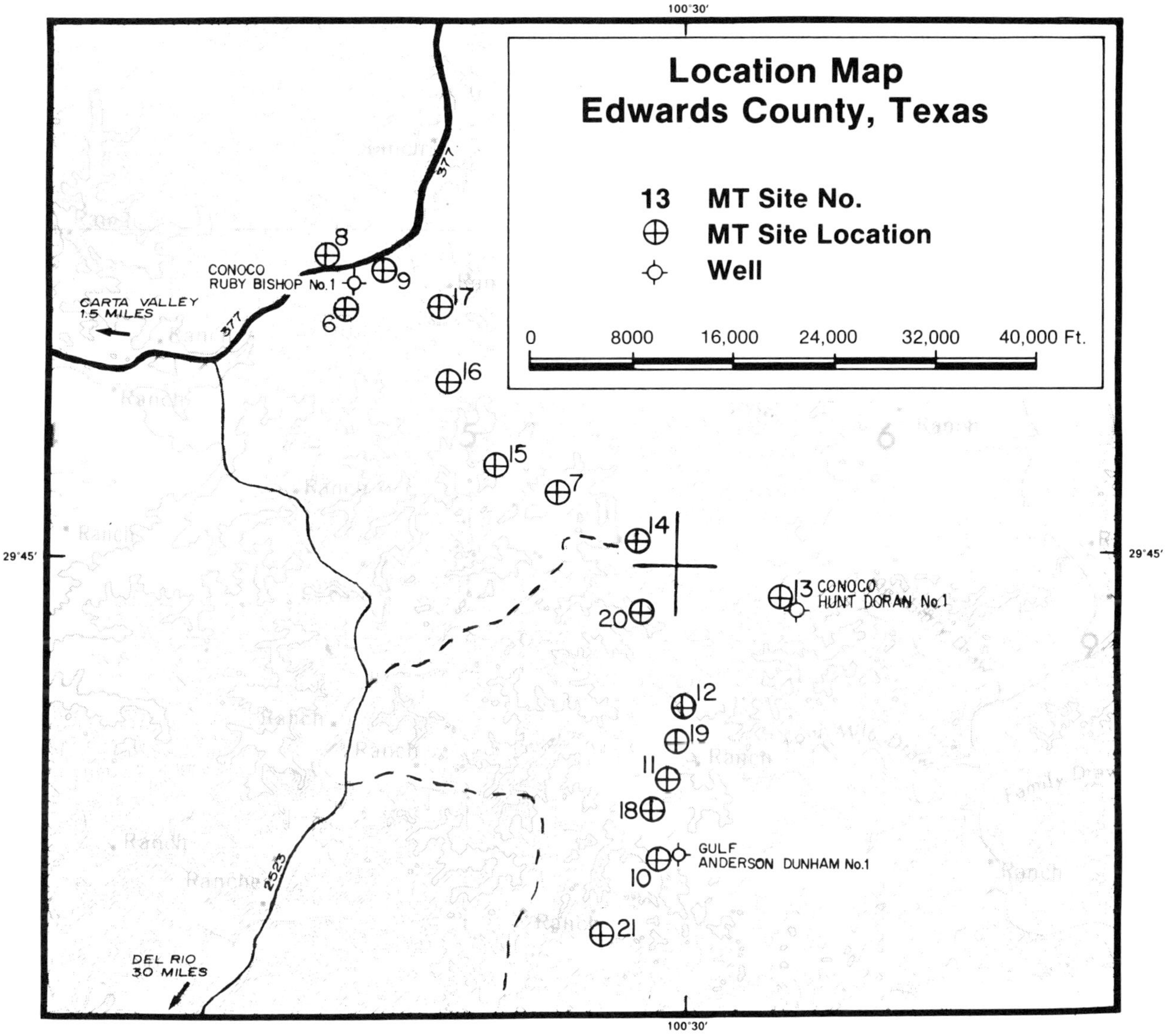
100°30′
Location Map
Edwards County, Texas
13 MT Site No.
MT Site Location
Well
0 8000 16,000 24,000 32,000 40,000 Ft.
CONOCO
RUBY BISHOP No.1
CARTA VALLEY
1.5 MILES
377
8
9
6
17
16
15
7
14
29°45′
13 CONOCO
HUNT DORAN No.1
20
12
19
11
18
GULF
ANDERSON DUNHAM No.1
10
21
2523
DEL RIO
30 MILES
100°30′

Fig. 15.

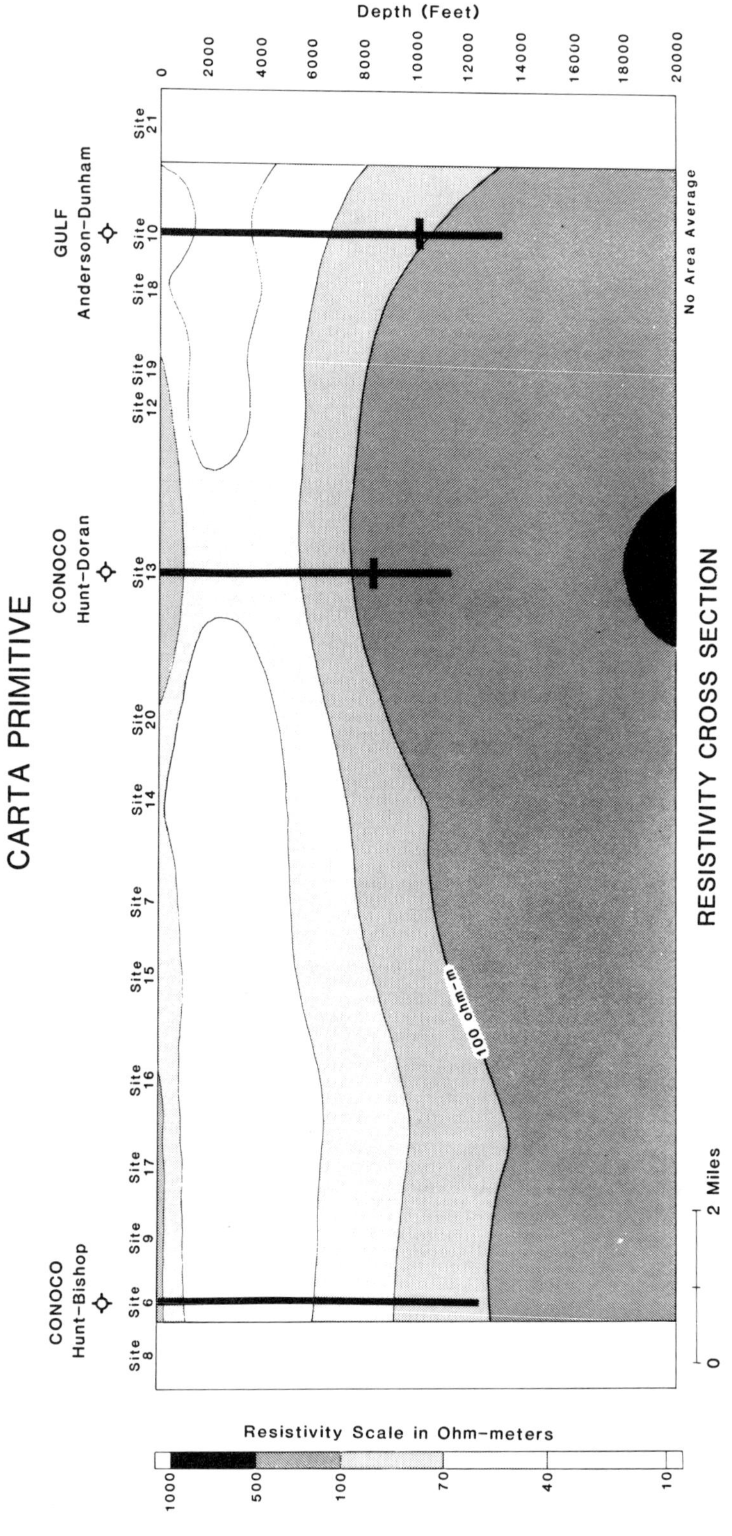
CARTA PRIMITIVE
CONOCO Hunt-Bishop
CONOCO Hunt-Doran
GULF Anderson-Dunham
Site 8
Site 6
Site 9
Site 17
Site 16
Site 15
Site 7
Site 14
Site 20
Site 13
Site 12
Site 19
Site 18
Site 10
Site 21
Depth (Feet)
0
2000
4000
6000
8000
10000
12000
14000
16000
18000
20000
100 ohm-m
No Area Average
0
2 Miles
RESISTIVITY CROSS SECTION
Resistivity Scale in Ohm-meters
1000
500
100
70
40
10

Fig. 16

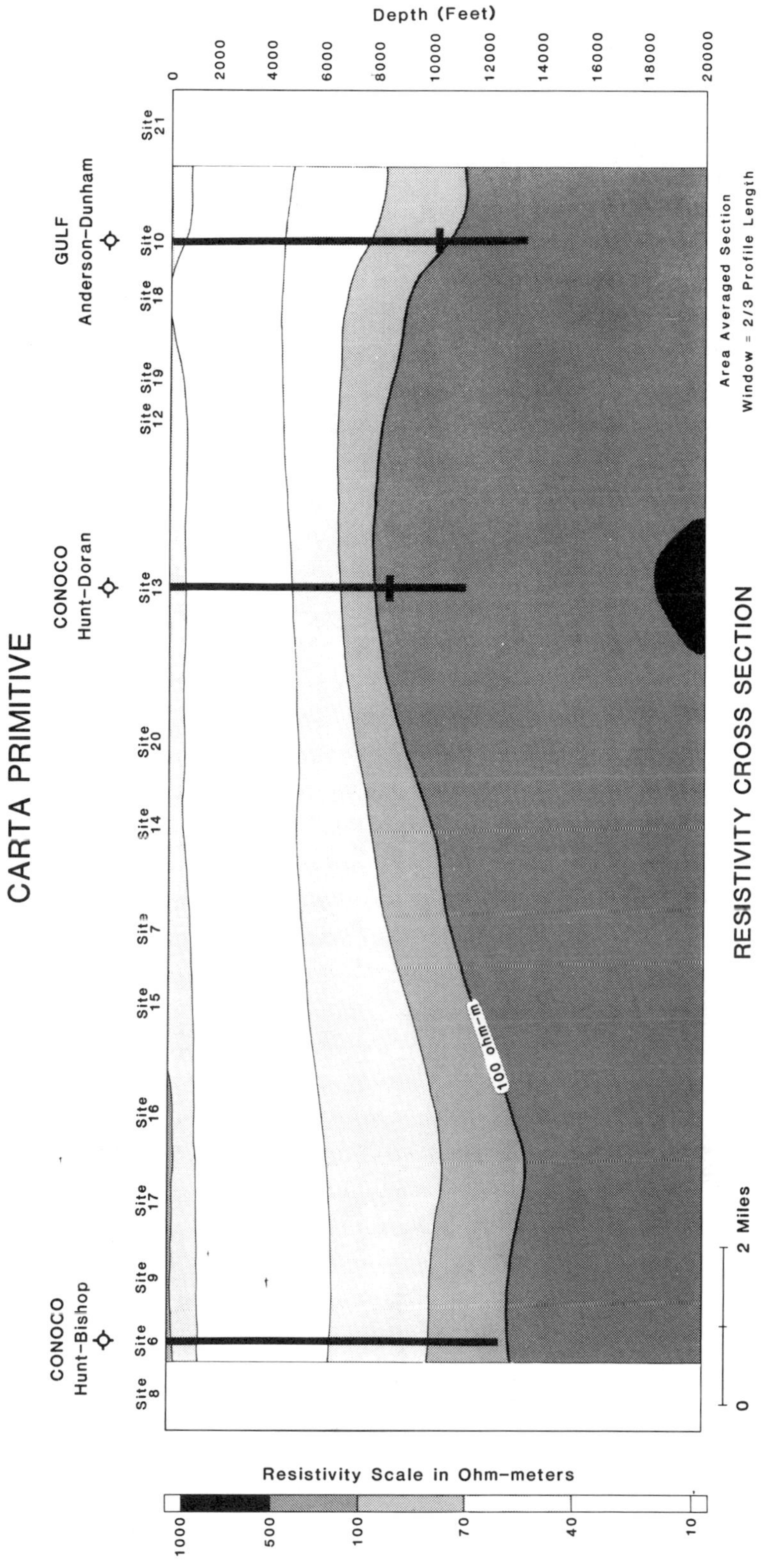
CARTA PRIMITIVE
CONOCO Hunt-Bishop
CONOCO Hunt-Doran
GULF Anderson-Dunham
Site 8
Site 6
Site 9
Site 17
Site 16
Site 15
Site 7
Site 14
Site 20
Site 13
Site 12
Site 19
Site 18
Site 10
Site 21
Depth (Feet)
0
2000
4000
6000
8000
10000
12000
14000
16000
18000
20000
100 ohm-m
0
2 Miles
RESISTIVITY CROSS SECTION
Area Averaged Section
Window = 2/3 Profile Length
Resistivity Scale in Ohm-meters
1000
500
100
70
40
10

Fig. 17

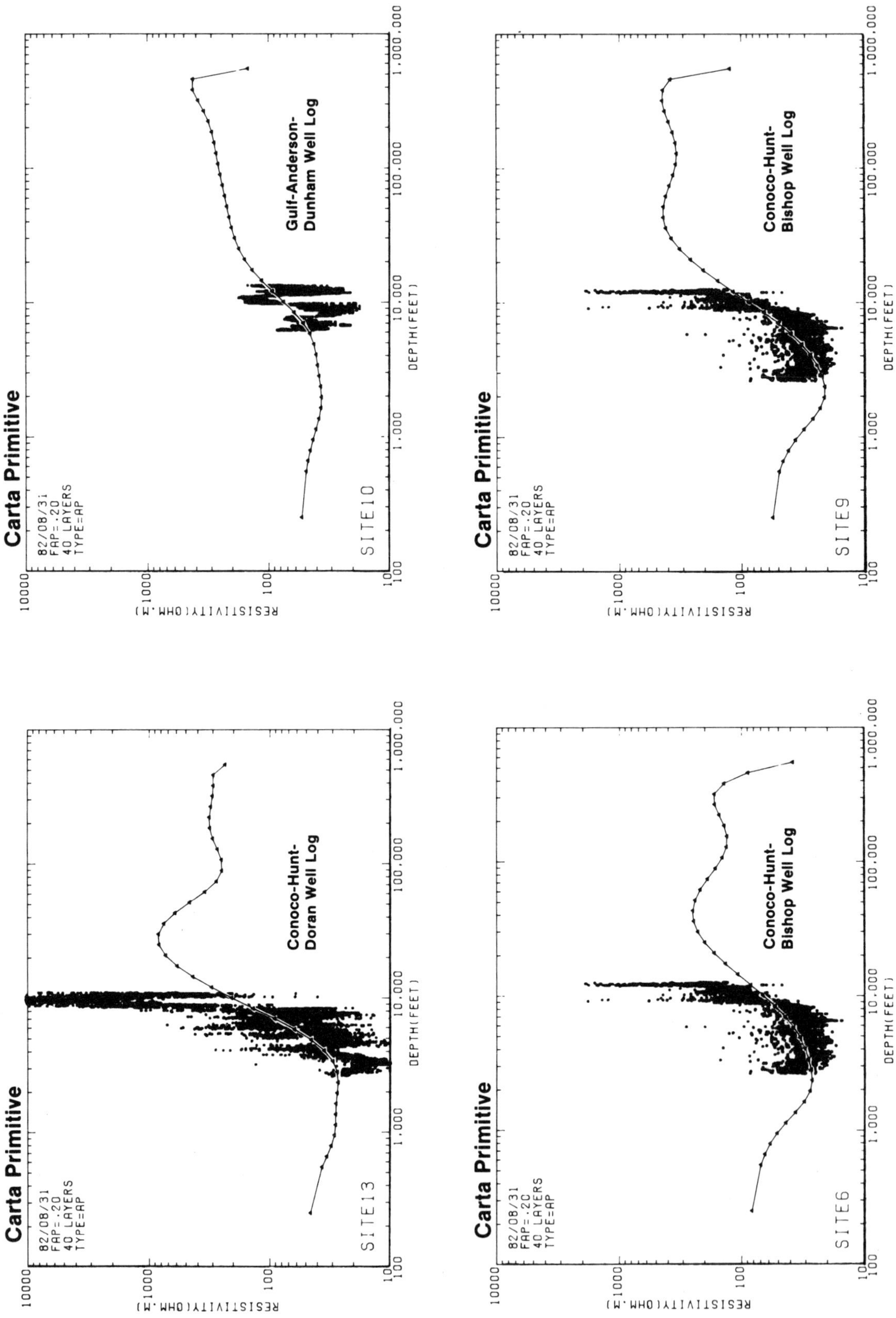

Fig. 18. Comparison of MT inversions with well logs.

UPDATE ON THE USE OF REMOTE SENSING IN OIL AND GAS EXPLORATION

G. Brian Bailey, J.L. Dwyer, J.R. Francica, and M.S. Feng

ASARCO, INC.

ABSTRACT

Satellite remote sensing is a rapidly evolving science that can play an increasingly important role in oil and gas exploration. As one of many exploration tools, remote sensing can contribute most significantly to the exploration process by providing regional geologic information and by targeting areas for more detailed exploration.

Explorationists have used satellite data with increasing regularity since the launch of LANDSAT 1 in 1972. Too often, however, their studies have lacked a systematic and comprehensive approach to analysis and interpretation; consequently, they have been incomplete and have not resulted in the extraction of all structural *and* lithologic information obtainable from LANDSAT data. In recent years, some companies have begun routinely to apply digital (computer) processing techniques to LANDSAT data. The resulting enhanced image products typically help geologists to extract more lithologic and structural information from the data. For example, improvements can be made in image products used to discriminate certain rock types, to identify zones of alteration, and to identify and map faults and other fractures.

New and future satellite systems, both U.S. and foreign, will provide data with improved spectral and spatial resolution that will be more useful in geologic studies than data from LANDSATS 1, 2, and 3. The 10 and 30 m spatial resolution of the French SPOT and the LANDSAT 4 thematic mapper, respectively, will permit identification of smaller geologically significant features. Stereoscopic data from the SPOT satellite will result in better topographic control for more accurate stratigraphic mapping. The thematic mapper will provide information with which to better discriminate rock types and identify rock alteration. Currently being conducted at the Earth Resources Observation Systems (EROS) Data Center and elsewhere are studies which use airborne thematic mapper simulator data and other prototype sensor systems to evaluate the types of geologic information that future satellites will provide.

Perhaps the greatest benefit to resource exploration from satellite data will come when these data are used routinely with other pertinent data in a digital database approach to exploration. Modern computer technology allows the rapid integration and synthesis of satellite, topographic, gravity, aeromagnetic, geochemical, and other data from a given area. After such data are geographically registered, they can be digitally processed, within the constraints of a defined geologic model, to identify and focus exploration efforts on target areas that have the greatest potential for success.

The evolutionary process that satellite remote sensing has undergone during the past ten years is shown by results of studies the authors have conducted in the hydrocarbon-rich Uinta and Piceance Basins in Utah and Colorado and the Qaidam Basin in China. Incremental benefits to geologic and exploration studies are documented, but more importantly, techniques and methodologies that can be most usefully applied on a routine and operational basis in resource exploration are illustrated.

INTRODUCTION

Following the launch of LANDSAT 1 in 1972, the perceived usefulness of satellite data as a tool in nonrenewable natural resource exploration suffered from some exaggerated claims of application successes. That perception has also suffered because inexperienced users of LANDSAT data have sometimes failed to extract most of the geologic information available from the data and have thereby drawn negative conclusions, regarding its usefullness. Nevertheless, during the ten years since the launch of the first LANDSAT satellite, LANDSAT data have become recognized throughout most of the petroleum industry as an important exploration tool. The principal role of these data has been in providing regional geologic information useful in targeting areas for more detailed exploration. In the decade ahead, the role of satellite remote sensing will be expanded. Land resource satellites of the 1980s will send back to earth data which will have spatial and spectral resolution, characteristics better suited for extracting geologic information than the data provided by satellites of the 1970s. These improved data, along with recent advances in the techniques and technology of digital data processing (particularly in the area of digital database compilation and manipulation), will have a positive impact on petroleum exploration in the 1980s.

GEOLOGIC ANALYSIS AND INTERPRETATION OF LANDSAT IMAGES

The synoptic perspective of LANDSAT images facilitates the rapid extraction of geologic information for large areas. In addition, as is discussed in the next section, digital (computer) processing of LANDSAT data further facilitates the extraction of geologic information by enhancing various spectral and textural features. However, the full extraction of geologic information contained on LANDSAT images, whether they are standard or enhanced, is primarily dependent upon the application of a systematic and comprehensive approach to analysis and interpretation (see Bailey et al., 1983; Francica, 1983).

Manual geologic analysis and interpretation techniques are most commonly applied to standard black-and-white and false-color composite images, such as those available from the Earth Resources Observations Systems (EROS) Data Center. The goals of geologic analysis and interpretation are (1) to detect variations in the reflected solar radiation represented on the image, (2) to attach geologic significance to those variations, and (3) to compile the geologic information from the image to produce a geologic interpretation which is meaningful in the context of the problem being addressed. The solution of most geologic problems is dependent upon structural information, lithologic information, or both. Extracting this information from LANDSAT images involves analyzing the image in terms of landforms, drainage, cover characteristics, and lineaments. All variations in spectral reflectance visible on LANDSAT images are related to one or more of these parameters. Once these parameters are described and correlated, a geologic interpretation can be derived. The detail and accuracy of such an interpretation are determined by factors such as the limitations imposed by the spatial and spectral resolution of the sensor system, the scale and format of the image used for interpretation, and the geologic knowledge and skill of the interpreter, as well as scene-dependent variables such as type and density of vegetative cover. Nevertheless, a systematic and comprehensive approach to geologic analysis and interpretation of LANDSAT images should result in the extraction of maximum geologic information and should permit meaningful geologic interpretations to be made.

Figure 1 is a six-scene LANDSAT image mosaic covering the western part of the Qaidam Basin, People's Republic of China. The Qaidam Basin is a topographic and structural sedimentary basin that covers approximately 100,000 km^2 in Western China. Petroleum is currently being produced from the nonmarine sedimentary rocks within the basin, but the region is still being actively explored. LANDSAT images provided information leading to a rather detailed and complete interpretation of the regional structural geology of this large area in a short period of time (Bailey and Anderson, 1982).

Landform, drainage, cover type, and lineament analyses were performed on the images which comprise the mosaic. From these analyses the structural interpretation shown on Figure 2 was derived. This interpretation was greatly facilitated by the negligible vegetative cover and good exposure of rocks in the Qaidam Basin. Soils, windblown sands, and alluvial deposits cover rocks in parts of the basin, but these materials were not a severe hindrance to geologic interpretations of bedrock in the western part of the basin.

The aspect of the basin geology that is most readily interpreted form the images is the folding in the sedimentary rocks. The basin sediments have been folded into a large number of generally northwest trending, commonly doubly plunging anticlines and synclines. On the images, the folds are defined by a series of semielliptical patterns which on the ground represent the traces of continuous sedimentary beds or associated weathered materials. The axes of the folds identified on the images are plotted on Figure 2. Most of the folds identified were determined to be anticlines on the bases of landform characteristics and apparent dips of the exposed stratigraphic units. Stereoscopic analysis of sidelap image areas was valuable in establishing some of these interpretations. Field observation and information supplied by Chinese geologists confirm that a few of the folded structures identified on the images have no associated surface outcrops and represent buried anticlines (eg. point A, Fig. 2).

Lineament analysis of the LANDSAT images provided information important to a geodynamic and structural interpretation of the western part of the Qaidam Basin. The distribution of interpreted faults and fractures occurring within the study area is also shown on Figure 2. Lineaments observed on the images were classified as belonging to one of four general groups: strike slip faults, thrust faults, normal or reverse faults, and fractures. Lineaments were not interpreted as faults unless apparent offset along them could be observed or inferred.

The geologic data derived from studies of LANDSAT imagery of the western Qaidam Basin are significant in several respects. They resulted in an accurate interpretation, based on comparison with Chinese geologic maps as well as our own field observations, of the structural geologic setting of an area greater than 50,000 km^2. The image interpretations were accomplished during a period of just a few months as compared to the many years required to derive similar interpretations from field-acquired data. Many major, exposed folded structures were precisely located, identifying potential targets for detailed exploration work. Of even greater significance to exploration efforts specific to the Qaidam Basin was the demonstration of LANDSAT's ability to detect subsurface folded structures that were not identified in detailed surface geologic mapping.

Interpretations of faulting within the basin have less direct significance for petroleum exploration than do interpretations related to folding, yet several observations can be made. The recognition of an apparent major thrust sheet (Fig. 2) is significant because there exists a potential for discovering favorable structural traps in the lower plate rocks even though there may be no indication of their presence on the surface. The existence of the proposed thrust sheet should be considered when interpreting seismic and other geophysical data collected from the area.

The recognition of high-angle faults which offset some of the major folds may be significant both in the locating of offset hydrocarbon reservoirs and because some of these faults could themselves be acting as hydrocarbon traps. The major significance of fractures that have been interpreted lies in their effects on the migration of hydrocarbons into or out of structural or stratigraphic traps, as well as in their effects on increased reservoir permeability.

The structural interpretations derived from these studies have been used to develop a predictive structural model for application in the eastern part of the Qaidam Basin (Bailey and Anderson, 1982). The eastern part of the basin has more unconsolidated surficial cover than does the western part of the basin. This material obscures surface indications of folded structures which nevertheless may be present. The model, which is based on global tectonic theory, may be useful in guiding exploration efforts in the eastern part of the basin.

While LANDSAT images are most commonly used to derive structural information, as in the Qaidam Basin study, application of systematic and comprehensive techniques of analysis and interpretation can often provide important regional lithologic information as well. Figure 3 is a band 7 LANDSAT image covering a portion of northwestern Colorado and northeastern Utah, including the Piceance basin. The geologic interpretation (Fig. 4) of this area contains considerable amounts of both lithologic and structural information. It was derived by applying the techniques of image analysis and interpretation described in this section, as well as some of the basic principles of geology. Lithologic unit labels designate the dominant rock type(s) and relative age interpreted for each unit. It should be stressed that this is *only an interpretation* which, in fact, could be further refined with additional time and effort. However, Colorado and Utah state geologic maps show that it provides a reasonably accurate protrayal of the regional geologic setting of this area. Finally, it is an interpetation which was derived with little bias due to previous geologic knowledge of the region.

Landforms and drainage were the parameters most important in the interpretation of lithologic variation occurring within the image area. As indicated by the descriptions in the explanation to Figure 4, most lithologic units were defined on the basis of topographic and drainage characteristics. The units interpreted on Figure 4 do not necessarily correspond to formal rock-stratigraphic units, although similarities in areal distributions of image-defined units and units mapped on Colorado and Utah state geologic maps do exist. Rather, the image-defined units were interpreted, from observable landform, drainage, and cover type characteristics, to be lithologically consistent groups of similar aged rocks and thus could define rock-stratigraphic units.

Some of the image-defined units have similar descriptions in the explanation to Figure 4. Consequently, it is logical to suspect confusion among units. To be sure, some misinterpretation has probably occurred. However, the subtleties of landform and drainage characteristics within the units, which are impossible to describe in an abbreviated legend, often allow the experienced interpreter to decide confidently which image-defined unit best describes the rocks in question. Furthermore, such confusions are often resolved by applying the basic geologic law of superposition.

Landform and drainage characteristics were also important in the interpretation of the folded structures represented on Figure 4. As a classic example, the Piceance Creek basin (Area A, Fig. 3) is clearly defined by a parallel and subparallel drainage network. The directions of drainage flow are largely controlled by the dips of the underlying sediments. Other examples include the Rangely and Skull Creek anticlines (Areas B and C, respectively, Fig. 3) which are recognized by arcuate hogback landforms as well as by the distributions of interpreted stratigraphic units, which themselves were defined mostly on the basis of landform and drainage characteristics.

Lineaments, the other major parameter of image analysis, were primarily important in the interpretation of faulting; that is, they focused the interpreter's attention on areas where faults may exist. A lineament overlay constructed for Figure 3 contained many more lineaments than are represented on Figure 4. However, if a lineament could not be inferred to be a fault based on some indication of offset, it was not represented on Figure 4. Lineaments that showed apparent offset were mapped as faults on Figure 4, and the inferred direction of offset was noted. Lineaments whose direction of offset could not be established were mapped as probable faults on Figure 4.

This systematic and comprehensive approach to analysis and interpretation of LANDSAT imagery resulted in the definition of the regional geologic setting of this large area in a period of just a few days. In the context of petroleum exploration, the work has shown that the region is dominantly comprised of sedimentary rocks (mostly clastic rocks) which could be conducive to the occurrence of petroleum. Folded structures which could form hydrocarbon traps and areas that might contain especially thick sequences of rocks favorable to the occurrence of hydrocarbons were identified and located. Areas were identified which, because of the presence of volcanic or metamorphic rocks, could be eliminated from initial consideration in an evaluation of the petroleum resource potential of the area. The work demonstrates the utility of LANDSAT data as a regional or reconnaissance geologic tool.

DIGITAL PROCESSING OF LANDSAT DATA

Reflectance data are acquired in four wavelength bands by the LANDSAT multispectral scanner (MSS) and converted on board the satellite to 6-bit digital numbers which range from 0 (darkest values) to 63 (brightest values). These data are transmitted back to earth where they are rescaled to 0-127 (7-bit data) or 0-255 (8-bit data). These digital data may be used to make photographic images, or they first may be computer processed to produce any of a variety of enhanced images.

The purpose of digital image enhancement is to produce an image from the original data such that certain or selected features in the image are more easily detected and interpreted. Depending upon overall objectives, geologic studies of LANDSAT data often employ one or more of a variety of digital processing techniques. While it is beyond the scope of this paper to discuss any of these techniques in detail, a few of the more commonly used digital enhancement techniques are illustrated and briefly described.

The digital numbers, or brightness values (BV's), as they are commonly called, transmitted by the satellite back to earth are oftentimes concentrated over only a small portion of the 0-63 BV range available. The result is a photographic image that has poor overall scene contrast. To overcome this situation, an image processing technique called contrast enhancement (contrast stretching) is applied to the data. The technique involves expanding the original range of brightness values over the full (e.g., 0-255) dynamic range available. Typically, the data are stretched in a linear manner, so that the improvement in contrast is equal throughout the original data. However, various nonlinear contrast enhancement techniques exist which allow selected ranges in the original data to be stretched more than others. The LANDSAT images in Figures 1 and 3 have been linearly contrast enhanced. In the case of Figure 3, the contrast enhancement did not provide a significant improvement because the contrast in the original data was very good. However, the contrast-enhanced images which comprise the mosaic in Figure 1 were much more useful for interpeting geologic information than were images produced from the unstretched original data. Contrast stretching is a scene-dependent enhancement technique. Each scene must be evaluated independently to determine first whether contrast stretching is required, and if so, to determine what range within the data should be streched and to what degree.

Contrast enhancement is a technique which facilitates the overall extraction of geologic information from LANDSAT data. Band ratioing, on the other hand, is oftentimes employed specifically to improve lithologic discrimination. In this technique, corresponding brightness values from one band are divided by the brightness values from another band, and the resulting numbers are scaled to produce an image. Three such ratios from the same area can be used to generate a color-ratio image. The result of ratioing is to minimize the effects on scene brightness that result from variations in steepness and direction of topographic slope. Thus, reflectance differences between rocks, soils, and other surface materials which may result from compositional differences are enhanced. Figure 5 is a LANDSAT MSS band 5 image showing the west plunging nose of the Split Mountain anticline near Vernal, Utah. It is presented for comparison with Figure 6, which is a LANDSAT MSS bands 7/4 ratio of the same area. Notice how the topographic detail is suppressed and the lithologic variation is enhanced in the ratio image.

Data acquired in the four LANDSAT MSS bands for any given scene are often highly correlated; that is, brightness values of a certain magnitude in one band will likely predict the corresponding value in another band. The large covariance associated with highly correlated channels implies redundant information between channels. Principal components analysis is a linear transformation technique used to minimize data redundancy, as well as to enhance areas of subtle spectral disparity (Jenson and Waltz, 1979). Statistical parameters of the original data, such as variance and covariance, are used to calculate the transformation, which generates four new axes (bands) that are uncorrelated and ordered. The variance (information) is concentrated in the higher order axes, or components. For example, the first principal component contains the greatest overall variance in the original data, and the image is similar in appearance to an albedo image produced from the four original bands. Succeeding components contain less variance and typically produce a grainy appearing image, but an image which often contains information that is significant from a geologic point of view.

Figure 7 is the first principal component image produced from the four-band LANDSAT MSS data over Split Mountain. Notice the similarity of this image and the band 5 image of the same area shown on Figure 5. Figure 8 is the third principal component image produced from the same data and at the same time as Figure 7. Compare the lithologic variation visible in Figure 8 with that in Figure 7. Color compositing of three of the principal component images allows representation of more information and thus generally produces a more useful image than any of the single component images alone.

Canonical analysis is another type of data transformation commonly used to enhance spectral information related to different cover types such as rocks, soils, and vegetation. Statistical parameters (mean, variance, covariance) are generated from the image data for representative samples of the cover types to be enhanced. The data are mathematically transformed so that the within-catagory covariance is minimized and the among-category covariance is maximized, thus providing increased spectral separability of the selected cover types (Jenson and Waltz, 1979). The output is similar to the results of principal components analysis in that the amount of variance from the original data decreases in the lower order components.

Spatial filtering is an image-enhancement technique which enhances two-dimensional spectral frequencies in the data. Low pass filters use averaging techniques to enhance low-frequency tonal differences. Their application produces an image which appears to be out of focus, but an image in which larger regions of generally dark and generally light scene brightness become more obvious. The main use of such filters in processing LANDSAT data is in the production of high-pass filtered and edge-enhanced images. A high-pass filtered image results from subtracting the low-pass filtered image from the original image. An edge-enhanced image is produced by adding the high-pass filtered image back to the original image. In an edge-enhanced image, closely spaced changes in scene brightness are enhanced. Thus, features such as lineaments and drainage are commonly accentuated. Figure 9 is an enlarged, edge-enhanced LANDSAT band 7 image of a portion (Area A, Fig. 1) of the Qaidam Basin. Even allowing for the difference in image scale, notice how much more clearly the drainage networks and lineaments are defined on Figure 9 than on Figure 1.

Sometimes it is desirable to enhance linear features of a specified trend. Directional filters can be employed to accomplish this objective. They apply directionally weighted averaging functions to the data, as opposed to the nonweighted functions used in the standard edge enhancement techniques discussed previously. Figure 10 is an example of a directionally filtered LANDSAT band 7 image of the same area illustrated in Figure 9. In this image the northeast trending linear features have been enhanced. It should be pointed out that while directional filters successfully enhance features exhibiting the selected trends, they may introduce some artifacts in the image.

CHARACTERISTICS AND APPLICATIONS OF NEW AND FUTURE SATELLITE SENSORS

Although LANDSAT data have proven to be a useful tool in exploration geology during the past ten years, the 80-m spatial and the four-band spectral resolution characteristics of LANDSAT MSS data do impose limitations on their overall usefulness. The French SPOT satellite, which is scheduled to be launched in 1984, will be capable of providing multispectral data with 10 m spatial resolution, and it will be capable of acquiring stereoscopic imagery. Both of these improvements will provide new and improved data useful in petroleum exploration.

The recently launched LANDSAT 4 carries a new sensor called the thematic mapper (TM), which is the most significant improvement in LANDSAT sensor technology in the ten-year history of the program. The thematic mapper is a multispectral scanning system that senses reflected energy in six bands and emitted thermal energy in one band. Table 1 lists the nominal spectral bandpasses of the seven TM bands. The instantaneous field of view (IFOV) of the thematic mapper is 30 m, which results in distinct improvement in spatial resolution of TM images over images produced from Landsat MSS data. Perhaps of more importance to geologists, however, is the improvement in the spectral characteristics of TM data which result in significantly improved capabilities for lithologic discrimination.

TABLE 1
LANDSAT 4 THEMATIC MAPPER
SPECTRAL BANDPASSES

	WAVELENGTHS
Band 1	0.45 - 0.52 μm
Band 2	0.52 - 0.61 μm
Band 3	0.63 - 0.69 μm
Band 4	0.76 - 0.90 μm
Band 5	1.55 - 1.75 μm
Band 6	10.40 - 12.50 μm
Band 7	2.08 - 2.35 μm

For example, the inclusion of a 2.08 - 2.35μm band on the thematic mapper was a direct result of pressure from the geologic exploration community. The theoretical significance of this band relates to the pronounced absorption features in the spectra of clays and other hydrous minerals caused by the presence of bound OH^- ions. Some of these minerals occur in certain rocks and alteration types in sufficient quantities to affect the rock's spectral response in the 2.08 - 2.35 μm band. In addition, the widths of the TM bands in the visible and near-infrared portions of the spectra are narrower than those of the LANDSAT MSS, and the radiometric sensitivity of each band is greater. Thus, for example, the Fe^{+++} absorption band at 0.83 μm will be relatively more noticeable in TM band 4 because its influence is proportionately greater than it is in the wider 0.8 to 1.10 μm LANDSAT MSS band 7.

Although the first LANDSAT 4 TM images have been produced, TM data will not be routinely available for much of the United States until at least mid-1983. However, since the mid-1970s, the National Aeronautics and Space Administration (NASA) has been acquiring data from airborne instruments which simulate TM data. Studies of these data have shown that they do indeed provide more geologic, especially lithologic, information than LANDSAT MSS data. Numerous studies have demonstrated the usefulness of thematic mapper simulator (TMS) data, and thus by inference TM data, in delineating zones of alteration associated with hydrothermal ore deposits.

However, much less information is available on the usefulness of TM data in mapping sedimentary rocks.

Studies are currently being carried out at the EROS Data Center to investigate and document potential benefits to petroleum exploration from the application of TM data. Emphasis is being placed on determining the capability of these data to discriminate various sedimentary rocks occurring in the Uinta and Piceance Basins of northeastern Utah and northwestern Colorado. TMS data acquired over five detailed study areas which are representative of the sedimentary geology in the region are being studied. Image data from one of the areas are presented here to illustrate the spatial and spectral resolution characteristics and information content of the data.

Figure 11 is the TMS band 3 (0.63-0.69 micrometers) image of the Split Mountain detailed study area. This image shows the better spatial resolution in the TMS data when compared with LANDSAT MSS data (Fig. 5). These images provide a realistic comparison of the spatial resolution improvement that is characteristic of LANDSAT 4 TM data.

Figure 12 is a TMS bands 4/2 ratio image of the Split Mountain area. There is a more obvious separation of certain lithologic units in this image as compared to that provided by a single TMS band (Fig. 11). Also, there is better clarity and lithologic separation in the TMS ratio image as compared with those of the corresponding LANDSAT MSS ratio image of the same area (Fig. 6).

Figure 13 is a principal components image produced from seven-band TMS data; that is, the thermal channel data were also used in the creation of this image. While care must be exercised in the interpretation of images which include both reflected and emitted data, this image provides excellent separation of several units and thus would be useful in a geologic interpretation of the area.

Other digital processing techniques have also resulted in TMS images that contain useful geologic information. However, the examples presented here serve to illustrate many of the important benefits that will be realized in studies of sedimentary terranes through the application of LANDSAT 4 TM data. Further, it is important to realize that in the case of both the TMS ratio and principal components examples, significantly more lithologic information is displayed when three single-band images are composited to produce a false-color image.

DEVELOPMENT AND IMPLEMENTATION OF A DIGITAL GEOLOGIC DATABASE

Database development is not a new concept in exploration geology or other natural resource disciplines. In recent years, however, research into the processing of remotely sensed data has resulted in the advancement of digital processing techniques for analysis, merging, and integration of various spatial data types.

Geologic and geophysical data are commonly used to establish important spatial relationships in two- and three-dimensional space for surface and subsurface analysis. Conventional methods of display and analysis, however, have relied upon producing appropriate base maps and interpretive overlays. Merging various data types has involved drafting new maps or combining overlays, often a time-consuming and tedious process. Digital computer techniques provide a means for rapid access, manipulation, and display of a wide range of spatially related data. This section reviews the approach currently being used in the development of a digital database for targeting areas of potential hydrocarbon occurrence in the Uinta Basin region.

The first stage in database design involves the identification of project goals and the relevant data needed to achieve them. The acquisition of these data is contingent upon availability, cost, level of detail (resolution), and capability to process the data digitally. Once all the necessary data have been acquired, procedures and priorities are established for the types of processing and interpretation that will be conducted to extract information which will be evaluated through application of a conceptual exploration model.

Data employed by geologists are spatial in nature and are usually compiled from a variety of sources. Therefore, the structure of a regional digital database can be described in terms of two levels which involve manual interpretation, statistical and spatial analysis, merging, and integration. The results of these procedures may then be selectively put into a model designed to target areas for more detailed exploration.

The establishment of a Level I database involves several steps; some tasks are performed concurrently, others must follow a prescribed sequence.

The overall purpose of Level I is to have all original data converted to digital format and geographically referenced. This process includes the following: digitizing information such as lithologic units, fold axes, lineaments, faults, and structural contours from geologic maps; reformatting existing digital data for system compatibility; correcting radiometric and geometric errors in multispectral image data; modifying geophysical data, such as upward or downward continuation; and generating interpolated arrays for continuous variables, such as aeromagnetic, gravity, geochemical, and radiometric data. The results of these processes are sets of discrete points, lines, and polygons and continuous surfaces that are geographically referenced to a common coordinate system. These products are the basic datasets for further processing and interpretation.

The level II database work involves the actual processing of data for information extraction. Multispectral image data are typically enhanced, as previously discussed, to facilitate interpretation of geologic information. Geologic information interpreted from image analysis can be digitized to supplement other geologic data previously digitized from maps. For areas where no published geologic information is available, data interpreted from image studies become particularly important. Proximity analysis is one way of evaluating discrete data such as faults, fractures, fold axes, etc. digitized from geologic maps or interpretive overlays. A continuous surface image is produced which shows the distance any pixel is from a given type of feature (fault, fracture, fold). Thus, for example, the potential of any area's being underlain by a structural trap can better be evaluated. Such evaluations should, of course, be made in conjunction with available subsurface data.

Regional geochemical data can be used in the database and processed to locate anomalies that may be related to accumulation of hydrocarbons. Although specific relationships for the Uinta Basin are not firmly established, areas of anomalously high or low values (for certain elements) may prove to be of significance. Evaluation of radiometric data (U, Th, K) can be conducted similarly. Ratio calculations, slope generation, and color contouring may aid in the interpretation of these data.

Once the various data have been processed and the appropriate products (Level II) generated, a conceptual model which relates variables in each dataset to the potential for hydrocarbon occurrence can be tested and modified in an iterative process. Although such a model has not yet been implemented for the Uinta Basin, one approach can be illustrated. The datasets reside in the database as continuous surfaces (images) that are spatially referenced. Each grid cell (pixel) has a numerical value for each data layer. These values can be weighted according to the relative importance attached by the explorationist. The output from a simple, yet effective, additive model will be an image where the pixel values are the sums of the weighted values from the input datasets. Areas postulated to be favorable for detailed exploration will have high overall values. By selectively changing the input datasets and respective weights, various iterations of the model can be run and refined.

Digital geologic databases are still in the developmental stage, but their potential is significant. As more and more disparate data become available for specific areas of exploration interest, and as advances continue to be made in computer technology, digital geologic databases will become effective operational petroleum exploration tools in the decade ahead.

SUMMARY AND CONCLUSIONS

In the ten years since the launch of LANDSAT 1, satellite remote sensing for geologic applications has evolved from an unknown and unaccepted, unconventional technique to a respected part of geologic and exploration science. Over the years, much advancement has been made in techniques for digitally enhancing LANDSAT data, and now we are at the threshhold of a new era in sensor technology—an era that promises to provide useful new geologic information. Yet, some things remain unchanged and have little prospect for change in the near future. The need for systematic and comprehensive techniques of analysis and interpretation to extract all available information from satellite data will not change, and they must not be overlooked.

The future for geologic remote sensing is bright. Satellite data will continue to serve primarily as sources of geologic information, but the promise of new and better information is exciting. The development of digital geologic databases for resource exploration is an important part of the future of remote sensing. Such development is a natural extension of satellite remote sensing not only because satellite data are ideal for inclusion in a database but also because of remote sensing's growth with computer technology. Indeed, satellite remote sensing has become and will continue to be an important part of exploration geology.

ACKNOWLEDGEMENTS

The authors wish to thank Charles M. Trautwein, W. Terry Lehman, William C. Draeger, and Susan K. Jenson, for their many helpful suggestions during preparation of this manuscript. We also wish to acknowledge the work of Cynthia A. Sheehan who performed initial image enhancements of LANDSAT data from the Qaidam Basin.

REFERENCES

Bailey, G.B., and Anderson, P., 1982, Applications of LANDSAT imagery to problems of petroleum exploration in Qaidam Basin, China: American Association of Petroleum Geologists Bulletin, v. 66, no. 9, p. 1348-1354.

Bailey, G.B., Francica, J.F., Dwyer, J.L., and Feng, M.S., 1983, Extraction of geologic information in LANDSAT MSS and thematic mapper simulator data from the Uinta and Piceance Basins, Utah and Colorado: International Symposium on Remote Sensing of Environment, Second Thematic Conference, Remote Sensing for Exploration Geology, Fort Worth, Texas, 1982, Proceedings [in press].

Francica, J.R., 1983, A systematic approach to manual analysis and interpretation of LANDSAT Imagery of the Ladakh Himalaya: International Symposium on Remote Sensing of Environment, Second Thematic Conference, Remote Sensing for Exploration Geology, Fort Worth, Texas, 1982, Proceedings [in press].

Jenson, S.K., and Waltz, F.A, 1979, Principal component analysis and canonical analysis in remote sensing: American Society of Photogrammetry, annual meeting, Proceedings, v. 1, p. 337-348.

Rohde, W.G., Lo, J.J., and Pohl, R.A., 1978, Earth Resources Observation Systems Data Center LANDSAT digital enhancement techniques and imagery availability: Canadian Journal of Remote Sensing, v. 4, no. 1, p. 63-76.

Taranik, J.V., 1978, Characteristics of the LANDSAT System: U.S. Geological Survey Open File Report 78-187, 76 p.

Taranik, J.F., 1978, Principles of computer processing of LANDSAT data for geologic applications: U.S. Geological Survey Open File Report 78-117, 50 p.

Trautwein, C.M., Greenlee, D.D., and Orr, D.G., 1982, Digital data base application to Porphyry Copper mineralization in Alaska: U.S. Geological Survey Open File Report 82-801, 14 p.

U.S. Geological Survey, 1975, Workshop on remote sensing and Earth Resources Technology Satellite [LANDSAT] Interpretation: U.S. Geological Survey Open File Report 75-196.

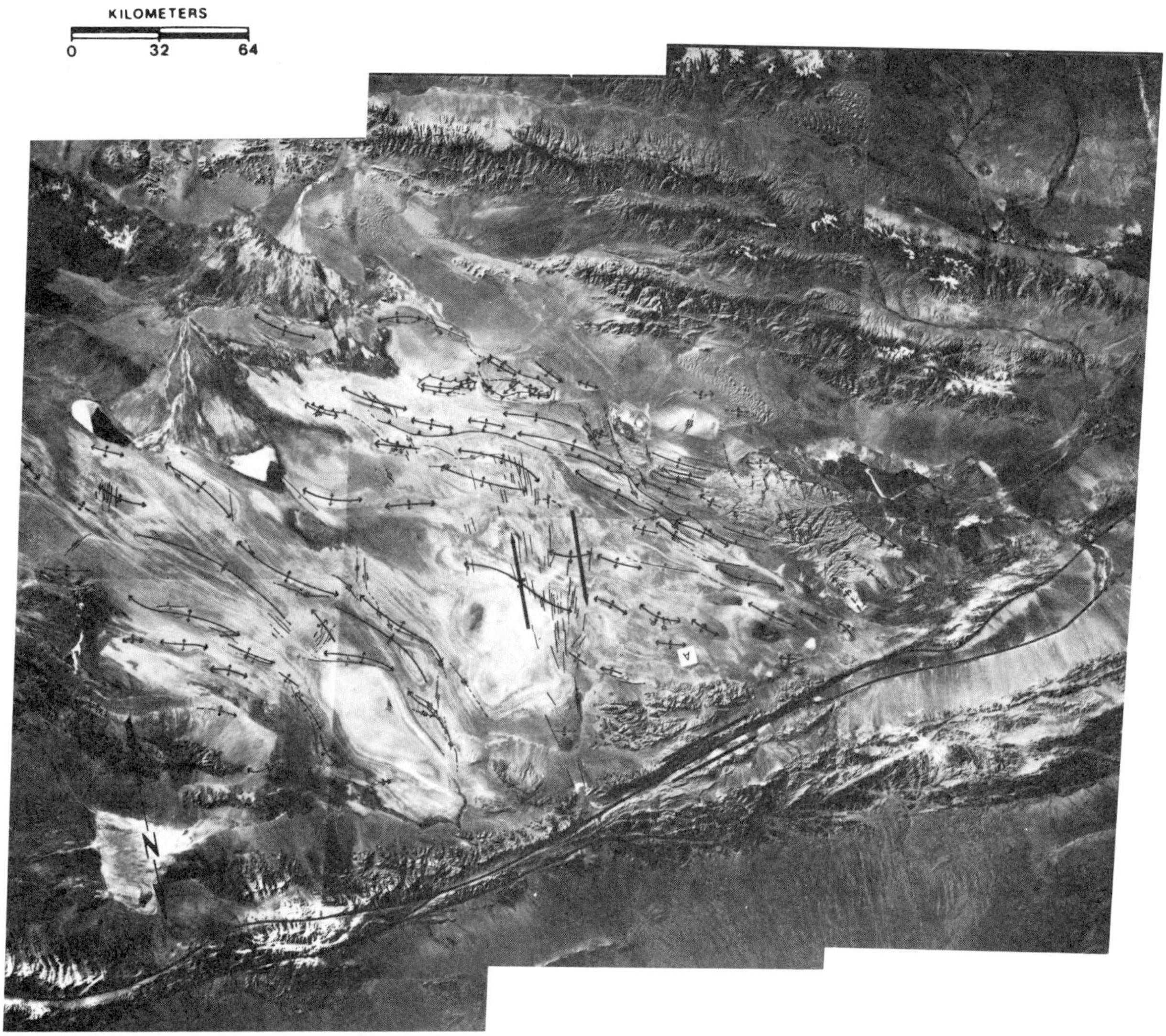

Fig. 1. Six-scene LANDSAT mosaic of the western part of the Qaidam Basin, People's Republic of China. (B&W copy of color original from Bailey and Anderson, 1982.) Note: Letter A denotes location of detailed study area.

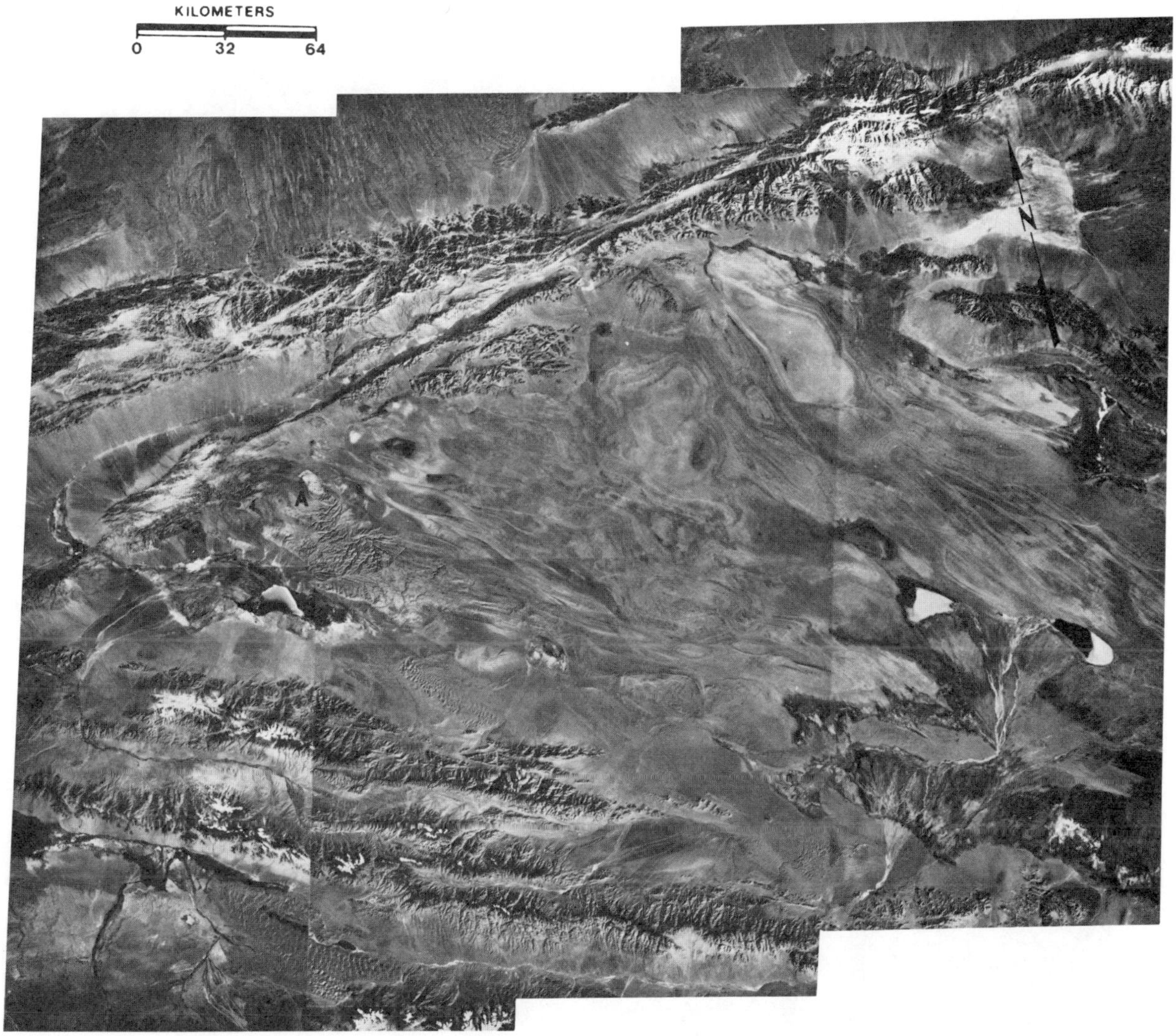

Fig. 2. Structural geologic interpretation of the western Qaidam Basin derived from manual analysis and interpretation of the six-scene LANDSAT mosaic (Fig. 1; from Bailey and Anderson, 1982). Standard structural notations are used. Note: Letter A denotes location of a buried anticline which was interpreted from image studies.

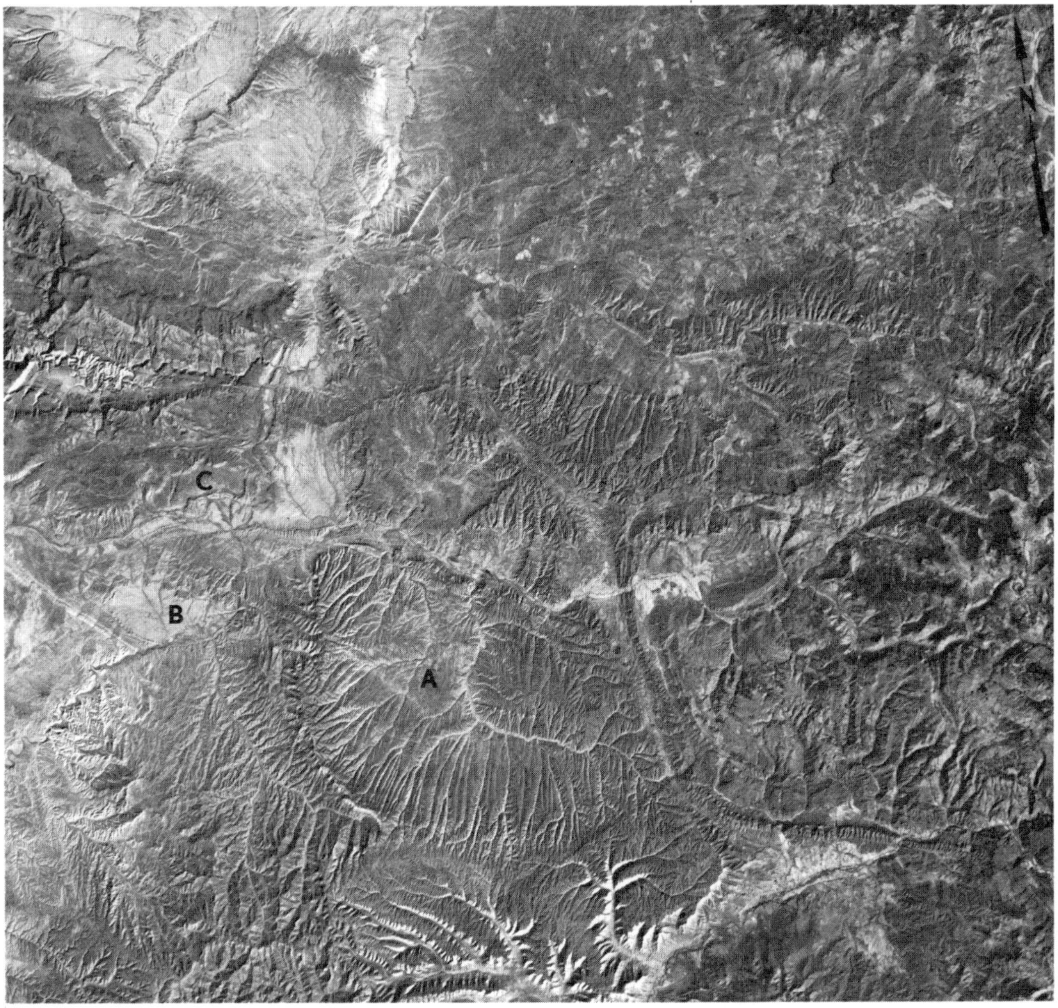

Fig. 3, Band 7 LANDSAT image of the Piceance Basin region, northwestern Colorado and northeastern Utah. Note: Letters A, B, and C denote locations of Piceance Creek Basin, Rangely anticline, and Skull Creek anticline, respectively.

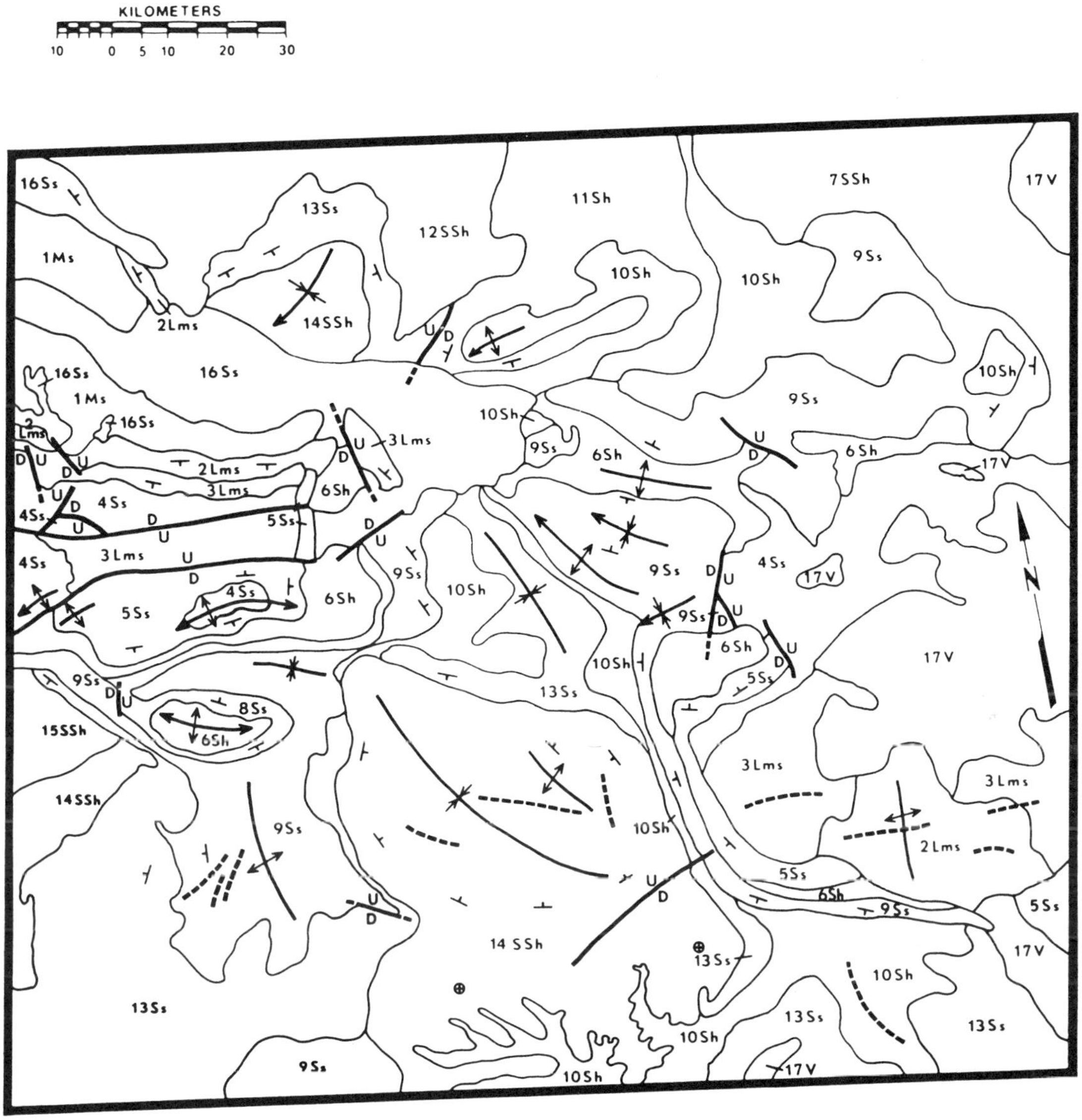

Fig. 4. Geologic map of Piceance basin region derived from a manual interpretation of the Landsat scene shown in Figure 3.

Figure 4 Explanation

Youngest

17V
Volcanic Rocks
Resistant rocks; form medium- to high-relief, subangular to subrounded topography. Course- to medium-textured, dendritic to subrectangular drainage. Very dark to medium brownish red. Subtle flow features in places.

16Ss
Sandstones (some shales)
Medium- to less-resistant rocks; form low-relief, rounded or smooth topography. Coarse- to medium-textured, dendritic to subparallel drainage. Light bluish grey to brownish-grey.

15SSh
Sandstones and Shales
Medium- to less-resistant rocks; form low- to medium-relief, subangular to rounded topography. Medium-textured, subparallel to dendritic drainage. Light yellowish tan where vegetation is sparse.

14SSh
Sandstones and Shales
Medium- to less-resistant rocks; form medium- to low-relief, subrounded to rounded topography. Generally medium-textured, parallel to dendritic drainage. Light to medium bluish grey where vegetation is sparse.

13Ss
Sandstones (some shales)
Mixed resistant and less-resistant rocks; form high- to low-relief, subangular to subrounded topography. Generally medium-textured, dendritic and subrectangular drainage. Commonly vegetated. Medium bluish grey to brownish grey where vegetation is sparse.

12SSh
Sandstones and Shales
Medium-resistant to nonresistant rocks; generally form medium- to low-relief, subangular to rounded topography. Medium- to fine-textured, dendritic and subrectangular drainage. Light bluish grey to brownish grey.

11Sh
Shales
Nonresistant rocks; form low-relief, subangular to subrounded topography. Fine-textured, pinnate drainage. Dark grey to brownish grey.

10Sh
Shale (with some sandstones)
Generally nonresistant rocks; form low-relief, subrounded topography. Fine-textured, dendritic and sub-parallel drainage. Sparse natural vegetation; medium to light bluish grey.

9Ss
Sandstones (with inter-bedded shales)
Mixed resistant and less-resistant rocks; form mostly medium-relief, angular and subangular to rounded topography. Prominent ridge former. Medium- to fine-textured, subdendritic and subparallel drainage. Commonly vegetated, medium to light bluish grey where vegetation is sparse.

8Ss
Sandstones
Medium-resistant rocks; form low- to medium-relief, subangular topography. Medium- to coarse-textured, subparallel and subrectangular drainage.

7SSh
Sandstones and Shales
Mixed resistant and less-resistant rocks; form high- to medium-relief, subangular to subrounded topography. Subrectangular to subdendritic drainage. Medium to dark brownish red or brownish grey depending on type and density of vegetation.

6Sh
Shales
Nonresistant rocks; form low-relief, subrounded topography. Fine-textured, dendritic drainage. Typically light bluish grey in color with sparse vegetation.

5Ss
Sandstones
Resistant rocks; form medium-relief, subrounded to subangular topography. Medium-textured, subparallel to dendritic drainage. Reddish brown where more densely vegetated, brownish grey where vegetation is less dense, light tan where vegetation is absent.

4Ss
Sandstones
Generally resistant rocks; form medium- to high-relief, subangular to subrounded topography. Coarse- to medium-textured, subparallel to subrectangular drainage. Brownish red and grey where vegetated. Very light tan where vegetation is absent.

3Lms
Limestones
Generally resistant rocks; form medium-relief (subdued), subrounded to rounded topography. Coarse-textured, subparallel to subrectangular drainage. Brownish red where vegetated, medium to dark grey where vegetation is sparse.

2Lms
Limestones
Generally resistant rocks; form high- to medium-relief, subangular to subrounded topography. Medium-textured, dendritic to subrectangular drainage. Brownish red where vegetated, light grey where vegetation is sparse.

1Ms
Meta-sedimentary Rocks
Mixed resistant and less-resistant rocks; form medium- to high-relief, angular to subrounded topography. Coarse-textured, parallel and dendritic drainage. Dark reddish brown where vegetated, dark grey where vegetation is sparse. Subtle parallel banding in places.

Oldest

INTERPRETED LITHOLOGIC CONTACT

DIRECTION OF STRIKE AND DIP OF SEDIMENTARY ROCKS

AXIS OF FOLD IN SEDIMENTARY ROCKS

FAULT U D

PROBABLE FAULT

NOTE: Colors refer to appearance on a false-color composite image and will vary with acquisition date and color reproduction.

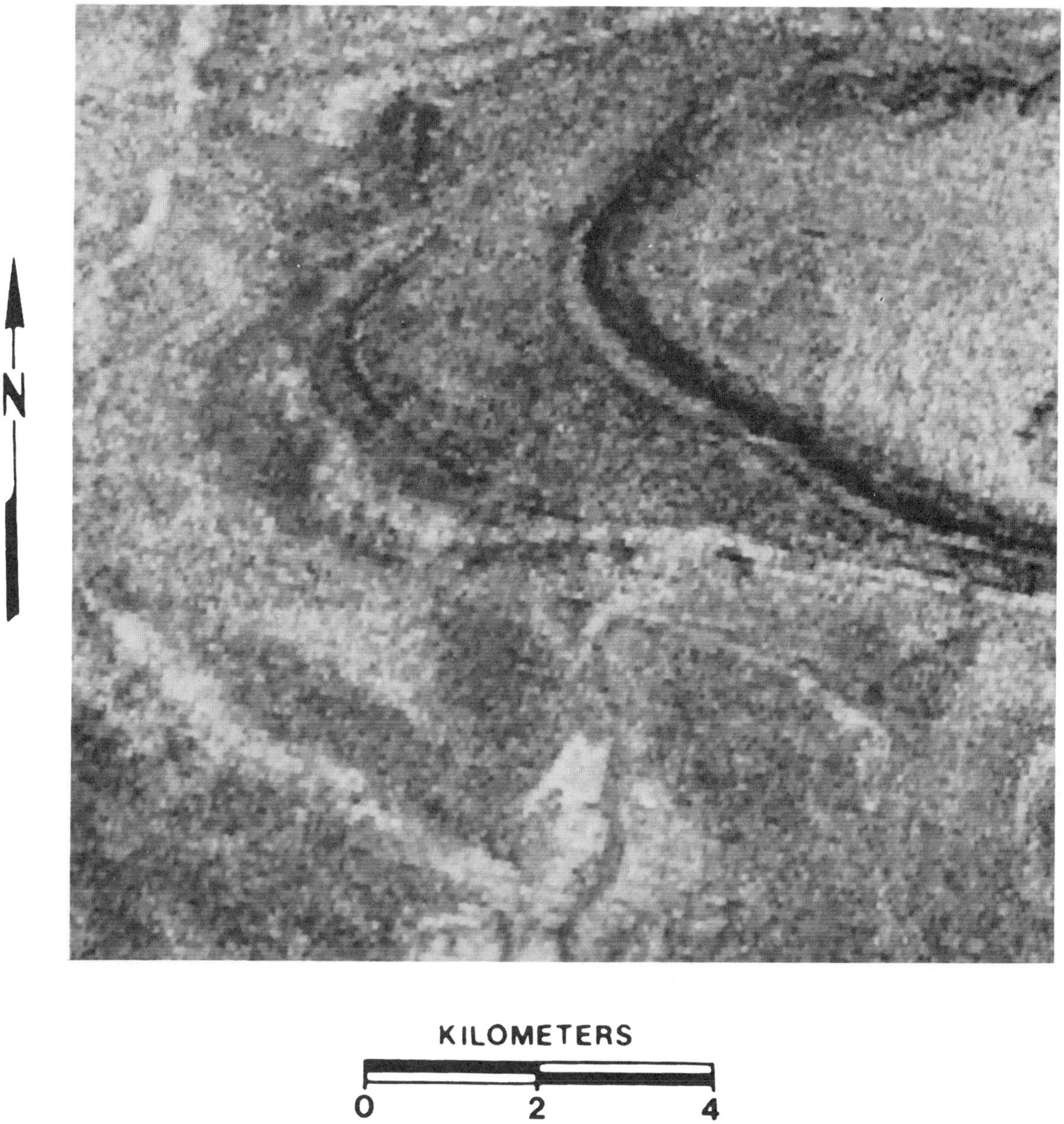

Fig. 5. LANDSAT band 5 image showing the western part of the Split Mountain anticline near Vernal, Utah. The image is an enlarged LANDSAT subscene from data acquired in June, 1973.

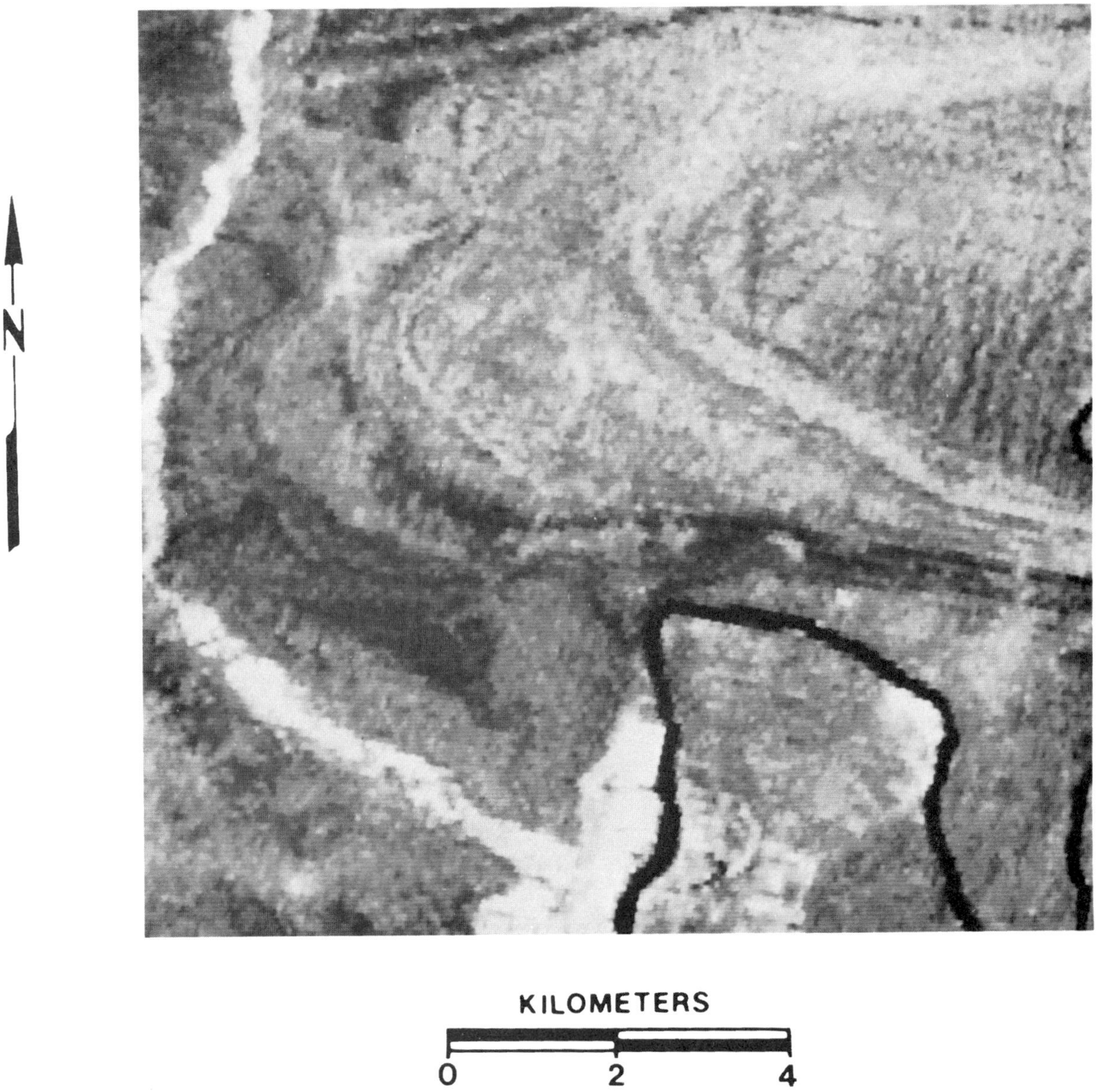

Fig. 6. LANDSAT bands 7/4 ratio image produced from data acquired over the Split Mountain anticline in June, 1973.

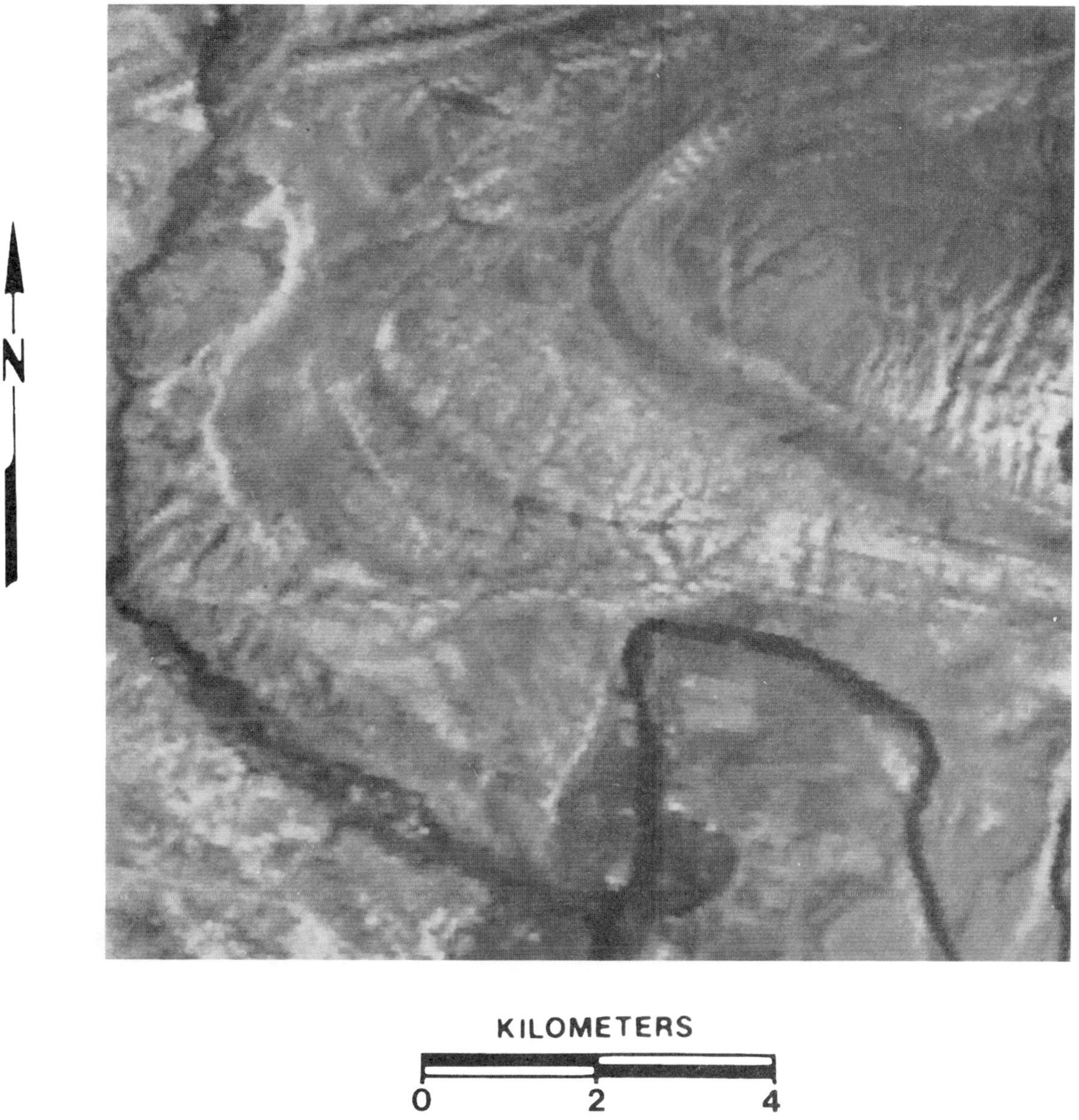

Fig. 7. LANDSAT principal components image of the Split Mountain anticline. The image is the first principal component and was produced from LANDSAT MSS data acquired in June, 1973.

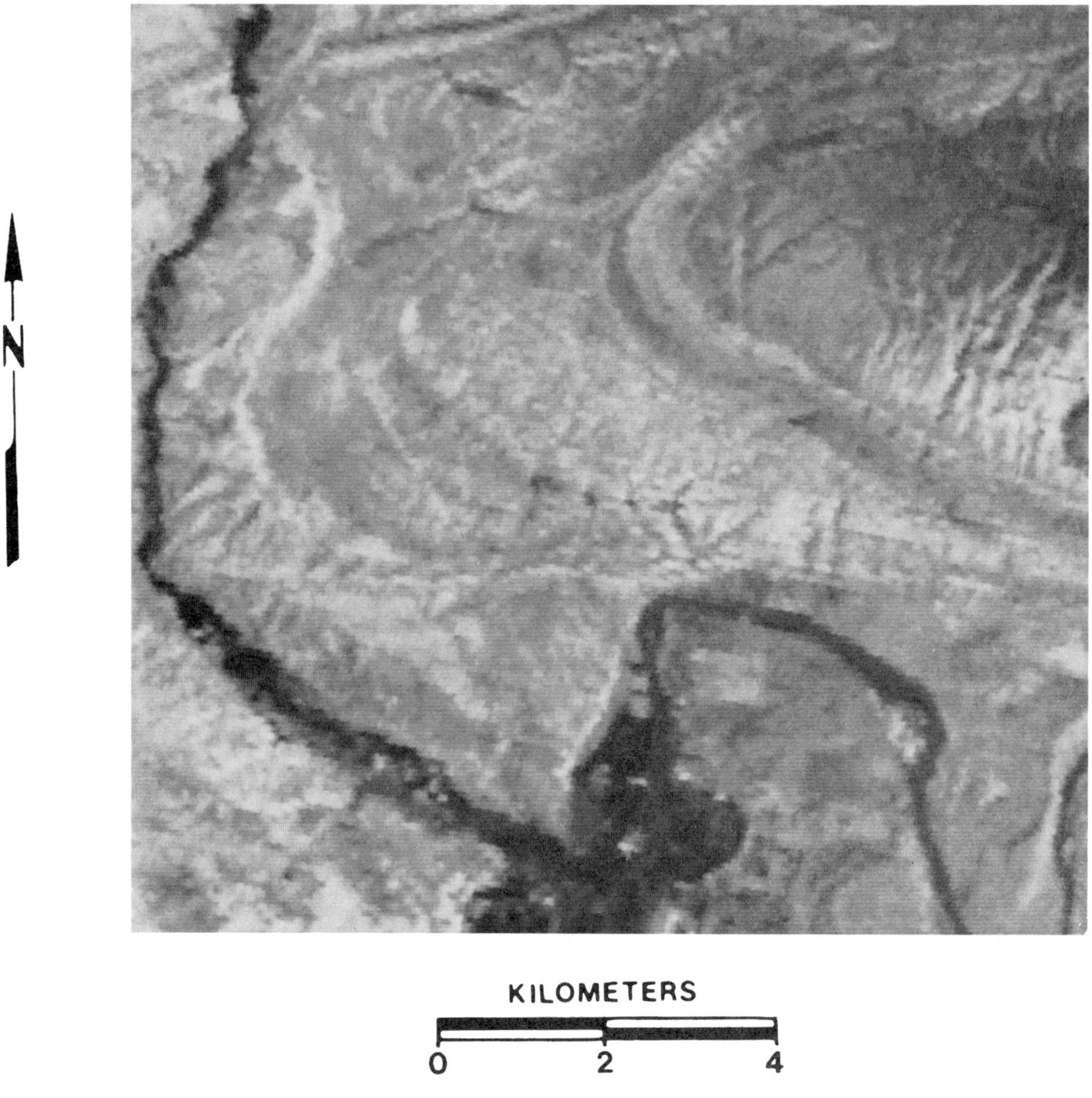

Fig. 8. Image of the third principal component produced from the same LANDSAT MSS data and at the same time as Figure 7.

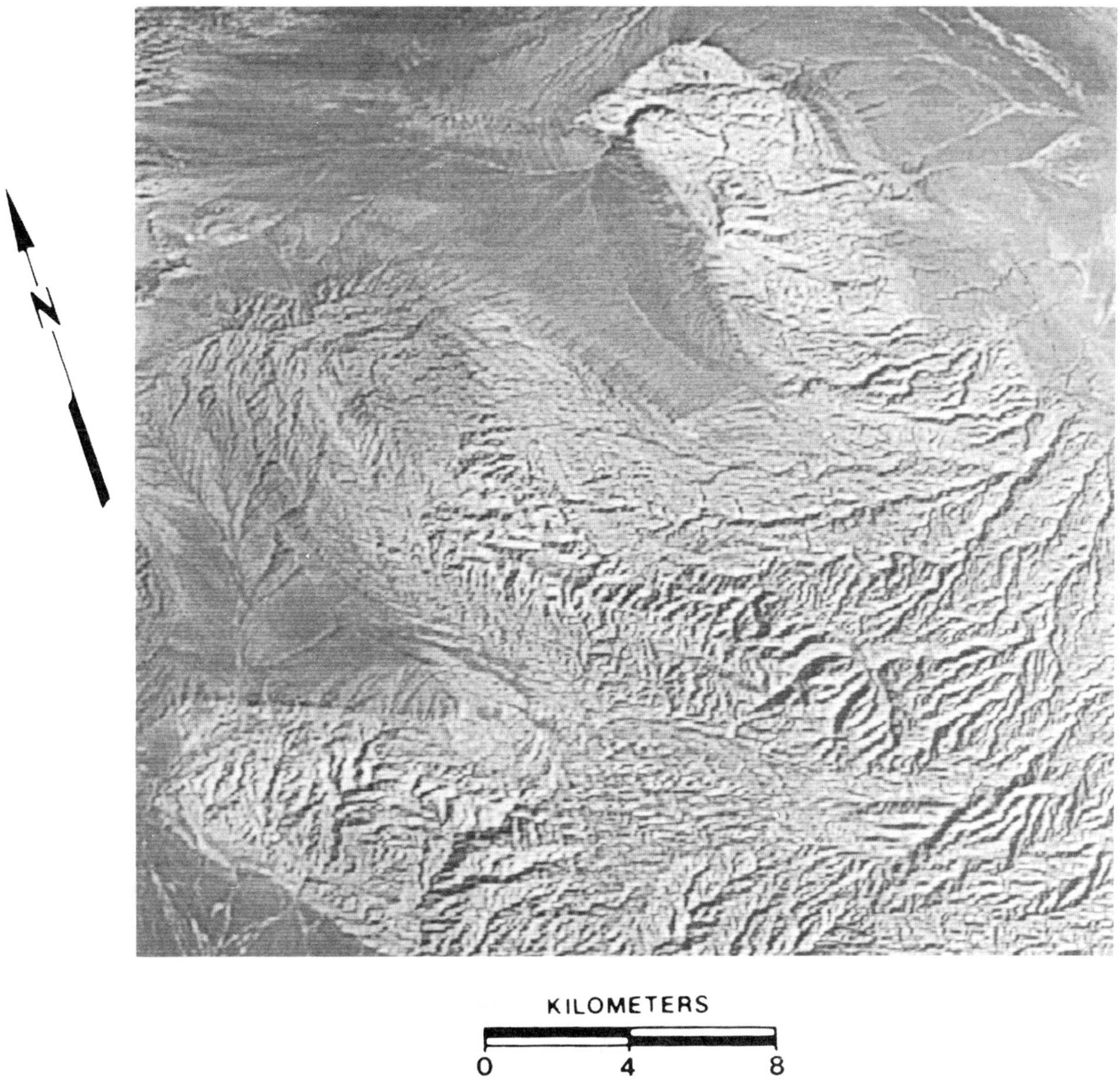

Fig. 9. Edge-enhanced LANDSAT band 7 subscene showing the northern part of the Youshashan area, Qaidam Basin, People's Republic of China (Area A, Fig. 1).

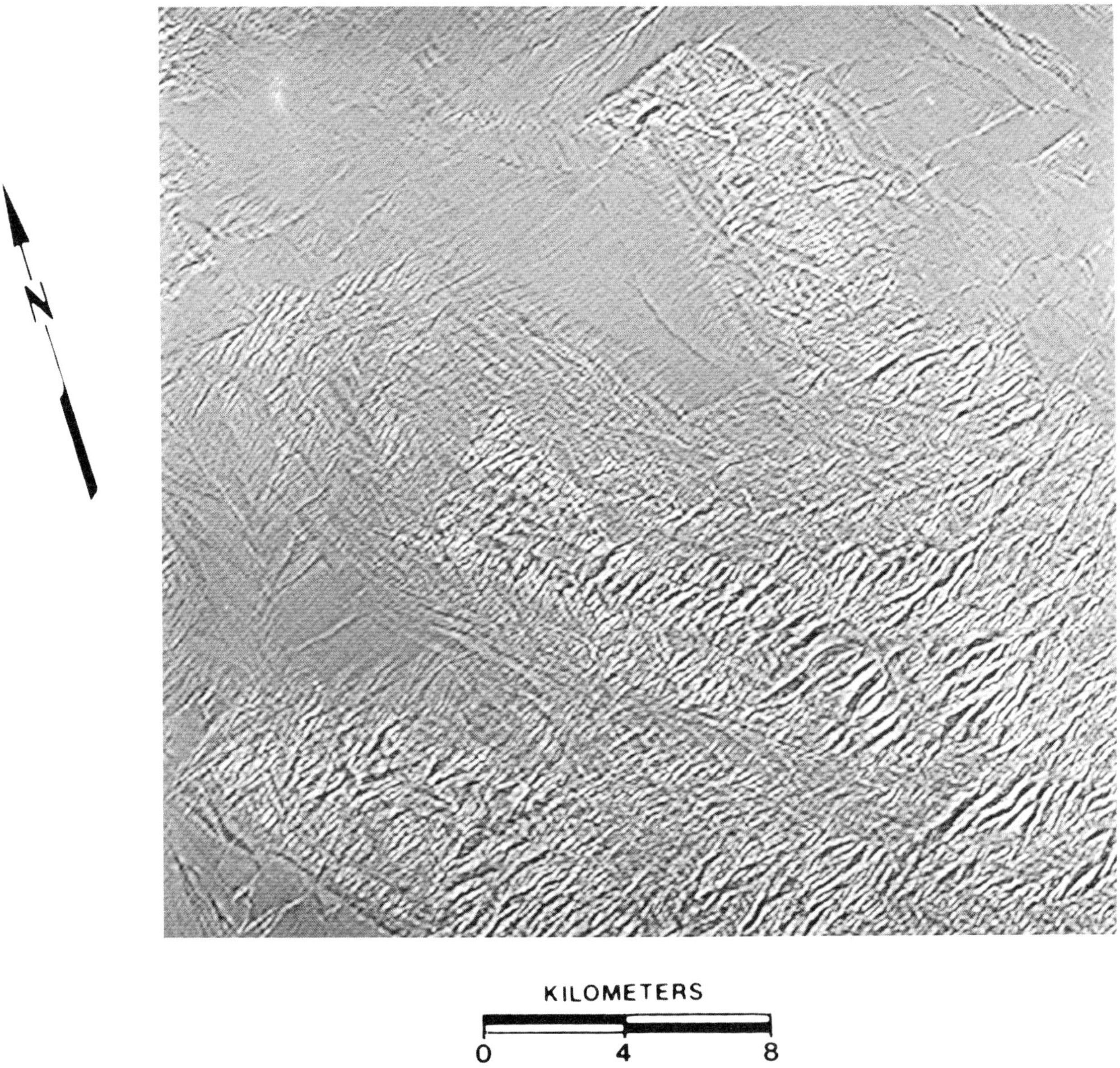

Fig. 10. Northeast directionally enhanced LANDSAT band 7 subscene showing the northern part of the Youshashan area, Qaidam Basin, People's Republic of China (Area A, Fig. 1).

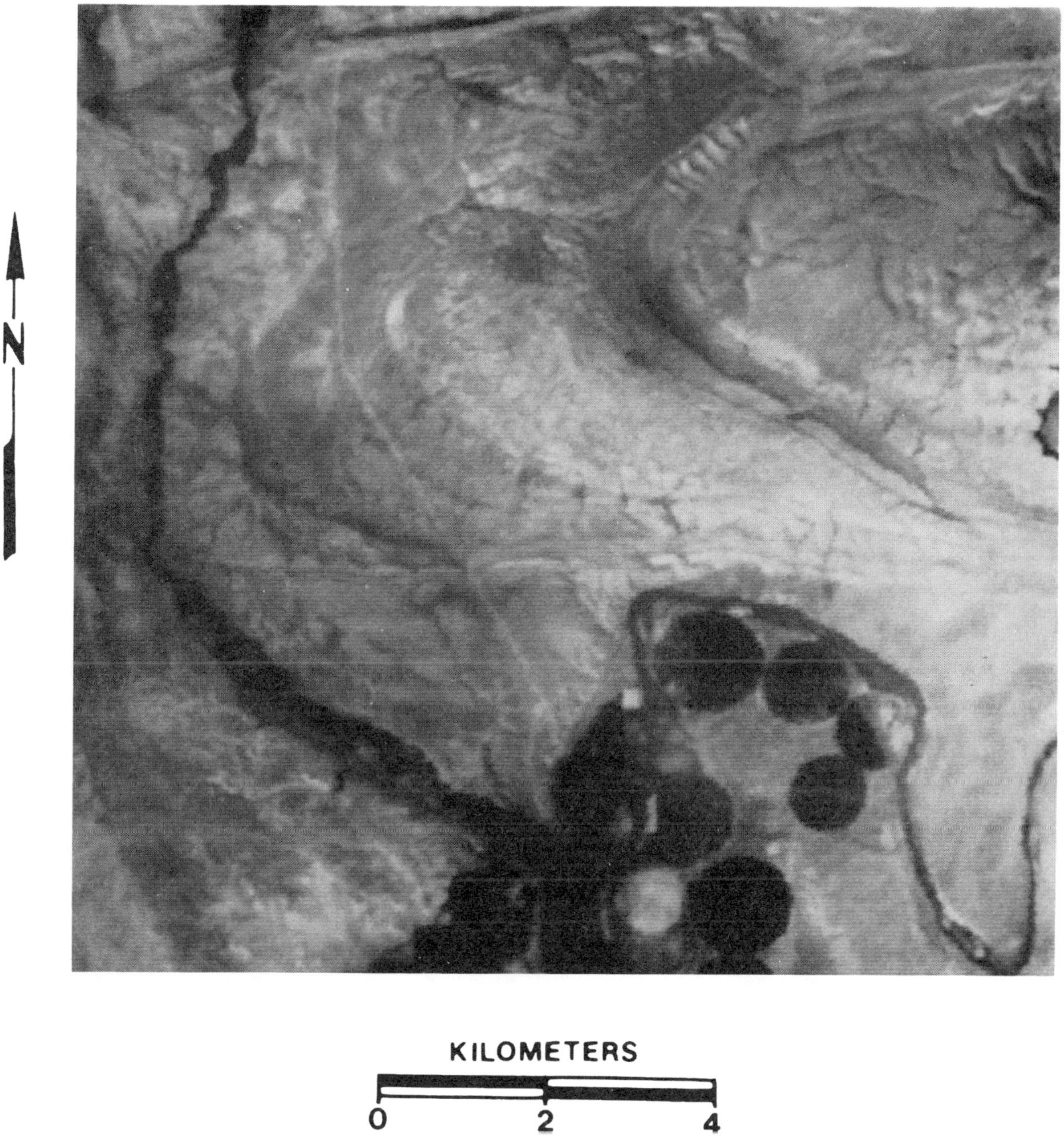

Fig. 11. TMS band 3 image of the Split Mountain anticline.

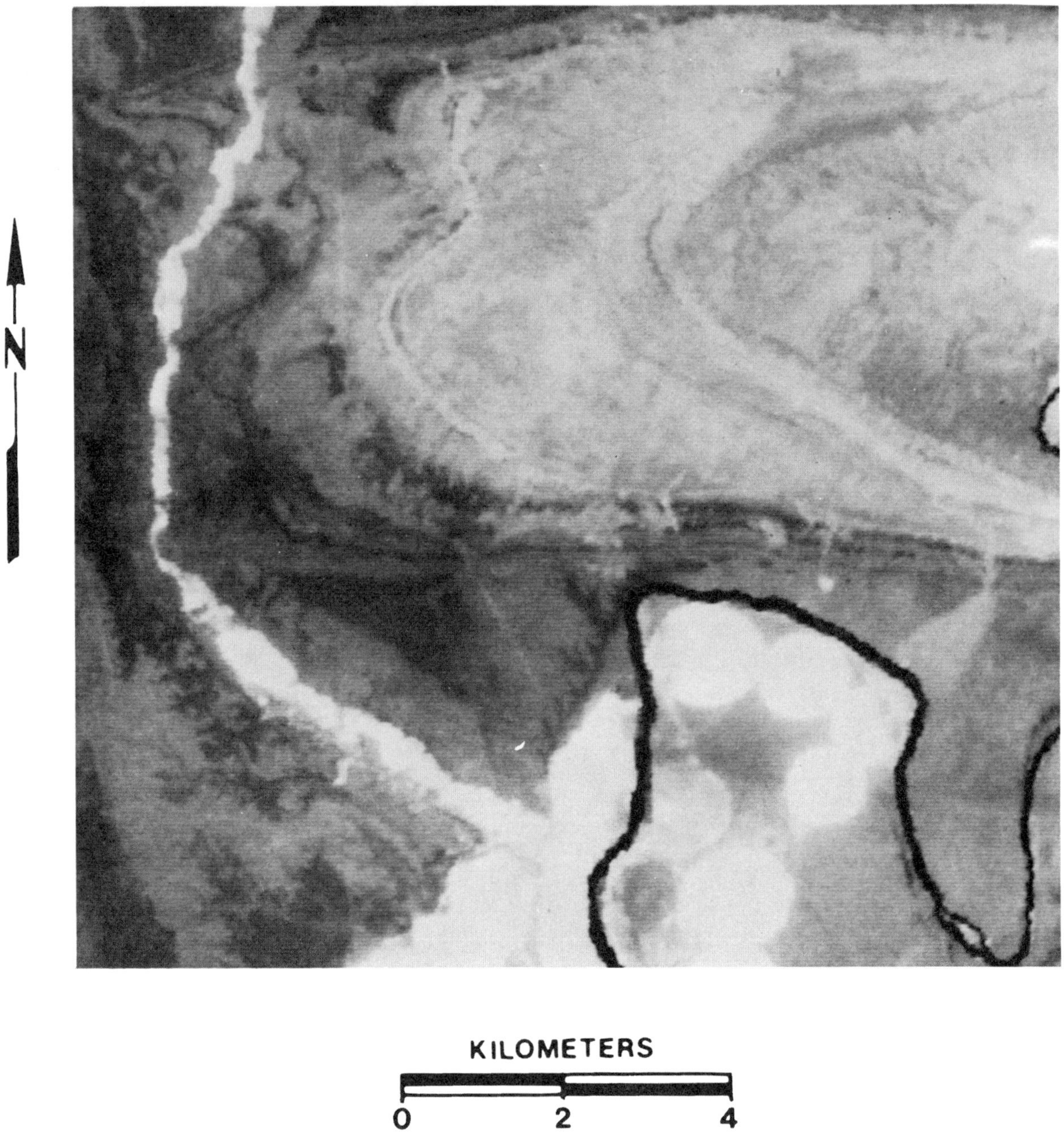

Fig. 12. TMS bands 4/2 ratio image of the Split Mountain anticline.

Fig. 13. TMS principal components image of the Split Mountain anticline. The image is the fourth principal component produced from seven-band TMS data.

DEEP WATER DISCHARGE: A MECHANISM FOR THE VERTICAL MIGRATION OF OIL AND GAS

Paul H. Jones

P. H. JONES HYDROGEOLOGY, INC.

INTRODUCTION

Upward movement of formation water in deep sedimentary basins has long been recognized as an important factor in geothermal heat flow and sediment mineral diagenesis. Its importance in the migration and accumulation of fluid hydrocarbons has been considered by many investigators, but only during the past thirty years or so has information been developed that enables description and explanation of a rational vertical migration model. Key elements in this model are the widespread occurrence of geopressure below some known depth in sedimentary basins, information regarding the depth-temperature, depth-fluid pressure, and depth-formation water salinity distributions; information on the sediment facies distribution and mineral diagenesis; and geologic structure. These factors define the basin hydrogeology and hydrology. Other key elements, for which information has been available only during the past ten years or so, are the aqueous solubility of whole petroleum and natural gas under the conditions of elevated temperature and pressure known to occur in the geopressure zone at great depth; the kinetics of organic matter diagenesis (the generation, preservation, and natural cracking of fluid hydrocarbons); and the effects of dissolved inorganic solids and gases on the solubility and stability of fluid hydrocarbon molecules.

Field and laboratory data regarding key elements of the fluid migration model, necessary to test the usefulness of the model as an exploration tool in the search for oil and gas, are available in the public record for the northern Gulf of Mexico basin. A study area in coastal southeastern Louisiana was selected because it provides an opportunity to test the model in deposits ranging in age from upper Oligocene to middle Miocene in an area where the top of the geopressure zone has been mapped, the geologic structure is well known, and both commercially productive and nonproductive structural traps occur. The study area (Fig. 1) is a belt some twenty miles wide and eighty miles long, extending from Hancock County, Mississippi, southwestward to Valentine dome in Lafourche Parish, Louisiana.

THE FLUID MIGRATION MODEL

The general concept of the fluid migration model (Fig. 2) asserts that movement of formation water in the geopressure zone in the general direction of fluid pressure release is reflected by changes in the temperature, interstitial fluid pressure, and salinity of formation water in the zone of fluid movement. The concept also identifies structural settings most likely to favor such fluid movement as those where the gravitational stress due to the sediment load is locally reduced by structural deformation. Such settings occur generally along growth faults, in fault zones adjacent to structural grabens, and where expansion of section occurs above diapiric structures such as salt domes. Although they are commonly referred to as "pull apart" features, such structures are in reality the result of gravity sliding on shear zones defined by interstitial fluid pressure gradients and sediment facies boundaries where changes in the shear strength of the rock matrix occur.

The structural setting of Figure 2 is the growth fault; downthrow has occurred along two primary shear zones and one secondary shear zone (antithetic fault). The dips of the shear planes decrease with depth as functions of interstitial fluid pressure (Bruce, 1973), and faults die out both upward and downward. The upward limit of growth faulting corresponds in general to the top of the geopressure zone; the lower corresponds to the depth at which interstitial fluid movement approaches zero.

In the fluid migration model, the upward movement of formation water from depths where the fluid pressure gradient exceeds 0.7 pounds per square inch per foot of depth (psi/ft) is focused upon the lower part of the main growth fault (F-1), from which it is diverted to the overlying growth fault (F-2) through sands 1, 2, 3, 4, 5, and 6. These are, or have been, hydraulically continuous with sands in the rollover sequence. Downthrow on faults F-1 and F-2 has reduced the horizontal component of compressional stress in deposits above them because of the curvature of the fault planes; compensation has resulted in movement on the antithetic fault F-3, the location and dip of which are also controlled by the fluid pressure regime.

Downwarp of the 0.7 psi/ft pressure gradient isopleth is the result of water movement along the paths indicated by the open arrows. The water that has moved upward is at a higher temperature than that of the water displaced, and the 250°F isotherm is upwarped in the deposits it occupies. As the salinity of the rising water is lower, the invaded aquifers are freshened. Viewed together, the isopleths of pressure, temperature, and water salinity reflect the paths of fluid movement and the bulk flow distribution by their displacements. It is apparent that, in the fluid migration model, maximum movement has occurred along the growth faults F-1 and F-2 in the deeper deposits. Flow was diverted up the zone of the antithetic fault F-3, resulting in the development of two "highs" on the 200°F isotherm, and the maximum drop in fluid pressure (shown by the downwarp of the 0.5 psi/ft isopleth) is in the deposits between the growth faults F-1 and F-2.

Oil and gas exsolved from the rising waters—as pressure and temperature drop occurred along the flow paths—accumulated in structural traps that were formed as movement on the faults tilted beds and closed avenues of fluid escape. As shown in the fluid migration model, continuing movement on faults may open up oil- or gas-filled traps, in a valvelike fashion, by placing sand beds across faults in positions previously occupied by shale. Oil and gas may or may not be released in such instances, depending upon such factors as capillarity, pressure differentials, and silica cementation.

THE STUDY AREA

The area selected for test of the usefulness of the fluid migration model as an exploration tool in the search for commercial deposits of oil and gas has been thoroughly explored by conventional techniques. There are at least fifty producing fields and several thousand drilled locations, and known structural traps either are barren or contain noncommercial accumulations of oil and gas. The test was designed not only to identify new prospects but also to explain why drilled prospects were not productive, and perhaps to indicate locations where production might be developed.

Subsurface conditions were analyzed and interpreted using electric logs for some 250 wells drilled in or near the study area, maps of geologic structure available from commercial sources, published reports, and data in the public record. Two regional geologic sections were prepared, roughly parallel and 2 to 8 mi apart, from northeast to southwest. (See Figs. 3 and 4.) A map of geologic structure, based upon marker beds a short distance from the top of the geopressure zone, is divided into three provinces (Fig. 5).

In Province 1, to the northeast, the marker is the *Heterostegina* limestone; in Province 2, the marker is the *Cristellaria "I"* sand, and in Province 3, to the southwest, the marker is the *Textularia "W" stapperi* sand.

GEOLOGY

The study area is underlain by a massive sequence of progradational deposits, mainly alluvial and deltaic sand and shale, which increases in thickness gulfward from a few thousand feet at the northeast to perhaps 50,000 ft at the southwest. Several regional growth fault systems cross the study area obliquely; the northernmost of these, downthrown gulfward, follows the north shore of Lake Pontchartrain and crosses the north end of Lake Borgne (Fig. 5). The vertical displacement across this fault system exceeds 10,000 ft. In Province 2, fault displacements are small except near the crests of diapiric structures. In Province 3, growth faults of large displacement flank the Valentine dome and form the Hollywood-Valentine graben. Here the massive Harang shale is overlain by a transitional series of interbedded sands and shales, which underlie the main sand series. (See Figs. 3 and 4.) In the transitional series there are several structural traps, some of which contain commercial accumulations of oil and gas.

FLUID MIGRATION

Each of the structural traps shown on the geologic maps and sections can be analyzed using the fluid migration model, by preparing detailed maps showing selected isothermal, isopressure, and isosalinity conditions at depth. For each well used as control, electric well-log heading data on maximum bottom-hole temperature for each logging run were corrected to equilibrium temperature using the correction curve for coastal Louisiana (Kehle, 1972), and the depths at which equilibrium temperature was 200°F, 250°F, and 300°F (if reached) were interpolated. Isothermal maps were made showing the depth of occurrence of the 200°F isotherm (Fig. 6) and the 250°F isotherm (Fig. 7) in the study area.

For each well used as control, mud weight data on the electric log heading for each logging run were expressed in terms of the depth-pressure gradient for the depth interval of the logging run. These calculations were verified using shale resistivity data on the electric log and the Schlumberger conversion chart (George, 1959). The depth of occurrence of the 0.5, 0.7, and 0.9 psi/ft depth-pressure gradients were interpolated for each well, and a map was made showing the depth at which the fluid pressure gradient reached 0.7 psi/ft in the study area (Fig. 8). The profiles of the 200°F, 250°F, and 300°F isotherms and the 0.5, 0.7, and 0.9 psi/ft pressure gradient isopleths are shown on the geologic sections, Figures 3 and 4.

Formation water salinity change with depth at each well location was estimated using the spontaneous potential curve values on the electric log and the Schlumberger equation. Because of the inaccuracy of the method, the salinity of water in sand-bed aquifers was expressed as multiples or fractions of sea water salinity (~35,000 mg/1). No maps of the formation water salinity/depth relations are included in this paper, but work maps were prepared and used in the study.

Using the guidelines developed in the fluid migration model (Fig. 2), the isothermal, isopressure, and isosalinity maps were analyzed. Anomalous upwarps of isotherms and salinity isopleths and downwarps of pressure gradient isopleths were investigated using map superposition and locality profiles. The locations of all apparent temperature (T), pressure (P), and salinity (S) anomalies in the study area are shown on Figure 9. Such anomalies do not always indicate fluid migration; salt and shale diapirs can produce similar anomalies, as is evident in Province 2. Of the some thirty anomalies identified in Figure 9, local geologic conditions support fluid migration in only twelve.

Of these twelve anomalous fluid migration localities, the locations of which are shown on Figure 10, eight are occupied by or are adjacent to oil and/or gas fields. All seven identified fluid migration localities in Province 3 are large scale anomalies. The four migration localities in Province 2 are smaller scale anomalies, probably because displacement on faults is generally less than in Province 3. Only one of the twelve identified fluid migration anomalies occurs in Province 1, and this is near its boundary with Province 2. Although it is areally the largest anomaly mapped, no existing oil or gas fields lie within or near it.

The fluid migration anomaly south of Valentine dome (Province 3, Fig. 5) was selected for detailed analysis. As shown on the north-south cross section, Figure 11, the Hollywood-Valentine graben is bounded on the north by three major faults along which stepwise downward displacement to the south exceeds 1,600 ft on the *Cristellaria "I"* marker. The graben is bounded on the south by a horst; down-to-the-north displacement of the *Bigenerina humblei* marker on the south boundary fault is almost 1,000 ft. On the south flank of the horst, down-to-the-south displacement of the *Bigenerina humblei* marker exceeds 1,400 ft on two closely spaced faults.

Upward movement of formation water from great depth is indicated by downwarp of the 0.7 and 0.9 psi/ft pressure gradient isopleths and upwarp of the 250°F isotherm. Upward flow to intermediate depth has been most pronounced along the northern margin of the graben. Flow was probably upward along the main fault to the massive sands that immediately underlie the *Cristellaria "I"* marker in the Texas Company, DeLaune No. 1 well. Thence, it was northward to the third fault and upward, in the vicinity of the McCarter, South Coast well, as indicated by the sharp upwarp of the 200°F isotherm. A divergence of the upward flow appears to have occurred southward across the graben in *Textularia "W" stapperi* sands, which rise steeply southward to a fault which branches off the main south boundary fault in the vicinity of the California Co., A. J. Hotard No. 1 well. The upwarp of the 200°F isotherm centered at the Hotard well could be the result of upward flow along the faults that bound the horst on the south, but this interpretation is not supported by the fluid-pressure profile; downwarp of the 0.5 psi/ft isopleth occurs only at the Hotard well.

Inferred major fluid migration paths are indicated by heavy arrows on Figure 11. The drop in temperature and fluid pressure and increase in formation water salinity along these flow paths reduce the aqueous solubility of fluid hydrocarbons in the moving water. Exsolution and accumulation of natural gas along these flow paths could have filled the commercial gas reservoirs tapped by the Hotard No. 1 well and the South Coast well. The graben-horst feature, redrawn in simplified form as Figure 12, shows how the area of expansion (AOE) tends to be the focus of deep water discharge, with entrapment of fluid hydrocarbons in the downthrown block or in either of the adjacent upthrown blocks, depending upon sediment facies distribution and the flow paths of ascending fluids. The AOE at the intersections of faults of large displacement on the southeast flank of the Lake Long structure (Fig. 13) is associated with a strong temperature anomaly identified as T-17 on Figure 9.

By far the most important locus of deep water discharge by vertical fluid migration in the northern Gulf of Mexico basin is the area of structural expansion formed by the intersection of two or more growth faults, as shown in Figure 14. How and where the oil and gas are likely to occur is shown on the fluid migration model, Figure 2. It is in this setting that most commercial oil and gas accumulations are found in the Gulf Coast.

CONCLUSIONS

It is evident that deep water discharge is an effective mechanism for the vertical migration of oil and gas, and that the fluid migration model can be a very useful exploration tool. In the study area, the model was most effective in deposits of middle Miocene age where displacements on gravitational faults generally exceeded 500 ft.

Structural features which focus deep water discharge and form traps suitable for the accumulation of oil and gas include growth faults, graben-horst systems, and domal upwarp from shale and/or salt diapirism.

REFERENCES

Bruce, C.H., 1973, Pressured shale and related sediment deformation: mechanism for development of regional contemporaneous faults: American Association of Petroleum Geologists Bulletin, v. 57, p. 878-886.

Gatenby, G. M., 1980, Exploration ramifications of subsurface fluid migrations in the Lake Borgne-Valentine area of southeastern Louisiana: Gulf Coast Association of Geological Societies, Transactions, v. 30, p. 91-104.

George, Rush, 1959, Approximate formation pressure from resistivity logs in south Louisiana: Schlumberger Well Surveying Corp., New Orleans, Louisiana (Log Interpretation Chart).

Kehle, R. O., 1971, Geothermal Survey of North America, 1971 annual progress report: unpublished duplicated report, Research Committee, American Association of Petroleum Geologists, Tulsa, Oklahoma, 31 p.

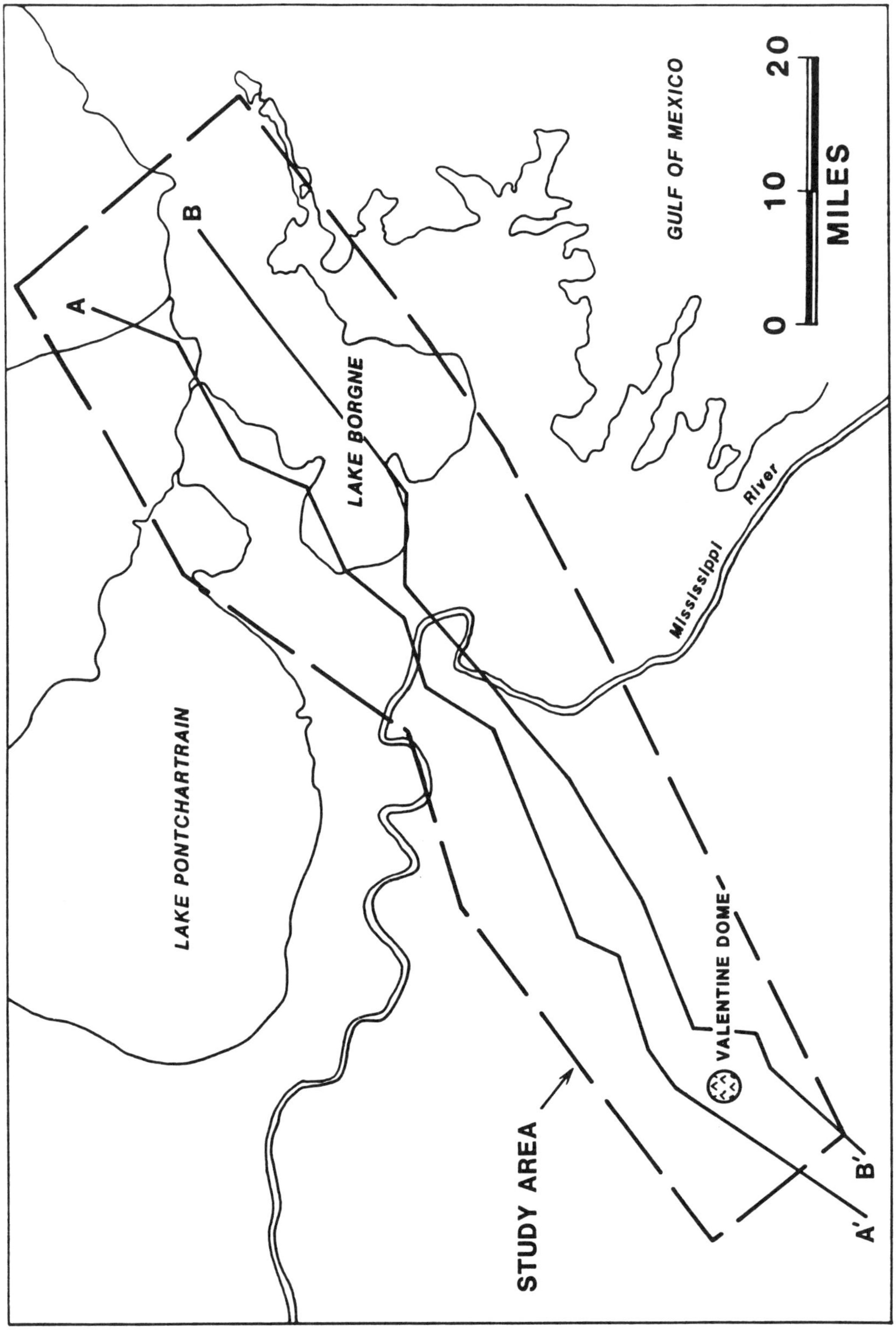

Fig. 1. Locations of the study area and lines of geologic cross sections A-A' and B-B' in coastal southeastern Louisiana.

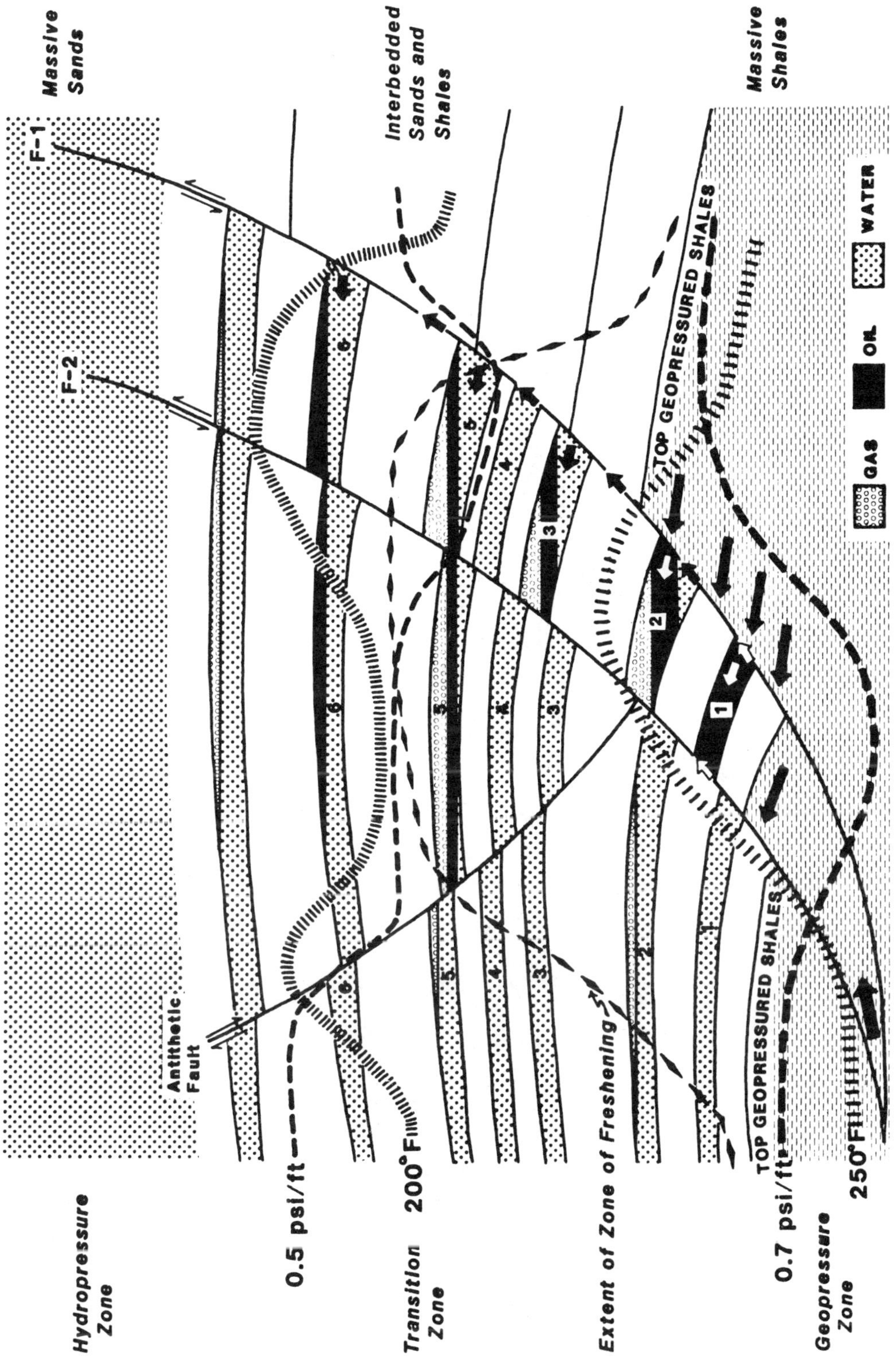

Fig. 2. Hypothetical fluid migration profile for a Gulf Coast Tertiary regressive depositional system.

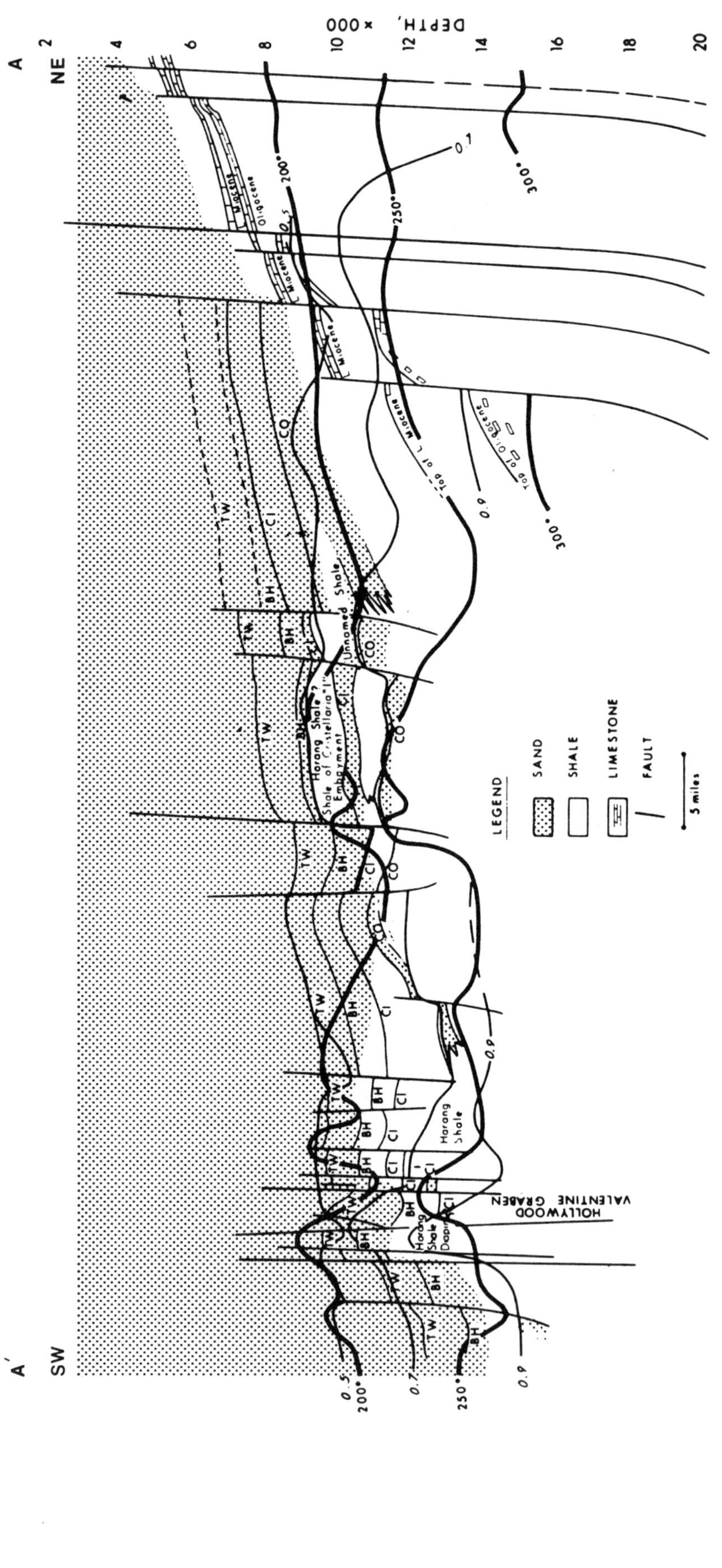

Fig. 3. Geologic cross section A-A' from Hancock County, Mississippi to the Valentine field in central Lafourche Parish, Louisiana, showing relations of temperature and pressure isopleths to sediment facies distribution and structural deformation. (after Gatenby, 1980).

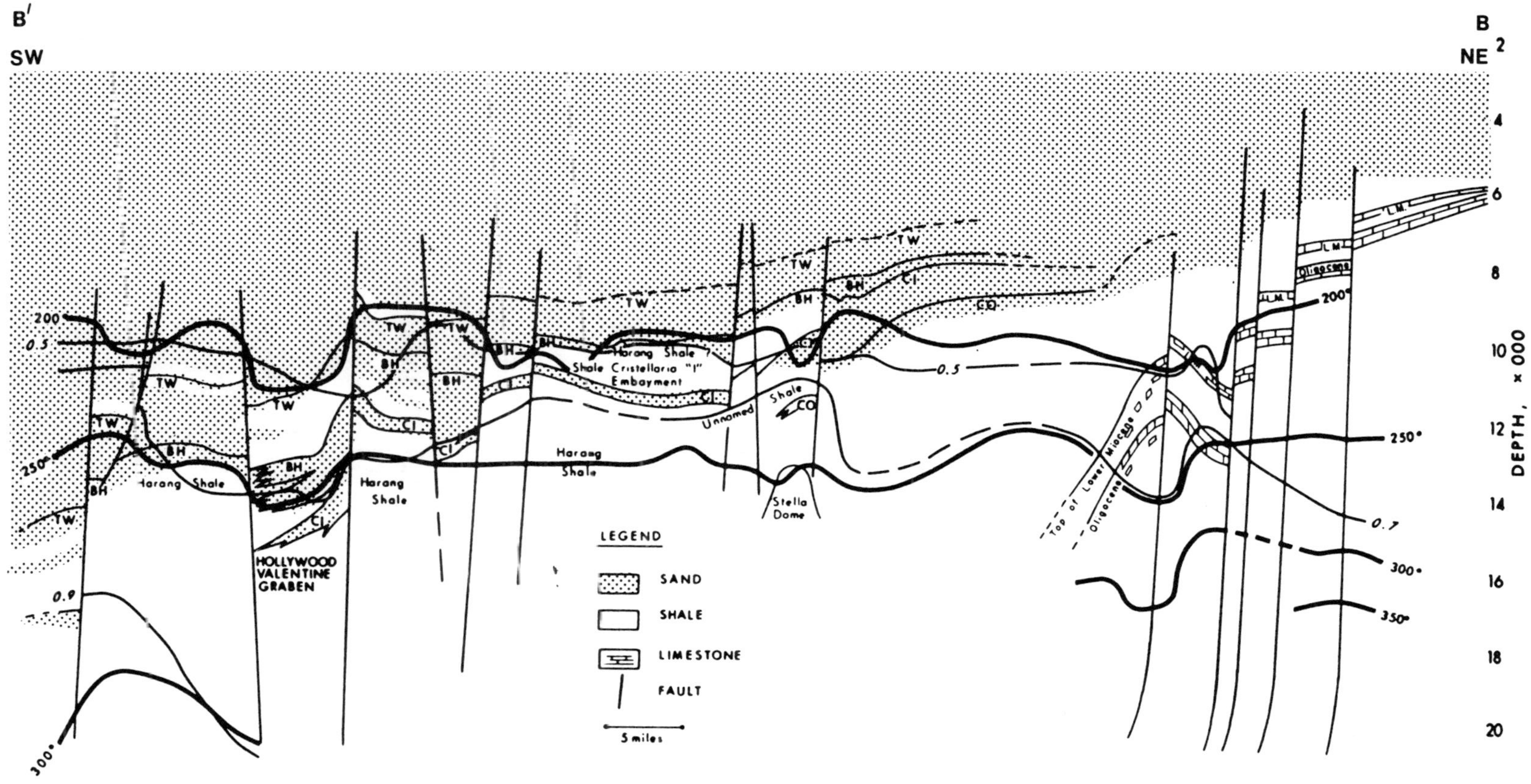

Fig. 4. Geologic cross section B-B' from Hancock County, Mississippi to the Valentine field in central Lafourche Parish, Louisiana, showing relations of temperature and pressure isopleths to sediment facies distribution and structural deformation (after Gatenby).

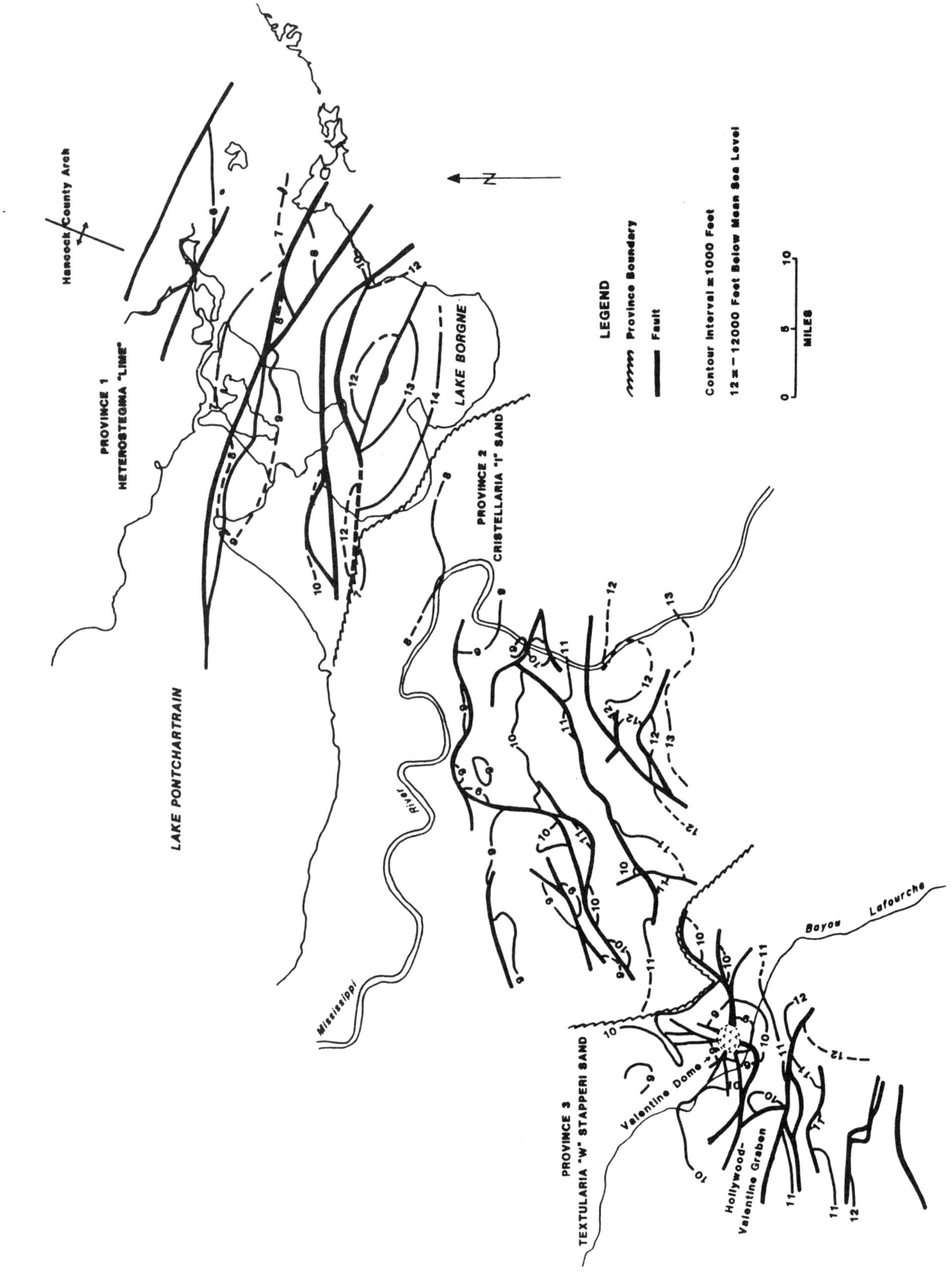

Fig. 5. Geologic structure of the study area mapped on marker beds at comparable depth in each of the three depositional provinces.

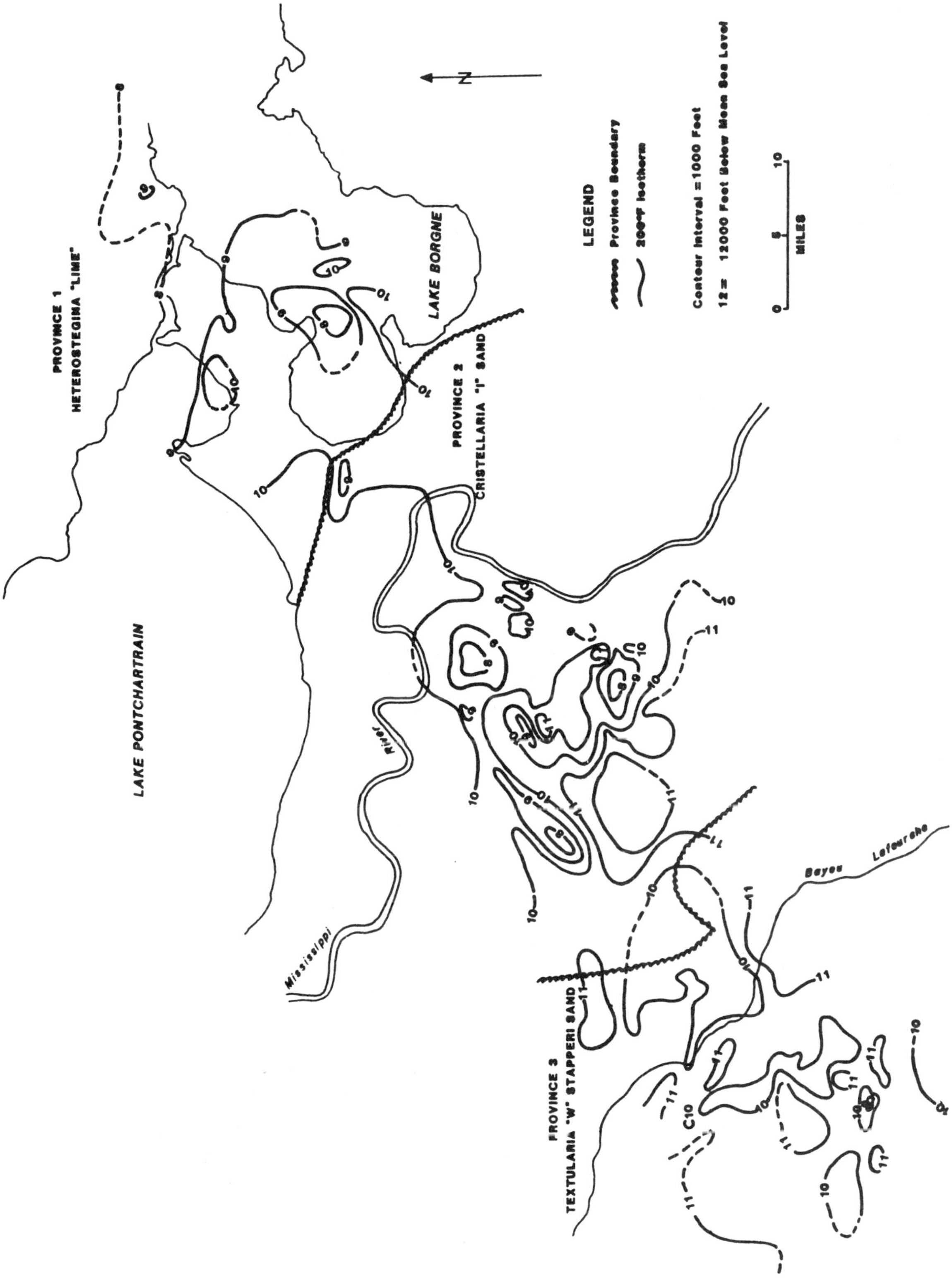

Fig. 6. Depth of occurrence of the 200°F isothermal surface in the study area.

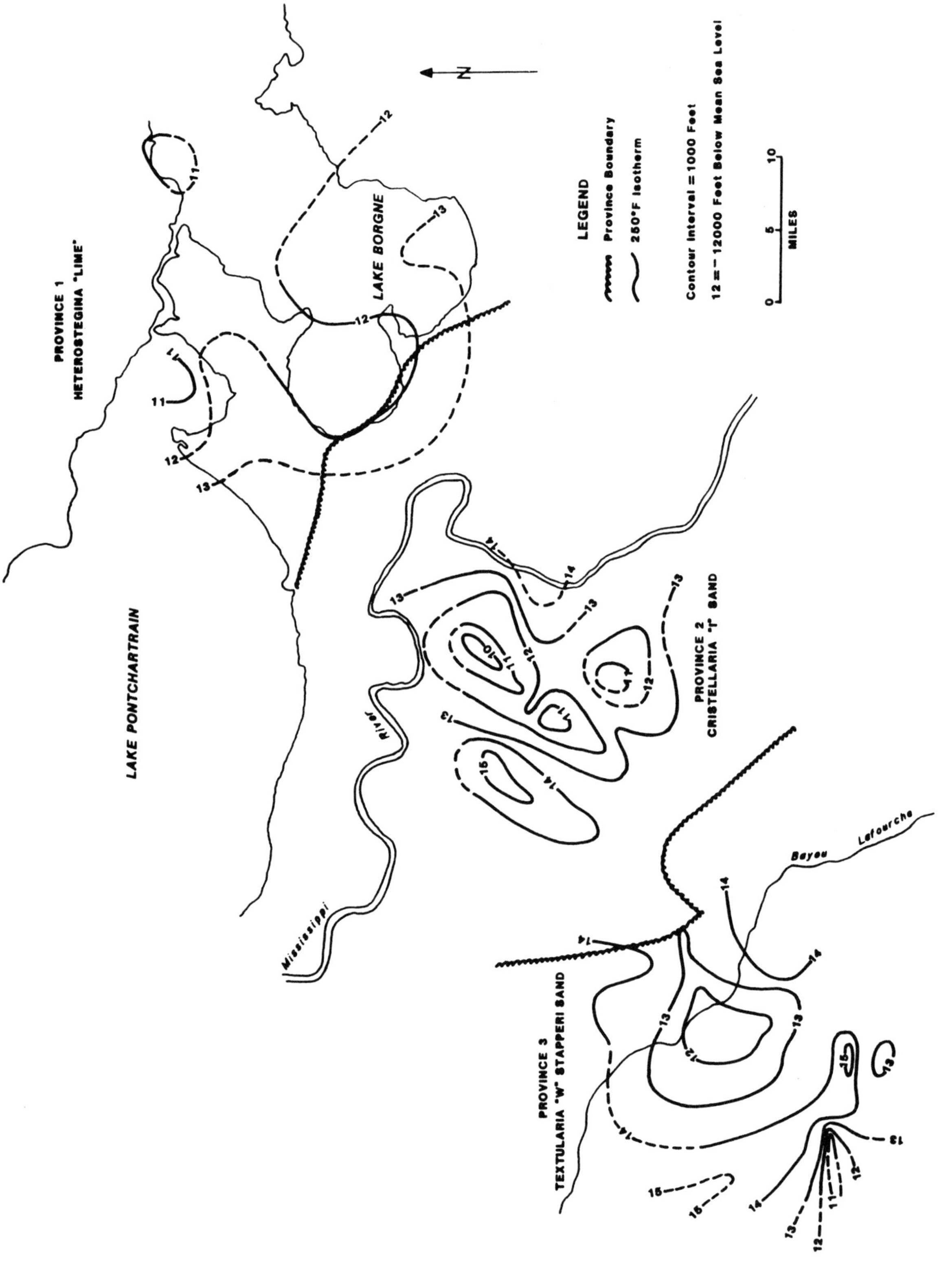

Fig. 7. Depth of occurrence of the 250°F isothermal surface in the study area.

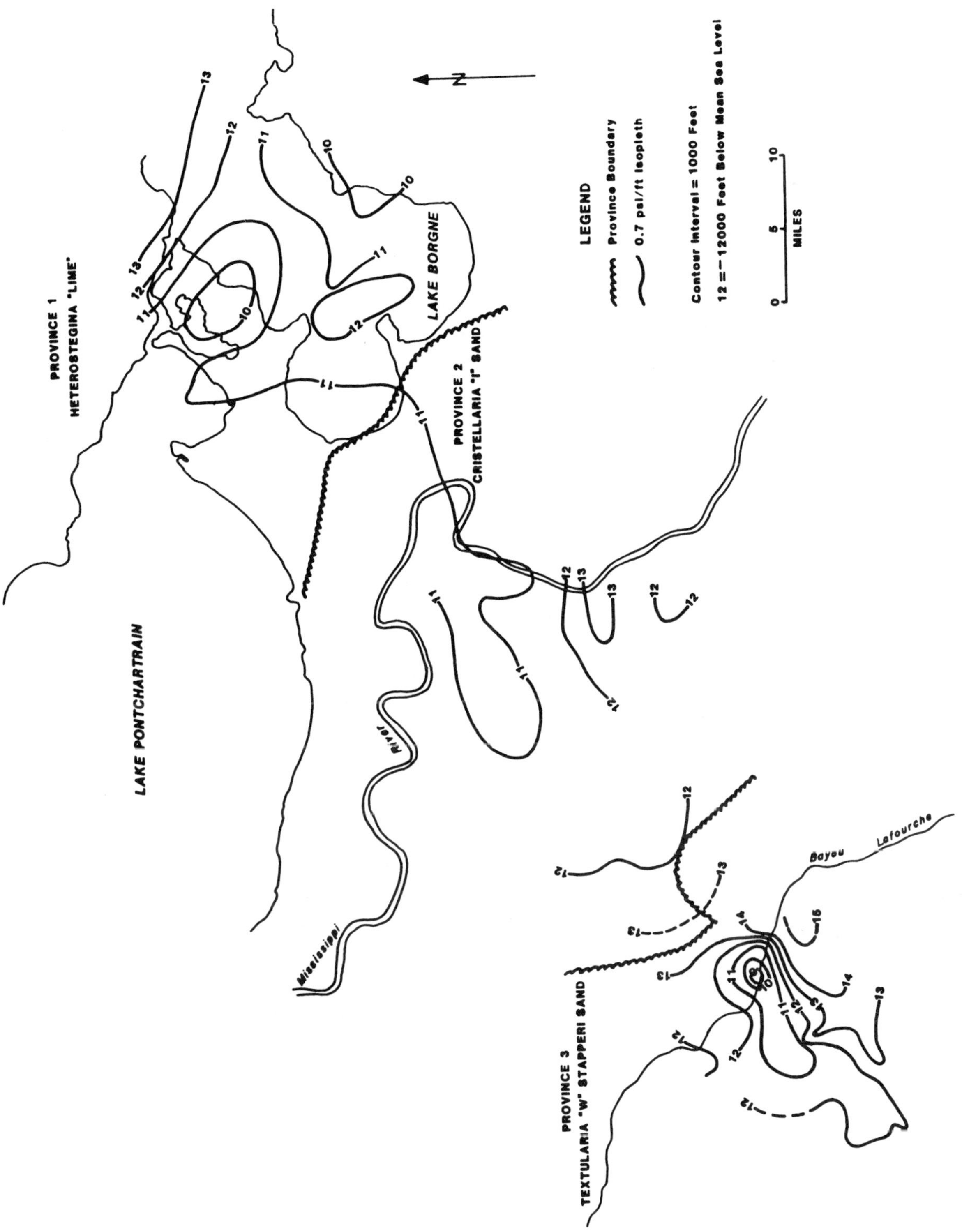

Fig. 8. Minimum depth at which the depth-fluid pressure gradient equals 0.7 pounds per square inch per foot in the study area.

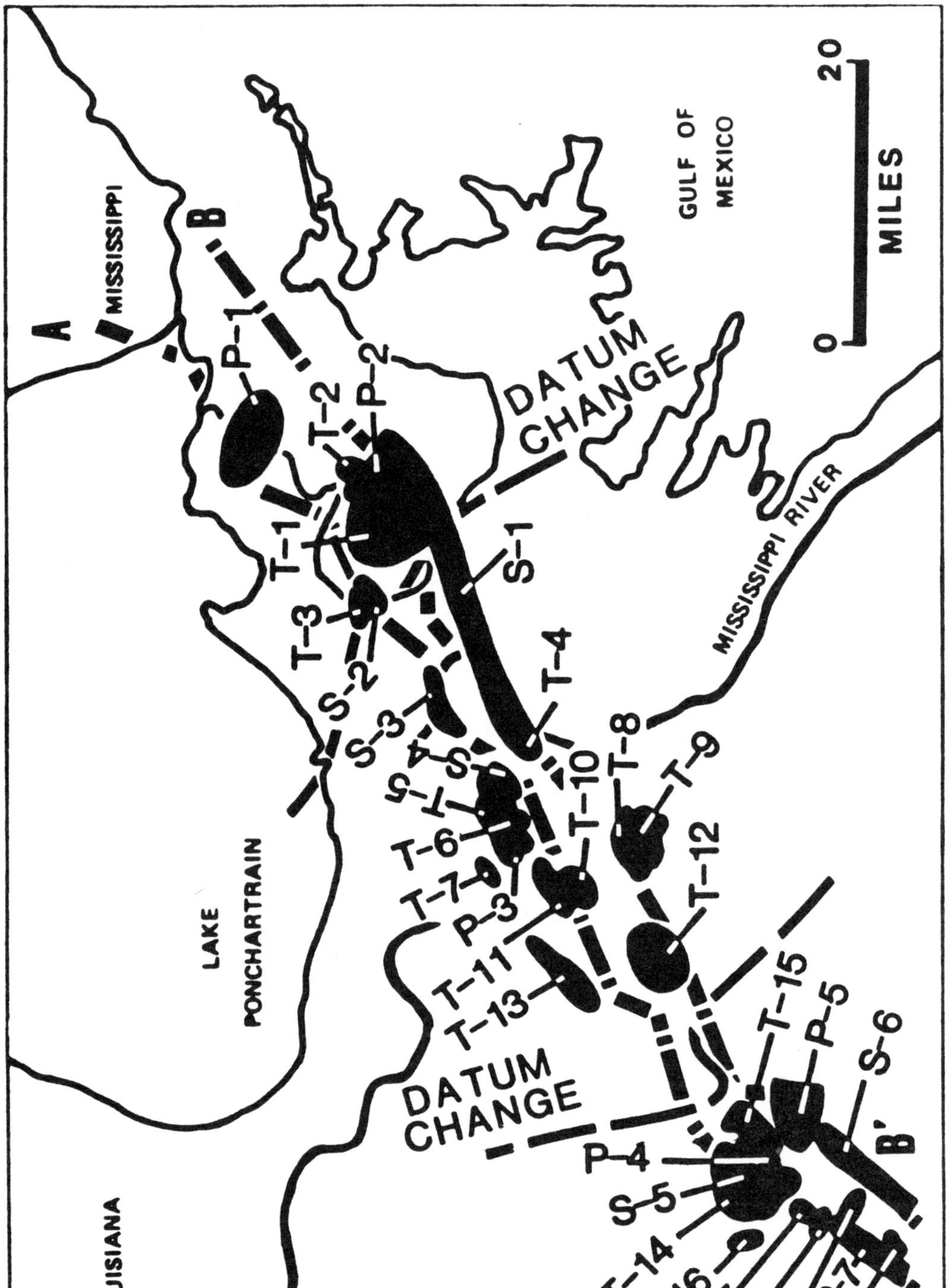

Fig. 9. Temperature, fluid pressure, and formation water salinity anomalies in the study area identified using the fluid migration model concept.

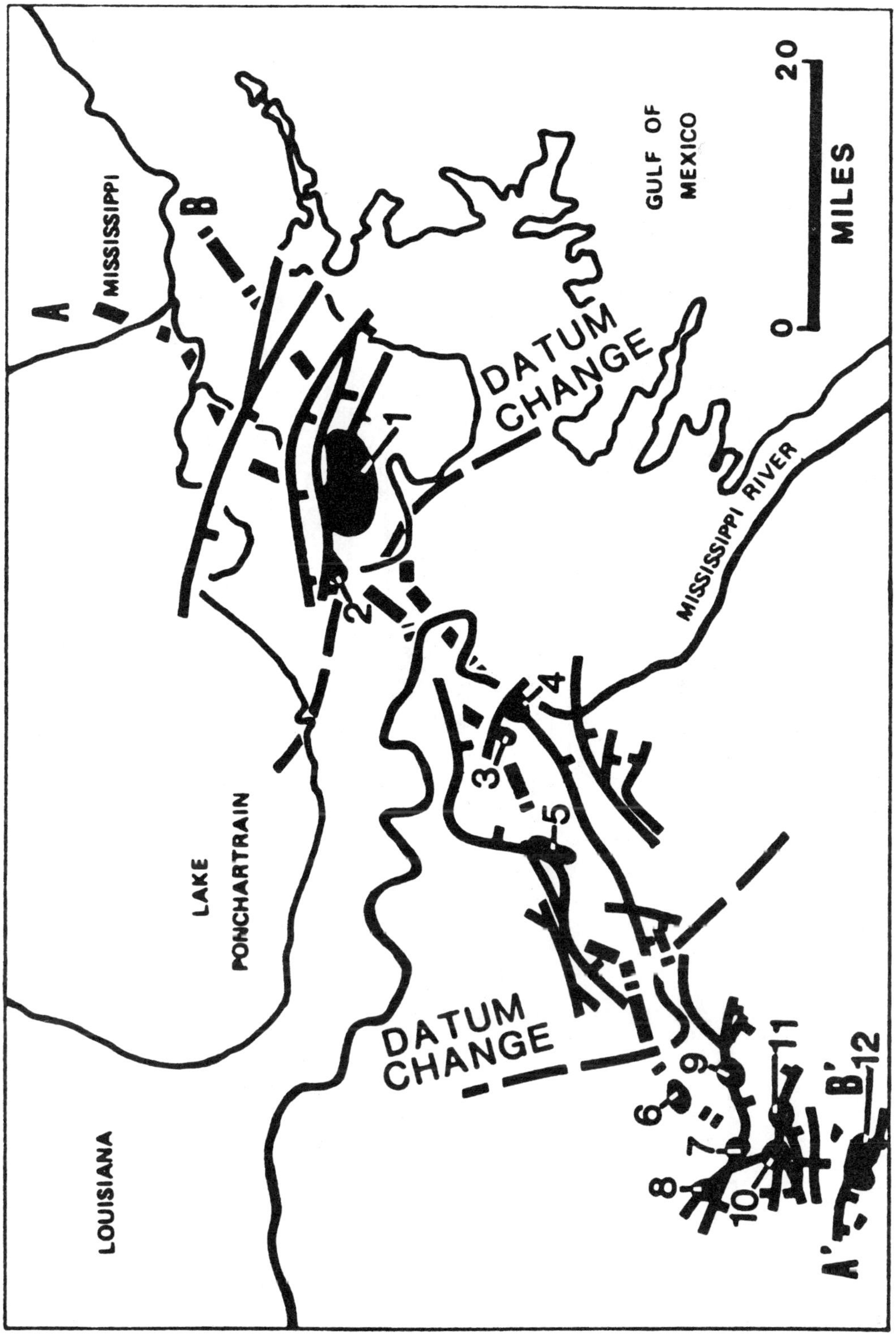

Fig. 10 Fluid migration anomalies on Figure 9 that are located on structure anomalies commonly associated with commercial accumulation of oil and gas.

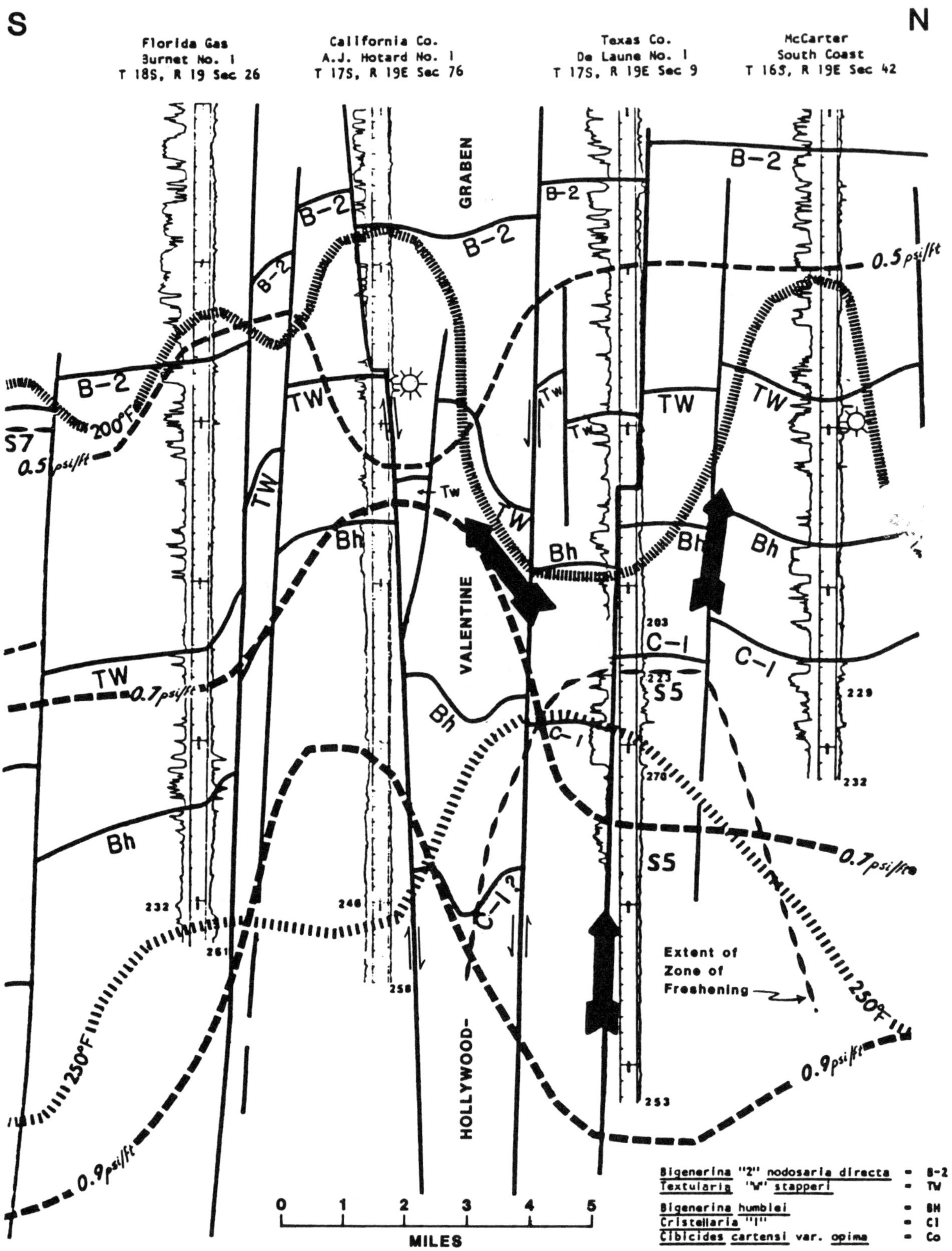

Fig. 11. Fluid migration paths in the Hollywood-Valentine graben indicated by upwarp of water-salinity and temperature isopleths, downwarp of fluid-pressure isopleths, and location of commercial reservoirs of natural gas.

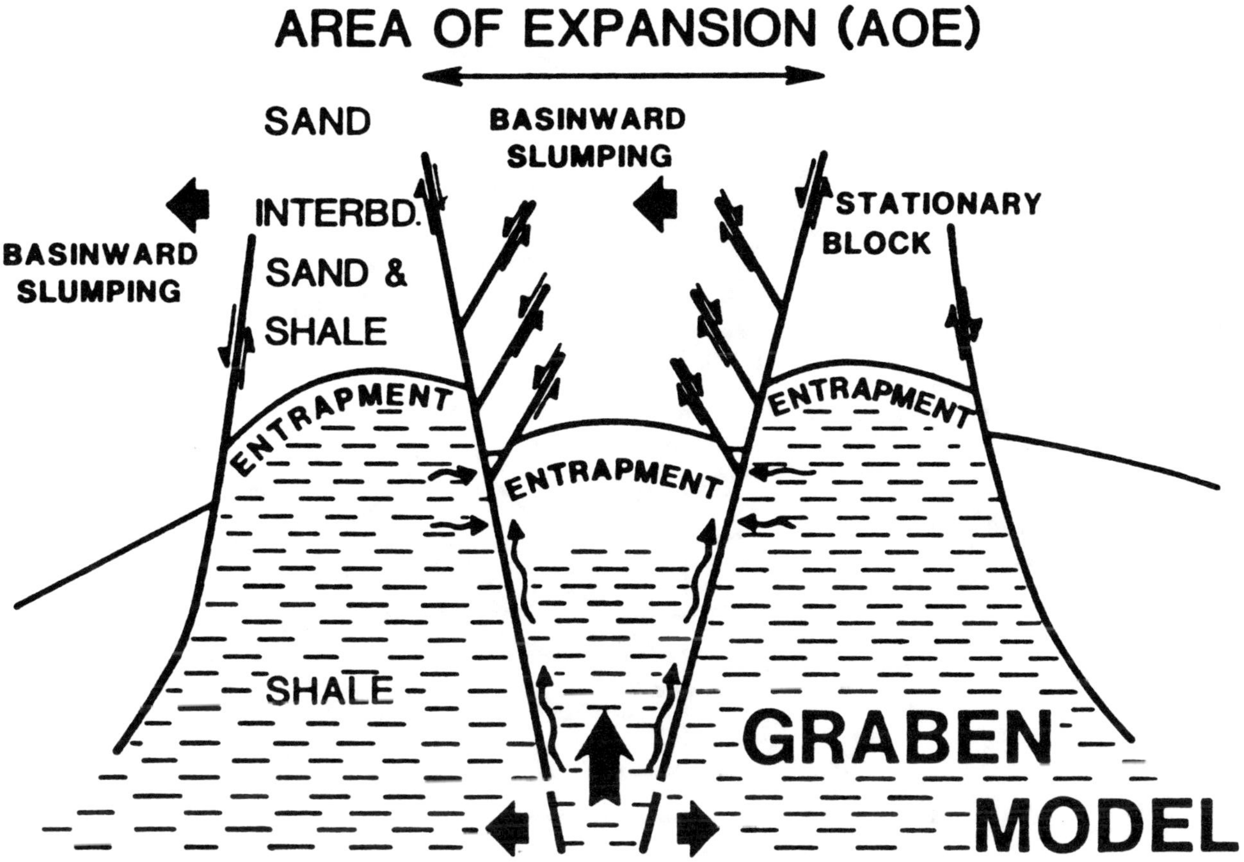

Fig. 12. Graben model indicating relation of structural deformation to vertical fluid migration paths.

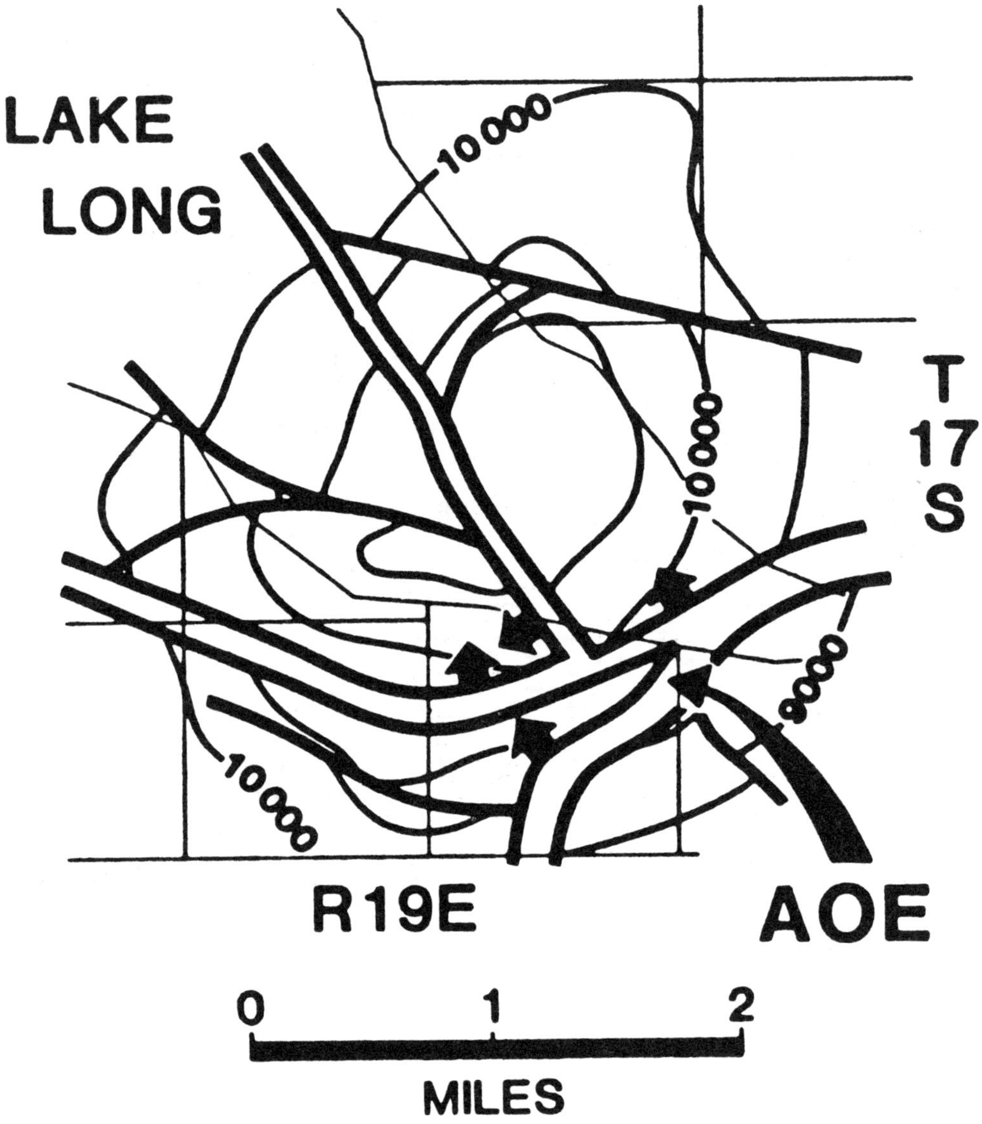

Fig. 13. Area of expansion on the Lake Long field structure which localizes upward fluid migration.

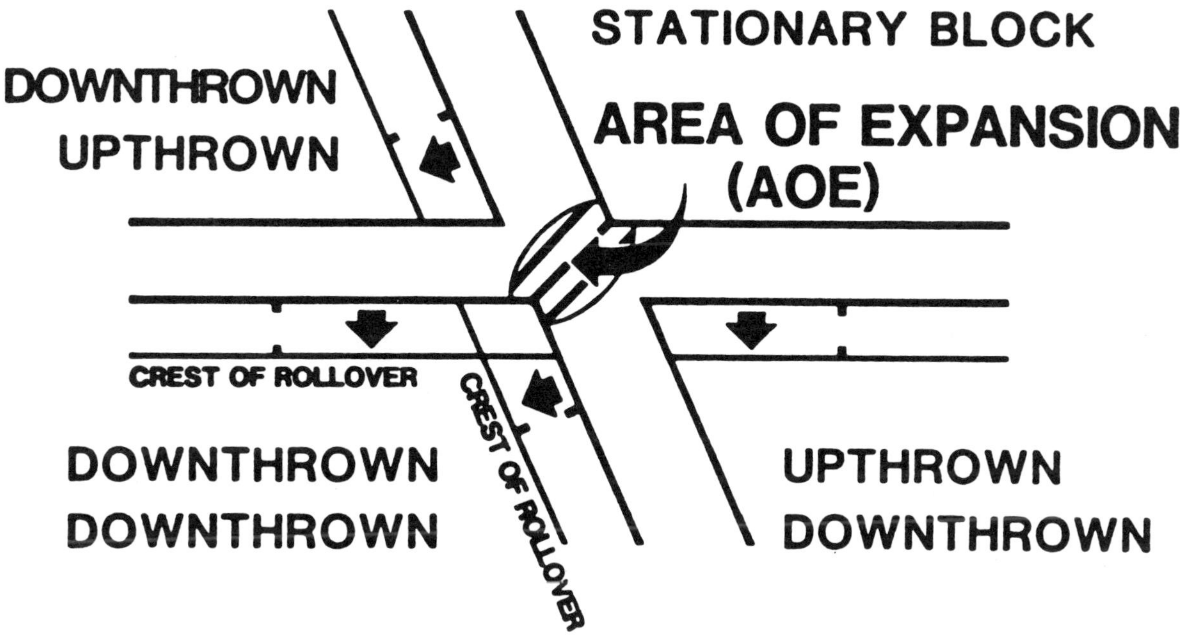

Fig. 14. Area of expansion which localizes upward fluid migration at growth fault intersections.